# Globalization and Sustainable Development

Environmental Agendas

Vladimir F. Krapivin and Costas A. Varotsos

# Globalization and Sustainable Development

## Environmental Agendas

Professor Dr. Vladimir F. Krapivin
Institute of Radioengineering and Electronics
Russian Academy of Sciences
Moscow
Russia

Associate Professor Costas A. Varotsos
University of Athens
Department of Applied Physics
Laboratory of Upper Air
Athens
Greece

SPRINGER–PRAXIS BOOKS IN ENVIRONMENTAL SCIENCES
SUBJECT *ADVISORY EDITOR*: John Mason B.Sc., M.Sc., Ph.D.

ISBN 978-3-540-70661-8 Springer Berlin Heidelberg New York

Springer is part of Springer-Science + Business Media (springer.com)

Library of Congress Control Number: 2007923412


Printed in Germany



Cover design: Jim Wilkie
Project management: Originator Publishing Services, Gt Yarmouth, Norfolk, UK

Printed on acid-free paper

# Contents

# Preface

A characteristic of the present global ecological situation is increasing instability or—put another way—a crisis in the civilization system, the global scale of which is expressed through a deterioration of human and animal habitats. The most substantial features of global ecodynamics of the late 20th and early 21st centuries include the rapid increase in world population (mainly in developing countries), increase in the size of the urban population (considerable growth in the number of megalopolises), and increase in the scales of such dangerous diseases as HIV/AIDS, hepatitis, tuberculosis, etc. With growing population size the problems of providing people with food and improving their living conditions in many regions will not only not be resolved but will become even more urgent. Any possible benefit from decrease in per capita consumption as a result of increased efficiency of technologies will be outweighed by the impact of such a growth in population size. Despite the predominant increase of population in developing countries, their contribution to the impact on the environment will not necessarily exceed that of developed countries. Key to ensuring sustainable development of the nature/society system (NSS) is the relationship between production and consumption, as mentioned at the World Summit on Sustainable Development in Johannesburg (2002).

As civilization has developed, so the problem of predicting the scale of expected climate change and associated change in human habitats has become more urgent. The matter primarily concerns the origin and propagation of natural phenomena causing the death of many creatures and large-scale economic damage. In the historical past, natural anomalies at various spatio-temporal scales are known to have played a certain role in the evolution of nature, causing and activating mechanisms for natural system regulation. With the development of industry and growing population density, these mechanisms have suffered considerable changes and acquired a life-threatening character. This is primarily connected with the growth and propagation in the level of anthropogenic disturbances in the environment. Numerous studies of resultant problems carried out in recent decades have shown that the frequency of catastrophic phenomena in nature and their scale have been continuously growing, leading to a growing risk to human life, financial losses, and to a breakdown of social infrastructure.

The search for a sustainable way for the NSS to develop depends on creating information technologies that are capable of forecasting the likely consequences of anthropogenic projects. In the absence of such technologies it will be difficult to improve predictions of the optimal conditions for human existence. One such approach has been developed by the authors. It consists in creating a new geoinformation-monitoring technology based on adaptively combining environmental observations with the results of numerical modeling of processes taking place in the NSS.

The book contains five chapters. Chapter 1 gives a general description of the notion of globalization and sets out the problems that need to be resolved to determine the direction of future trends in the development of civilization. Estimates are given of the state of energy resources of the biosphere and likely scenarios of how global ecosystems might evolve. The growth of globalization in practically all spheres of human activity is connected with numerous problems arising from the interaction between man and nature. Biogeochemical cycles are given top priority in Chapter 2. As a constructive approach to the solution of NSS contradictions, a simulation model of the NSS is proposed and described in Chapter 3. Chapter 4 is dedicated to technologies involved in collecting and processing information about the environment, which serve as the basis for assessing the simulation model's parameters and the consequences of various anthropogenic scenarios. Chapter 5 describes approaches that assess the risk of decision-making for its impact on the environment.

By analyzing present trends in civilization development and assessing global ecodynamics, the book considers the problem of global environmental change as an interdisciplinary problem facing many scientific fields of study. The authors' approach to solving this problem is based on GIMS technology, which can be used to obtain information about the NSS and select the appropriate criteria for assessment of its state.

The problems facing civilization and the prospects for its future development are so broad and comprise so many components that the aspects considered here are only a small part of a vast field of scientific and methodical studies of NSS interaction processes. The proposed adaptive–evolutionary scheme of combining monitoring data with the results of numerical modeling raises the hope that a mechanism that fully understands this interaction has been found, one that will ensure the transition to sustainable development.

The book will be useful to specialists dealing with the development of information technologies for nature conservation, global modeling, study of climate change, problems of NSS interaction, as well as geopolitics.

# Acknowledgment and dedication

The motivation to write this book came from the well-known Russian scientist Academician Kirill Ya. Kondratyev (1920–2006). In the process of preparing this book he sadly passed away, having left unique material and multiple proposals that were used in this book. The authors would like to express albeit posthumously their deep gratitude to him and wish to dedicate this book to his memory.

The area of scientific interests of Professor Kirill Kondratyev was extremely broad encompassing the theory of transfer of thermal radiation through the atmosphere, the greenhouse effect, natural and human-induced disasters and catastrophes, remote-sensing of the environment, and global climate change. He was the initiator and participant of numerous scientific projects that were related to the study of global climate change as well as to the understanding of interactive processes in the nature/society system (NSS).

Kirill Ya. Kondratyev received his PhD in 1956 at the University of Leningrad, where he also became Professor in 1956 and was appointed Head of the Department of Atmospheric Physics in 1958. He was Head of the Department of the Main Geophysical Observatory (1978–1982) and Head of the Institute for Lake Research (1982–1991), both in Leningrad. Since 1992, he was a Councilor of the Russian Academy of Sciences. He published more than 100 books in the fields of atmospheric physics and chemistry, remote-sensing, planetary atmospheres, and global change.

Kirill Kondratyev was elected an honorary member of the American Meteorological Society, the Royal Meteorological Society of Great Britain, the Academy of Natural Sciences "Leopoldina" (Germany), a foreign member of the American Academy of Arts and Sciences, a member of the International Astronautic Academy, honorary Doctor of Sciences of the Universities of Lille (France), Budapest (Hungary), and Athens (Greece). He was a long-time Editor-in-Chief of the Russian journal *Earth Observations and Remote Sensing*, member of the editorial board of a number of scientific journals: *Optics of the Atmosphere and Ocean*, *Proceedings of the Russian Geographical Society*, *Meteorology and Atmospheric Physics* (Austria),

*Időjárás* (Hungary), *Il Nuovo Cimento C* (Italy), *Atmósfera* (Mexico), *Energy and Environment* (Great Britain).

In recent years, he studied the Arctic environment from the viewpoint of global change. He and his scientific group (members of many Russian institutes) assumed the Arctic Basin to be a unique environment, playing a major role in the dynamics and evolution of the Earth system. Changes in the Arctic environment are considered one of a number of problems arising from the interaction between humankind and nature. He also took into account the consequences of the growth of globalization in almost all spheres of human activity.

Kirill Ya. Kondratyev was an encyclopedic and interdisciplinary broad-minded scientist. He struggled for justice and cleanness in scientific investigations. His numerous reports presented within the framework of international conferences were always directed at in-depth understanding of and plausible explanations for the processes in the global climate system. He advanced new ideas focused on the consideration that various environmental and anthropogenic problems are the main components of the nature/society system. His latest books were *Global Ecodynamics: A Multidimensional Analysis*, *Arctic Environment Variability in the Context of Global Change*, *Natural Disasters as Interactive Components of Global Ecodynamics*, and *Global Carbon Cycle and Climate Change*.

Kirill Ya. Kondratyev was a very communicative person. He loved life and always helped young scientists from Russia and other countries to investigate the unexplored borders of science. He was a defender of scientific truth. In many of his papers and books he criticized the Kyoto Protocol for not being based on scientific fact. His long life (86 years) was spent serving world science.

# Figures

# Tables

# Abbreviations and acronyms

| | |
|---|---|
| AAHIS | Advanced Airborne Hyperspectral Imaging Spectrometer |
| ACE | Aerosol Characterization Experiment |
| ACPI | Advanced Configuration and Power Interface |
| ACRS | Asian Conference on Remote Sensing |
| ADEOS | Adaptive Domain Environment for Operating System |
| AFO | AtmosphärenFOrschung |
| AGCM | Atmosphere General Circulation Model |
| AGU | American Geophysical Union |
| AI | Aerosol Index |
| AIDS | Acquired Immune Deficiency Syndrome |
| AIMR | Airborne Imaging Microwave Radiometer |
| AIRS | Atmospheric InfraRed Sounder |
| AIS | Airborne Imaging Spectrometer |
| AMMR | Airborne Multichannel Microwave Radiometer |
| AMPR | Advanced Microwave Precipitation Radiometer |
| AMSR | Advanced Microwave Scanning Radiometer |
| AMSR-E | Advanced Microwave Scanning Radiometer for EOS |
| AMSU | Advanced Microwave Sounding Unit |
| AOCI | Airborne Ocean Color Imager Spectrometer |
| AOGCM | Coupled Atmosphere–Ocean General Circulation Model |
| APS | Atomic Power Station |
| APT | Almost Plain Text |
| ARISTI | All Russian Institute for Scientific and Technical Information |
| ATMOFAST | ATMOsphere FASTness |
| AVNIR | Advanced Visible and Near Infrared Radiometer |
| AVSS | Atmosphere–Vegetation–Soil System |
| AWSR | Airborne Water Substance Radiometer |

| | |
|---|---|
| BC | Brundtland Commission |
| BEN | Base of Environmental Knowledge |
| BIO | BIOlogical block |
| BR | Brundtland Report |
| CASE | Computer Aided Software Engineering |
| CASI | Compact Airborne Spectrographic Imager |
| CBD | Convention on Biological Diversity |
| CCRS | Chicago Center for Religion and Science |
| CCSM | Community Climate System Model; Complex Climate System Model |
| CDCP | Center for Disease Control and Prevention |
| CERES | Clouds and the Earth's Radiant Energy System |
| CF | Cloud Forcing |
| CFC | ChloroFluoroorganic Compound |
| CF | Cloud Feedback |
| CGCM | Certified Graphic Communications Manager |
| CHRISS | Compact High-Resolution Imaging Spectrograph Sensor |
| CLD | Modeling the effect of Land Degradation on Climate |
| CLIMBER | CLIMate and BiosphERe |
| CLI | CLImate block |
| CLIMPACT | Regional CLimate modeling and integrated global change IMPACT studies in the European Arctic |
| CLIVAR | CLImate VARiability and predictability |
| CMIP | Coupled Model Intercomparison Project |
| CMS | Convention on Migratory Species |
| COADS | Comprehensive Ocean–Atmosphere Data Set |
| COARE | Coupled Ocean–Atmosphere Response Experiment |
| COLA | Centre for Ocean–Land–Atmosphere Studies |
| CON | CONtrol block |
| COP | Conference Of the Parties |
| CPS | Commerce and Population Service |
| CR | Club of Rome |
| CRL | Communication Research Laboratory |
| CSI | Canopy Structure Index |
| CSIRO | Commonwealth Scientific and Industrial Research Organization |
| CSSR | Earth Institute's Center for the Study of Science and Religion |
| CSTE | Council of State and Territorial Epidemiologists |
| CTM | Chemical Transport Model |
| DAT | DATabase control block |
| DB | DataBase |
| DEM | DEMographic block |
| DEMETER | Development of a European Multimodel Ensemble system for seasonal to inTERannual prediction |

| | |
|---|---|
| DO-SAR | DOrnier SAR |
| DOE | Department Of Energy |
| DPIR | Drivers, Pressures, Impact, Response model |
| EBM | Energy Balance Model |
| ECDO | Economic Cooperation and Development Organization |
| ECLAT-2 | European CLimate Action Towards the Improved Understanding and Application of Results from Climate Model Experiments in European Climate Change Impact Research |
| EEC | European Economic Community |
| EFIEA | European Forum on Integrated Environmental Assessment |
| EIA | Energy Information Administration |
| EJ | ExaJoule |
| EMIRAD | ElectroMagnetic Institute RADiometer |
| EMISAR | ElectroMagnetic Institute SAR |
| EMW | ElectroMagnetic Waves |
| ENSO | El Niño/Southern Oscillation |
| EOF | Empirical Orthogonal Function |
| EOS | Earth Observing System |
| EPA | Environmental Protection Agency |
| EPHI | Environmental Public Health Indicator |
| ERB | Earth's Radiation Budget |
| ERBS | Earth Radiation Budget Satellite |
| ERIC | Educational Resources Information Center |
| EROC | Ecological Rates Of Change |
| ERS | Earth Remote Sensing |
| ES | Ecosystem Service |
| ESA | European Space Agency |
| ESTAR | Electronically Steered Thinned Array Radiometer |
| ETEMA | European Terrestrial Ecosystem Modeling Activity |
| EU | European Union |
| EUROTAS | EURopean river flood Occurrence and Total risk Assessment System |
| EWI | Ecosystem Wellbeing Index |
| FAO | Food and Agriculture Organization |
| FDI | Fund for Direct Investments |
| GAIA | Global Action in the Interest of Animals |
| GCC | Global Cycle of Carbon |
| GCM | General Circulation Model |
| GCP | Global Carbon Project |
| GD | Greenhouse Gas Dynamics |
| GDP | Gross Domestic Product |
| GEF | Global Ecological Fund |
| GEO | Global Environment Outlook |
| GEP | Global Ecological Perspective |

| | |
|---|---|
| GGD | Greenhouse Gas Dynamics |
| GGP | Global Gross Product |
| GHG | GreenHouse Gas |
| GIMS | GeoInformation Monitoring System |
| GIS | Geographic Information System |
| GISS | Goddard Institute for Space Studies |
| GLOBEC | GLOBal ocean ECosystem Dynamics |
| GMNSS | Global Model of Nature/Society System |
| GMS | Global Meteorological Satellite |
| GNESD | Global Network on Energy for Sustainable Development |
| GOCART | GOddard Chemistry Aerosol Radiation and Transport |
| GOES | Geostationary Operational Environmental Satellite |
| GOES-EAST | GOES Eastern |
| GOMS | Geostationary Operational Meteorological Satellite |
| GSC | Global Sulfur Cycle |
| GWP | Global Warming Potential |
| HDI | Human Development Index |
| HERA | Human and Environmental Risk Assessment |
| HIV | Human Immunodeficiency Virus |
| HPS | Hydro Power Stations |
| HRPT | High-Resolution Picture Transmission |
| HSB | Humidity Sounder for Brazil |
| HWI | Human Wellbeing Index |
| HYD | HYDrodynamic process |
| IA | Integral Assessment |
| ICARTT | International Consortium for Atmospheric Research on Transport and Transformation |
| ICDES | International Center of Dataware for Ecological Studies |
| ICGGM | International Center on Global Geoinformation Monitoring |
| ICLIPS | Integrated assessment of CLImate Protection Strategies |
| ICPD | International Conference on Population and Development |
| ICPES | International Center of Information Provision of Ecological Studies |
| ICS | International Council for Science (see *http://www.icsu.org/index.php*) |
| IEA | International Energy Agency |
| IEEE | Institute of Electrical and Electronics Engineers |
| IFC | International Forum on Globalization |
| IGAC | International Global Atmospheric Chemistry |
| IGBP | International Geosphere–Biosphere Program |
| IHDP | International Human Dimensions Program |
| IIASA | International Institute for Applied System Analysis |
| IIED | International Institute for Environment and Development |
| *IJSD* | *International Journal of Sustainable Development* |
| ILO | International Labour Office |

| | |
|---|---|
| IMF | International Monetary Fund |
| InSAR | Interferometric Synthetic-Aperture Radar |
| INSAT | INdian SATellite |
| INTEX-NA | INTercontinental Chemical Transport EXperiment-North America |
| IPCC | Intergovernmental Panel on Climate Change |
| IR | InfraRed radiation |
| IREE | Institute of Radio Engineering and Electronics |
| ISA | Integrated Systems Approach |
| ISGGM | Integrated System for Global Geoinformation Monitoring |
| ITCT | Intercontinental Transport and Chemical Transformation |
| ITOP | International Transport of Ozone and Precursors |
| JAMSTEC | Japan Agency for Marine–earth Science and TEChnology |
| JERS | Japanese Earth Resources Satellite |
| JPL | Jet Propulsion Laboratory |
| KLINO | KLImaszenarien in Nord- und Ostsee |
| KP | Kyoto Protocol |
| LAI | Leaf Area Index |
| LP | Logical and Probabilistic methods |
| LRTP | Long-Range Transport Potential |
| LTREF | Long-Term Regional effects of climate change on European Forests |
| LWRF | Low-Wave Radiation Forcing; Long-Wave Radiative Forcing |
| MAC | Manufactures And Construction |
| MAE | Model of Equatorial Upwelling |
| MAGEC | Modelling Agroecosystems under Global Environmental Change |
| MAHB | Millennium Assessment of Human Behavior |
| MAI | Multilateral Agreement on Investments |
| MAX-DOAS | Multiple Axis Differential Optical Absorption Spectroscopy |
| MCBS | Magnetosphere/Climate/Biosphere System |
| MDC | Movement for Democratic Change |
| MEA | Millennium Ecosystem Assessment |
| MECW | Marx and Engels Collected Works |
| MEM | Microwave Emission Model |
| MGC | Minor Gas Component |
| MGNC | Model of Global Nitrogen Cycle |
| MGOC | Model of Global Oxygen Cycle |
| MGPC | Model of Global Phosphorus Cycle |
| MIGCM | Moscow Institute for Governmental and Cooperative Management |
| MISR | Multi-angle Imaging Spectro-Radiometer |
| MMIA | Methods and Models for Integrated Assessment |

| | |
|---|---|
| MMSD | Mining, Minerals, and Sustainable Development |
| MOCAGE | Multiscale Off-line/semi-on-line Chemistry And transport Grid-point modEl |
| MODIS | MODerate-resolution Imaging Spectroradiometer |
| MOPITT | Measurements Of Pollution In The Troposphere |
| MOS | Marine Observation Satellite |
| MOSE | Model of the Okhotsk Sea Ecosystem |
| MOUNTAIN | Environmental and social changes in MOUNTAIN regions |
| MOZART | Model for OZone And Related Tracers |
| MPCE | Model of Peruvian CurrEnt |
| MRE | Mineral REsource control |
| MSR | Microwave Scanning Radiometer |
| MSSA | Multi-channel Singular Spectral Analysis |
| NAFTA | North American Free Trade Agreement |
| NASA | National Aeronautics and Space Administration |
| NASDA | NAtional Space Development Agency of Japan |
| NCAR | National Center for Atmospheric Research |
| NDVI | Normalized Difference Vegetation Index |
| NH | Northern Hemisphere |
| NHC | National Hurricane Center |
| NOAA | National Oceanic and Atmospheric Administration |
| NPS | Nuclear Power Station |
| NRC | National Research Council |
| NSI | North–South Institute |
| NSF | National Science Foundation |
| NSS | Nature–Society System |
| OCTS | Ocean Color and Temperature Scanner |
| OECD | Organization for Economic Cooperation and Development |
| OES | Other Energy Source |
| OGCM | Ocean General Circulation Model |
| OPEC | Organization of Petroleum Exporting Countries |
| ORCD | Organization for Relief and Community Development |
| PCL | Permissible Concentration Level |
| PCM | Parallel Climate Model |
| PCMDI | Program for Climate Model Diagnosis and Intercomparison |
| PDU | Protocol Data Unit |
| PEHP | Production of Electric power and Heat for the Population |
| PEL | Petroleum Economics Limited |
| PIC | Potsdam Institute for Climate Impact Research |
| PIRA | Petroleum Industry Research Associates |
| PISDR | Partnership for Information Society Development in Russia |
| POEM | POtsdam Earth system Modelling |
| POL | Global cycle of general POLlutants |
| POP | Persistent Organic Pollutant |
| PRECIS | Providing REgional Climates for Impact Studies |

| | |
|---|---|
| PWI | Plant–Water Index |
| RAL | Rutherford Appleton Laboratory |
| RAMS | Regional Atmospheric Modeling System |
| RAS | Russian Academy of Sciences |
| RCM | Regional Climate Model |
| REP | REPort formation block |
| RF | Radiative Forcing |
| RFF | Resources For the Future |
| RFFS | Russian Fund of Fundamental Studies |
| RGGI | Regional Greenhouse Gas Initiative |
| RGS | Russian Geographical Society |
| RIBAMOD | RIver BAsin MODeling project |
| RICC | Randwick Information and Community Centre |
| RISTI | Russian Institute for Scientific and Technical Information |
| SAMUM | SAharan Mineral dUst experiMent |
| SAR | Synthetic Aperture Radar |
| SAT | Surface Air Temperature |
| SC | Solar Constant |
| SD | Sustainable Development |
| SDS | Scott Data System, a semi-empirical method of scaling |
| SH | Southern Hemisphere |
| SHADE | SaHAran Dust Experiment |
| SiB2 | Simple Biosphere model-2 |
| SIDC | Small Island Developing Country |
| SKB | Swedish Nuclear Fuel and Waste Management Company |
| SMAHF | Simulation Model for the Aral Hydrophysical Fields |
| SMMR | Scanning Multichannel Microwave Radiometer |
| SMS | Standard Monitoring System |
| SP | Strategic Plan |
| SPF | Soil–Plant Formation |
| SRCES | Scientific Research Centre for Ecological Safety |
| SRES | Special Report on Emissions Scenarios |
| SRVI | Simple Ratio Vegetation Index |
| SSM/I | Special Sensor Microwave/Imager |
| SST | Sea Surface Temperature |
| STI | Scientific–Technical Information |
| STP | Scientific–Technical Progress |
| SWR | Short-Wave Radiation |
| SWRF | Short-Wave Radiation Forcing |
| TEPA | Taiwan Environment Protection Administration |
| TERRAPUB | TERRA Scientific PUBlishing Company, Tokyo |
| TIPT | The Irish Papers Today |
| TMI | TRMM Microwave Imager |
| TNC | TransNational Corporation |
| TOGA | Tropical Ocean and Global Atmosphere |

| | |
|---|---|
| TOMS | Total Ozone Mapping Spectrometer |
| TRMM | Tropical Rainfall Measuring Mission |
| TW | TeraWatt (= $10^{12}$ watts) |
| UAE | United Arab Emirates |
| UARS | Upper Atmosphere Research Satellite |
| UHF | Ultra-High Frequency |
| UN | United Nations |
| UNCCD | United Nations Convention to Combat Desertification |
| UNCED | United Nations Conference on Environment and Development |
| UNDP | United Nations Development Program |
| UNEP | United Nations Environment Program |
| UNFPA | United Nations Population Fund |
| UV-B | UltraViolet-B |
| VNIR | Visible Near-IR radiometer |
| VOC | Volatile Organic Compound |
| WB | World Bank |
| WCC | World Church Council |
| WCDR | World Conference on Disaster Reduction |
| WCED | World Commission on Environment and Development. |
| WCRP | World Climate Research Program |
| WCSDG | World Commission on the Social Dimension of Globalization |
| WCW4 | Fourth World Conference on Women's Problems |
| WEFAX | The International Analog Image Transmission System |
| WEO | World Energy Outlook |
| WFS | World Food Summit |
| WFUNA | World Federation of United Nations Associations |
| WG | Working Group |
| WHiRL | Wide-angle High-Resolution Line-imager |
| WI | Wellbeing Index |
| WISE | World Information Service on Energy |
| WIT | Wessex Institute Transactions |
| WMO | World Meteorological Organization |
| WSSD | World Summit for Social Development |
| WTF | Wet Tropical Forest |
| WTO | World Trade Organization |
| WTP | Willingness To Pay |
| WUF3 | Urban Problems |
| WWI | Worldwatch Institute |

# About the authors

Vladimir F. Krapivin was educated at the Moscow State University as a mathematician. He received his PhD in geophysics from the Moscow Institute of Oceanology in 1973. He became Professor of Radiophysics in 1987 and Head of the Applied Mathematics Department at the Moscow Institute of Radioengineering and Electronics in 1972. He was appointed Grand Professor in 2003 at the World University for Development of Science, Education and Society. Vladimir Krapivin is a full member of the Russian Academy of Natural Sciences and Balkan Academy of Sciences, New Culture and Sustainable Development. He has specialized in investigating global environmental change by the application of modeling technology. He has published 19 books in the fields of ecoinformatics, theory of games and global modeling.

Costas A. Varotsos received his BSc in Physics at the Athens University in 1980 and his PhD in Atmospheric Physics in 1984. He was appointed Assistant Professor in 1989 at the Laboratory of Meteorology in the Physics Department of the Athens University, where he also established the Laboratory of the Middle and Upper Atmosphere. In 1999 he became Associate Professor of the Department of Applied Physics at Athens University. He has published more than 250 papers and 20 books in the fields of atmospheric physics, atmospheric chemistry, and environmental change. In particular, he documented that a major sudden stratospheric warming, which was considered to be a unknown event for the Southern Hemisphere, actually occurred in September 2002 and caused the Antarctic ozone hole to split in two.

# Summary

The basic global problems of NSS dynamics are considered and the key problems of ensuring sustainable development are discussed. The present trend in changing ecological systems and characteristics of today's global ecodynamics are both analyzed and estimated. Emphasis is placed on how best to organize global geoinformation monitoring to provide reliable control over the development of environmental processes by further obtaining prognostic estimates of the consequences of anthropogenic projects being realized. A new approach to NSS numerical modeling is proposed and demonstrative results are given of modeling the dynamics of this system's characteristics on realization of some scenarios of anthropogenic impact on the environment. Emphasis is given to the importance of and the need for development of adaptive algorithms of data-processing monitoring to reduce the cost of such monitoring and raise the reliability of the estimates obtained about global ecodynamics characteristics. Suggestions are given on how to develop technology to estimate the risk of realization of decisions on ecosystem management.

# 1

# Problems of globalization and sustainable development

## 1.1 INTRODUCTION

Nowadays, the problems of globalization and sustainable development have been discussed in a vast scientific literature (the list of references beginning on p. 271 illustrates only a small number of existing publications). Even more numerous are the concerns raised in the mass media. However, discussions of such problems contain many contradictions and disagreements. First of all, globalization and sustainable development (SD) problems are treated, as a rule, as independent and separate. Nevertheless, the priority of the globalization problem and subordinate importance of SD problems as one of the most important aspects of globalization processes raises few doubts.

A constructive view on the globalization and sustainable development problem was first broached by Russian scientists Kirill Ya. Kondratyev and Nikita N. Moiseyev (Kondratyev *et al.*, 1992, 1994, 1997; Moiseyev, 1999). The development of globalization theory took a number of different directions that were often contradictory and colored by socio-political souls (Friedman, 2005; Mander and Goldsmith, 2006; Cheru and Bradford, 2005; Chistobayev, 2001; Stanton, 2003; Wen *et al.*, 2002; Wijen *et al.*, 2005). Global socio-political issues have a long history beginning with the era of colonialism, then independence. After 1917 there existed two periods: the pre- and post-Cold War eras. The post-Cold War era led to the increasing influence of what some people these days call quasi-governments (such as the International Monetary Fund or the World Bank). This delimitation served for a certain time as the cause of principal differences between the existing concepts of the globalization process. At its most basic, there is nothing mysterious about globalization in spite of these differences. Really there only exist two main conceptions of globalization:

(i) Globalization that brings prosperity to certain countries while impoverishing other countries.
(ii) Globalization as the objective process behind development of the nature–society system (NSS), taking into account existing global population dynamics.

In the first case, many experts around the world identify globalization with "Americanization" of economics and culture (Friedman, 2005). The rest of the world seems to be following the U.S.A. and leaving behind their own ways of life. "Americanization" is the contemporary term used for the influence that the U.S.A. has on the culture and economics of other countries, substituting their way of life by the American way. This process is managed by the many international structures under the control of the U.S.A. Significant social networks and international forums have been organized with the purpose of proliferating capitalist interests. Some of these forums date from early in the 20th century, such as the International Chamber of Commerce, the International Organization of Employers, the Center for Environmental Diplomacy, and the Bretton Woods Committee. They were created as a specific response to questions about globalization (Furth, 1965; Brewer *et al.*, 2002; Devkota, 2005; Ehrloch and Ehrlich, 2004). These and other organizations and forums continue to attend to the interests of developed countries in the 21st century as well.

Reformists of global social democracy put forward the theory of neoliberal globalization which proposes the transformation of social ambience in the world. The countries of the former Soviet Union have felt the negative consequences of this theory fully. Therefore, the question "what is globalization?" is becoming the principal question in recent times. Many experts see it as a primary economic phenomenon involving the increasing integration of national economic systems through the growth in international trade, investment, and capital flows. Other experts define globalization as social, cultural, political, and technological exchanges between countries. As a result two alternatives were formed: anti-globalization and pro-globalization.

The alternatives to globalization promoted by anti-globalists include different approaches to the dynamics of world structure:

- recognition of the importance of free trade and investment for economic growth and development, but there also exist data that trade and investment have in fact increased poverty and inequality;
- sustainable development is best achieved through a single framework, integrating environmental protection and the promotion of economic growth and social equity;
- innovation is necessary for sustainable living;
- support for accumulation of power at the local level;
- implementation of trade barriers to protect local production and renunciation of international trade unless goods or services cannot be produced locally;
- free trade usually benefits wealthy countries at the expense of poor countries; and
- environmental protection is the key to sustainable development and is imperative for economic development.

The argument of pro-globalization groups is based on the promotion of free trade as the key to eliminating poverty and ensuring effective development and that market instruments such as intellectual property rights are necessary to protect the environment and promote development. Supporters of globalization argue that it can be rolled back and point to the period between the First and Second World Wars as evidence. The increase in world trade as a proportion of world GDP was proportionately greater between 1870 and 1914 than it has been since 1975. That expansion was stopped, not just by the First World War, but by the loss of support for free trade that followed. This led directly to the 1930 depression and indirectly to the Second World War. All these changes took place when the world population did not exceed 2 billion. At the present time the situation is rapidly changing with the exponential growth of world population exceeding 6.6 billion. Therefore, the globalization process is becoming an inevitable phenomenon of world life. It is for this reason that the list of key questions on globalization can be extended:

- Is there any alternative to globalization in the coming decades?
- What are the environmental impacts of globalization?
- How does globalization affect culture and religion?
- What is globalization and when did it start and how does it depend on population size?
- Who are the players in the interactions between nature and society?
- Why is there global inequality, and is it getting worse?
- What are the costs and benefits of free trade?
- What is the role of the internet and communications technology in globalization?
- Is globalization shifting power from nation states to undemocratic organizations?
- Is globalization resulting in industries in developed countries being undermined by industries in developing countries with inferior labour standards?
- What does globalization mean to isolated countries like Australia?

Nevertheless, when considering the prospects for life on Earth, it is necessary to proceed from the human criteria of assessing the level of environmental degradation, since over time local and regional changes in the environment become global ones. The amplitude of these changes is determined by mechanisms that manage the NSS dynamics and structure (Gorshkov *et al.*, 2000, 2004; Drabek, 2001; Ehlers and Kraft, 2005; Guista and Kambhampati, 2006). Humankind deviates more and more from this optimality in the way it interacts with the surrounding inert environment. At the same time, human society as an element of the NSS tries to understand the character of large-scale relationships with nature, directing the efforts of many sciences to study cause-and-effect feedbacks from this system. The basic item of global human society is the country; that is considered as an NSS component (Kondratyev and Krapivin, 2005e; Olenyev and Fedotov, 2003).

National security in present-day conditions is estimated on the basis of many criteria, mainly military, economic, ecological, and social in character. Development of an efficient method of objective analysis of the problem of national security requires the use of the latest methods to collect and process data on various aspects of the functioning of the world system. Such methods are provided by GIMS technology (Kondratyev *et al.*, 2002; Krapivin and Marenkin, 1999; Mkrtchian, 1992).

The development and realization of an efficient technology to assess ecological safety on a global scale may be possible through the International Center on Global Geoinformation Monitoring (ICGGM). This may make it possible to understand the mechanisms behind NSS co-evolution. The basic mechanism will be geared towards new technologies of data processing based on the progress of evolutionary informatics and global modeling. Here the point is about realization of an approach developed by some authors (Bukatova and Rogozhnikov, 2002; Krapivin *et al.*, 1998a; Nitu *et al.*, 2000a, b) to modeling processes in conditions of inadequate *a priori* information about their parameters and the presence of principally unavoidable information gaps.

## 1.2 GLOBALIZATION AND TRENDS IN CIVILIZATION DEVELOPMENT

Though the notion of globalization appeared in the works of French and American scientists only in the 1960s, strictly speaking globalization processes began in the early Holocene (10,000–15,000 years ago), when the modern form of *Homo sapiens* evolved, which was immediately followed by territorial expansion and aspiration to go beyond its ecological niche (these processes have been convincingly analyzed by Gorshkov *et al.*, 1998, 2000, 2004). Held *et al.* (2004) have justly noted that so far, "there has been neither convincing theory of globalization nor even a systematic analysis of its major features—which is rather surprising." Held *et al.* (2004) planned to answer the following questions:

(i) Is the present globalization a new phenomenon?
(ii) What is globalization associated with: death, resurrection, or change of the government?
(iii) Does the present globalization establish new limits to policy? How can globalization be introduced into the framework of civilization and democracy?

It has been noted that "to begin with, globalization can be presented as the process of broadening, deepening and acceleration of the global-scale cooperation affecting all aspects of the present social life—from cultural to criminal, from financial to spiritual." Held *et al.* (2004) have selected and analyzed the results of developments carried out by three schools called *hyperglobalists*, *skeptics*, and *transformists*. The data in Table 1.1 illustrate special features of these three conceptual approaches to analysis of globalization processes. Held *et al.* (2004) have emphasized that the main sources of discussions involve five principal questions:

(1) general conception of globalization;
(2) its causes;

(3) periodicity;
(4) effects; and
(5) ways of globalization development.

The basic parameters of the historical forms of globalization include (Shutt, 2001; Shuffelton, 1995; Rotmans and Rothman, 2003; Raven, 2005; Morgenstern and Portuey, 2005):

*Spatial–temporal parameters:*
(1) Vast global structures.
(2) Intensity of global interconnections.
(3) Rate of global fluxes.
(4) Dominating effects of global interconnections.

**Table 1.1.** Three conceptual approaches to globalization.

| | Hyperglobalists | Skeptics | Transformists |
|---|---|---|---|
| *What is new?* | Global century | Trade blocs, weaker management | Historically unprecedented levels of global interconnectivity |
| *Dominating features* | Global capitalism, global government, global civil society | The world is less interdependent than in the 1990s | "Deep" (intensive and extensive) globalization |
| *Power of national governments* | Lowers or collapses | Intensifies | Changes its methods |
| *Driving forces of globalization* | Capitalism and technology | States and markets | Combined forces of modern life |
| *Stratification models* | Destruction of old hierarchies | Increased marginalization of the South | New structure of the world order |
| *Dominating motive* | McDonalds, Madonna, etc. | National interest | Transformation of political community |
| *Conceptual comprehension of globalization* | As a change of human activity | As internationalization and regioning | As a change of interregional relationships and actions at a distance |
| *Historical development* | Global civilization | Regional blocs/ conflict of civilizations | Uncertain: global integration and fragmentation |
| *Final argument* | End of nation states | Internationalization depends on governmental agreement and support | Globalization transforms the government and the world's policy |

*Organizational parameters:*
(1) Globalization infrastructure.
(2) Institutionalization of global structures and the use of force.
(3) Model of global stratification.
(4) Dominating types of global interaction.

Held *et al.* (2004) have emphasized that "Globalization, as we tried to prove, is neither unique nor a linear process. Moreover, it is best presented as an extremely differentiated phenomenon that includes different spheres of activity and cooperation, such as political, military, economic, cultural, migratory, and ecological. Political, military, economic, and cultural areas and spheres of labour, migratory movements and ecology—these are basic zones of governmental actions which will be the subject of our further study."

Finally, we will use a rather lengthy quotation to describe the approach to globalization analysis which forms the basis of the monograph by Held *et al.* (2004):

- "Globalization is best understood when presented as the process or a series of processes, but not as some unique state. It is neither a result of a simplified logic of linear development nor the prototype of the global society or global community. Rather, globalization means an appearance of interregional structures and systems of interaction and exchange. In this respect, inclusion of national and social systems into wider global processes should be distinguished from the global unification, no matter how it was understood."
- "The spatial coverage and density of global and international interconnections form complicated webs and structures of relations between communities, States, international institutions, non-governmental organizations and multi-national corporations, which form the global order. These partially coinciding and interacting structures determine the ways of development of communities, States and social forces, which simultaneously compel them and impart authorities. In this respect, globalization is akin to the process of structuring, since it is the product of both individual actions and joint interactions between numerous executers and organizations scattered over the globe. Globalization is connected with the developing dynamic global structure giving possibilities and imposing limitations. But also, it is a strictly stratified structure, since the globalization process takes place rather non-uniformly: it takes into account both the existing models of inequality and hierarchy and at the same time it creates new rules of inclusion and exclusion, new ideas of winners and losers. Thus globalization can be understood as processes of structuring and stratification."
- "Few spheres of public life have remained unaffected by globalization processes. These processes have been reflected in all social spheres: cultural, economic, political, legislative, military—up to ecological. Most adequate is understanding of globalization as a multi-aspect and differential social phenomenon, which should be understood not as some unique state, but as something connected with models of intensified global interconnections in all key spheres of social activity. Hence, to understand the dynamics and consequences of globalization,

it is necessary to have some knowledge about different models of global relationships in each of these spheres. For instance, the models of global ecological relationships differ radically from those of cultural or military global interaction. Any general conception of globalization processes should proceed from the fact that globalization is presented best not as some unique state but as a differentiated and many-sided process."

- "Since globalization liquidates and breaks political boundaries, it is associated with destruction and re-distribution of the area of social-economic and political space. And since the economic, social, and political activity 'propagates' more and more over the globe, it cannot be organized by only the territorial principle. Each activity can be practiced in some regions but it is not secured territorially. Under globalization conditions, local, national, and even continental space (political, social, and economic) is re-distributed so that it not necessarily coincides with the legally determined territorial boundaries. On the other hand, with the intensifying globalization, it stimulates more and more the socio-economic activity not connected with a certain territory by creating sub-national, regional and extra-national economic zones, mechanisms of management and cultural complexes, which can also intensify the 'localization' and 'nationalization' of societies. Hence, globalization includes the elements of destruction and re-distribution of the territorial boundaries of political and economic power. In this respect, it can be best described as an extra-territorial process."
- "Globalization is connected with the increasing scale of power interference, that is, it increases the spatial extent of government bodies and structures. Really, power is one of the main globalization attributes. In the global system whose elements get more and more interdependent, an accomplishment of power authorities by making decisions, activity or inactivity of executive bodies on one continent can have substantial consequences for people, communities and households on other continents. Power relationships are deeply rooted in the globalization process itself. In fact, an increase of the extent of power relationship means that power centers and the realization of the power itself are more and more removed from subjects and regions suffering from their consequences. In this respect, globalization means the structuring and re-structuring of power relationships affecting at some distance. Models of global stratification play the role of intermediaries that provide an access to power centers despite the fact that in this respect the results of globalization are distributed not uniformly. Political and economic elites of the main world metropolises are much more strongly integrated into global structures and have more possibilities to control them than the farmers in Burundi worried about their survival."

In correspondence with the four-stage periodicity of the process of globalization consisting of pre-modern, early modern, modern, and contemporary stages of development, Tables 1.2 and 1.3 characterize the basic global fluxes and structures as well as the main properties of the historical forms of globalization.

**Table 1.2.** Basic global fluxes and structures.

| | Pre-modern period | Early modern period | Modern period | Contemporary period |
|---|---|---|---|---|
| *Basic global fluxes and structures* | Early imperial systems<br><br>World religions<br><br>Key empires and propagation of agrarian culture<br><br>Plague epidemics and pandemics<br><br>Long-range trade | Political and military expansion<br><br>Europe and the New World: demographic, ecological, epidemiological fluxes<br><br>Development of European global empires<br><br>New transatlantic economic exchanges | European global empires, military, political, cultural fluxes—geopolitics<br><br>Global circulation of Western secular teachings and ideologies<br><br>Transatlantic migrations, Asian diasporas<br><br>Global economy | Formation of global military relationships of the Cold War period and subsequent period, systems of regional and global government and international rights<br><br>Economic globalization before and after the Bretton Woods system: trade, financial markets, multi-national production and investment, transfer of technologies<br><br>Global ecological threats: "drawn components" of the environment, interdependency, trans-boundary pollution<br><br>New models of global migration<br><br>Global propagation of multi-national mass-media corporations, Western mass culture and discourses<br><br>New global communication and transport structures |

| | | | | |
|---|---|---|---|---|
| *Countries, boundaries, and territories* | All boundaries are fuzzy and uncertain<br><br>Migrations and other movements take place, as a rule, not across the boundaries but re-form political space, which continued during the following two epochs | Proto-nations and the appearance in Europe of absolute and constitutional states of the modern epoch<br><br>In the end of the period—first national revolutions (France, U.S.A.)<br><br>Kingdoms, empires and overbroad "sovereignty" still remains a dominating form in Europe, something similar is observed beyond Europe | National states are developing in the West and in Latin America<br><br>European and other national empires and kingdoms dominate everywhere | National states are the prevailing political unification<br><br>Appearing regional units<br><br>Multi-layer government |

**Table 1.3.** Basic properties of the historical forms of globalization (generalized).

| | Pre-modern epoch | Early modern epoch | Modern epoch | Contemporary epoch |
|---|---|---|---|---|
| *Extent* | Most fluxes are only trans-Euro-Asian, some are inter-regional and inter-civilizational | In some respects more global as America and Oceania become connected with Eurasia<br><br>Closer relationships between Europe, the Middle East, and Africa | The growing global scale: European empires intensively include eastern Asia and Africa in wide-area networks. The East is awakening<br><br>Global economy is propagating beyond the Atlantic zone | Most basic spheres are global in character. Some structures and relations, such as problems connected with warming, are of true global character, others such as trade are almost global |
| *Intensity* | Intensive but transient fluxes<br><br>Stable relationships, low intensity decreasing with distance | Intensive but transient fluxes<br><br>Stable political relationships of low intensity: components of transatlantic empires are loosely connected<br><br>The intensity of transatlantic economy and other global economic ties increases | New political and military empires and structures of long duration are being established<br><br>The intensity of economic and cultural influence sharply increases | Very high in economy and ecology<br><br>High in the sphere of culture and in the public sphere<br><br>Migration is less intensive than in earlier epochs, but it intensifies and approaches it |
| *Rate* | Low | Low | Mean | High for transport and communications, but in some cases instant and takes place in real time (TV, financial markets) |
| *Infrastructure* | Written language<br><br>In some regions book-printing | European innovations and intensification of marine technologies and navigation<br><br>Art | Railways<br><br>Transportation of cargo in steam vessels | Aviation<br><br>Telephone, computer, and digital communication network |

| | | | | |
|---|---|---|---|---|
| | Domestication of animals and their use as transport<br><br>Imperial world—some road networks<br><br>Some improvements in marine technologies and cartography | Book-printing propagates to Europe and then to its colonies<br><br>Organization of the European mail service | Telegraph and telephone communication<br><br>Appearance but still small use of combustion engines, radio, TV | Global systems of cable communication<br><br>Internet<br><br>Propagation of radio and TV |
| *Institutionalization* | Very low—only oecumenical (oecumenical church) and tribal system | Very low—only diplomacy, etc. | Mean but growing in some spheres, especially in economy and migration | High—all spheres, many forms |
| *Stratification* | High level of stratification only within civilization. Mother countries control key trade routes<br><br>Between civilizations—Islamic, Chinese, Indian—the main points of intersection of global networks; Europe, Africa, South-Eastern Asia remain on the periphery | Transatlantic structures in which European countries and colonies prevail<br><br>Control by Europeans of the structures on the Indian coastline intensifies being heavily opposed by local and Islamic commercial structures<br><br>Europe can neither penetrate the inner regions of Asia and Africa nor control them<br><br>Centers of mother country and its elite at key trade routes are, as always, closely | Undoubted prevalence of the European–American global institutions and global structures<br><br>British Empire is most powerful of all European empires<br><br>Some non-European countries can more strongly resist than others, some can be locally modernized—Japan, Turkey | Various models of stratification in different spheres of globalization<br><br>Political/military—most of the Cold War epoch with U.S.A.–U.S.S.R. dominating; it creates a very clear hierarchy of countries in these spheres. The post-Cold War period has led to a more distinct multi-polar stratification<br><br>Economic stratification of countries and regions. ECDO countries prevail in the |

*(continued)*

**Table 1.3** (*cont.*)

| | Pre-modern epoch | Early modern epoch | Modern epoch | Contemporary epoch |
|---|---|---|---|---|
| | | connected with global fluxes and structures, but in some spheres of economy, politics, etc., their influence increases | | process of globalization and its control. A balance of forces is biased from the U.S.A. in favor of other countries. Appearance of countries with recently industrialized economy |
| | | | | Cultural stratification is still unfair; U.S. cultural industry and English language prevail in the global propagation of mass culture—though it raises doubts |
| | | | | The North–South ecological stratification is very sharp regarding levels of consumption, but in creation of dangers and threats, and exposure to them their rate of propagation is non-uniform |
| *Methods* | Forced | Forced | Forced | Competition and discussions |
| | Religious/ideological | Imperial | Ideological (more and more secular) | Cooperation |
| | | Religious/ideological | Competition and discussions | Ideological/cultural |

Summarizing, Held *et al.* (2004) noted: "The contemporary globalization cannot be reduced to any one cause-and-effect process; it is only explained by a complicated interweaving of cause-and-effect connections. They include expansive aspirations of political, military, economic, migratory, cultural, and economic systems. Each of them has suffered from consequences of revolution taking place in the sphere of communication and transport in the late XX century, which has favored the process of globalization in all spheres of public activity and extremely broadened—due to an appearance of global infrastructures, providing the transportation of people, goods and symbols—possibilities of global interactions. Besides, this set of factors has raised the intensity and the rate of global interactions changing simultaneously the effect of globalization, especially with respect to its consequences for power distribution."

However, in connection with the contemporary stage of globalization, the combination of globalization trends observed in all main spheres of social interaction is of particular interest. This means that each of these spheres—political, military, economic, migratory, cultural, and ecological—has a special situation of its own, and all of them are in a system of complicated interconnections, which determines the peculiarity of contemporary globalization and stimulates its development. Hence, an explanation of contemporary globalization simply as the product of the expansive logic of capitalism development, global propagation of mass culture, or military expansion is inevitably one-sided and reductive. The fact that the development of capitalism, in its different forms, is subject to the expansionism logic and that this logic is also present in global aspirations of the military system and certain cultural complexes raises no doubts. Though, from the analytical point of view, it is necessary to evaluate the historical models of globalization propagated in each sphere of activity and distinguish between them, which can lead to a preconceived and fragmentary point of view. To conceive the contemporary stage of globalization, the state of globalization factors and the dynamics of their interaction should be studied.

Particularly active in publishing works on globalization problems has recently been the international publishing house Zed Books (London) which has issued a series of books on the problems of global socio-economic development.

In particular, the following books have been published by Gupta (2001), Guyatt (2000), Khor (2001), Madeley (2000), Rivero (2001), and Sibruk (2006). Though, there is still no common definition of the notion of globalization, it is clear that the matter concerns global-scale changes of the state of human–society relationships, dynamics of the scientific–technical progress (especially in the sphere of informatics and communication), socio-economic development and its aspects such as international trade, finance, etc. The contemporary estimates of the present state of civilization are rather contradictory (Andreev, 2003; Boiko, 2005; Borisov, 2005; BP, 2005; Craik, 2005; Kondratyev *et al.*, 2004; NRC, 1999; NSF, 2003; Sachs, 2005; WMO, 2005; WWI, 2003–2006). One of the indicators of this contradiction is, for instance, the statement of 108 Nobel Prize winners that it is not terrorism but the following three problems—poverty, global climate warming, and weapon propagation—that should attract public attention in the context of the global change problem. Obviously these problems do not cover all problems of global change

(Kondratyev, 1998, 2004a; Chambers, 2005; Corning, 2003; Fedotov, 2002; Kresl and Gallais, 2002; Marfenin and Stepanov, 2005).

Characterizing the present socio-economic situation, Khor (2001), Director of the Third World Network (a non-governmental organization on the problems of global politics), has justly noted that "while the adherents of the process of globalization have supported its possibilities and advantages, many politicians of the South, specialists and scientists as well as the community of non-governmental organizations both in the South and in the North, have become disappointed in the development of this process. One of the signals of this disappointment is the failure of the Conference of Ministers in Seattle held in December 1999 under the aegis of the World Trade Organization (WTO).

There are many reasons for disappointment in the process of globalization, and first of all, the absence of real benefit to most of the developing countries from their open economies despite the widely advertised statement about progress in export and profits. The fact consists in economic and social losses in many developing countries caused by a rapid financial and commercial liberalization. Under the influence of globalization, the material inequality has intensified, and the functioning of global economy by the laws of free market results in negative prospects in solving the problems of ecology, social development, and culture"

The book by Khor (2001):

- explains why in most countries globalization has not favored economic growth or reduction of social inequality and poverty;
- criticizes the governments of Western countries for their aspiration for non-democratic prevalence in international politics;
- rejects the propositions of the World Bank (WB), International Monetary Fund (IMF), and World Trade Organization (WTO) for developing countries to follow the principle of "one size suits everybody";
- substantiates the right of developing countries to decide when and how to open their economies; and
- explains that, eventually, the matter boils down to the fundamental problems of a rapid, fair, and diversified socio-economic development of the South, the resolution of which determines people's wellbeing and the good state of ecology.

The main goal for further discussion of the monograph by Khor (2001) is an analysis of economic globalization and some of its key aspects (financial, trade, and investment liberalization) in the context of the socio-economic development of the South, especially from the viewpoint of substantiation of such suggestions and decisions which can weaken the manifestation of negative aspects of globalization in the interaction of national and global economies. Though the development of the globalization process is mainly determined by the progress of present technologies (first of all, information and communication), its major driving force has been political decisions at national and global levels, which have recently led to rapid financial, trade, and investment liberalization. Though, of course, developing countries have been an important part of this process, making political decisions was mainly determined by developed countries as well as by international organizations under their control. Events connected with the developing financial crisis, which started in 1997 in South-East Asia, have brought forth doubts about the acceptability of standard measures recommended by international financial institutions to overcome critical situations. These doubts have been intensified by the failure of the WTO Conference in Seattle.

Economic globalization is not a new process, since for more than five centuries developed countries have more and more extended their industrial and commercial activity far beyond their own borders. However, during the last two to three decades the process of economic globalization has accelerated due to several factors. Mainly this has been favored by technical progress, especially the global-scale liberalization policy. The main feature of economic globalization is the destruction of economic barriers, international expansion of trade, finance, and industry, the increasing power of TNC and international financial institutions. Though economic globalization is developing very non-uniformly and is characterized by enhanced trade and investments in several countries in particular, practically all countries happen to be involved in it. So, for instance, the relative contribution of countries with low incomes to the development of global trade can be insignificant, but their participation in this process can have important socio-economic consequences from the viewpoint of exported goods prices or customs policy.

Of the three aspects of liberalization (finance, trade, and investment), most intensive was the development of financial liberalization, which manifested itself through a deeper relaxation of control of financial fluxes and markets (Khor, 2001). The daily mean global level of financial operations has grown from $15 billion in 1973 to more than $900 billion in 1992, and now it exceeds $1 trillion, with most of these operations being speculative: only <2% involves payment for trade operations. Interconnections between financial markets and systems and gigantic amounts of money flow are the cause of anxiety about the stability of the whole financial system and the possibility of some of its components failing or even the system as a whole. The reality of this anxiety has been illustrated by the east Asian crisis originating in the second half of 1992 and extending to such countries as Russia, Brazil, and others.

The process of trade liberalization has been developing as well, but at not as high a rate as in the sphere of finance. Global export increased from $61 billion in 1950 to $315 billion and $3447 billion in 1970 and 1990, respectively. The share of global export to GDP increased, on average, from 6% in 1950 to 12% and 16% in 1970 and 1990. An increase in the scale of trade was followed by a decrease in tariff barriers, both (on the whole) in developed and developing countries. However, high tariffs still prevail in developed countries in such sectors of the economy as agriculture, textile industries, and in some other sectors, in which the developing countries have made progress. Moreover, the practice of extra-tariff barriers has intensified, which hinders the access of developing countries to the markets of developed countries. All this has taken place against a background of intensive rhetoric about a "free market" (McMichael *et al.*, 2003; Michie, 2003; Morris, 2002; Oldfield, 2005).

The fact is that even now the colonial structure of trade has been preserved, within which the developing countries (which refers to Russia as well) supply raw materials in exchange for products—with the raw materials priced too low—which has catastrophic consequences for the economies of different countries. For instance, in 1989 the financial losses of the countries of sub-Saharan Africa constituted $16 billion, and in the period 1986–1989 they reached $56 billion (about 16–17% of GDP in 1987–1989). Such financial losses (which increased still further during the 1990s) are, apparently, the main mechanism behind the "pumping" of economic resources from the South to the North.

During recent years, the level of funds for direct investments (FDI) has been increasing, though more weakly than international financial fluxes. Most of these investments and their growth constituted the share of developed countries. From the early 1990s, there was also an increase in relative FDI fluxes in developing countries—up to 32% during the period 1991–1995 (compared with 17% in 1981–1990), which coincided with the liberalization of politics with respect to foreign investments in most developing countries, but this increase was confined only to a few developing countries. The share of the least developed countries constituted a very small part of FDI fluxes, and, respectively, this source of financing was insignificant for most developing countries—apparently, this situation will remain for the foreseeable future.

The main feature of the globalization process consisted in the growth of concentration and monopolization of financial resources and power of transnational corporations as well as global financial firms and funds. This process called "trans-nationalization"—manifested through a decrease in the number and enlargement of TNCs—began playing a growing role in the regulation of economic resources and markets. Instead of the earlier existing transnational companies controlling the production of a certain product, gigantic multi-national companies have appeared which participate in the production or trade of a wide range of products, services, and economic sectors. Through an amalgamation of TNCs, their decreasing number controls the growing share of the global market of goods and services. The sales volume of the 200 largest TNCs—constituting $3046 billion in 1982 (24% of global GDP)—reached $5862 billion in 1992 (26.8% of global GDP).

Perhaps the most important and unique feature of the present globalization process is the globalization of national politics and mechanisms of making political decisions. National politics (including its economic, social, cultural, and technological aspects), which has recently been under the jurisdiction of states and people within a given country, has become more and more affected by international agencies and processes or large private corporations. This has led to erosion of national sovereignty and reduced the possibilities of governments and people to make decisions in the sphere of socio-economic and cultural policies. Even in developed countries, large corporations assumed the right to make many important decisions at the expense of infringement of national power or the role of political and social leaders.

Erosion of the ability to make political decisions on a national level has partially resulted from liberalization of market technology development. Transnational companies and private financial institutions control huge resources which exceed the resources of many (or even most) states and thus can now affect the politics of many countries. Moreover, countries have to carry out a policy that agrees with the directives of international institutions. These key institutions are the World Bank (WB), International Monetary Fund (IMF), and World Trade Organization (WTO), as well as a number of powerful international organizations (especially the U.N. and their agencies), and others. As for the U.N., it should be stated that their role and importance have recently decreased. On the other hand, the importance of such institutions as the WB, IMF, WTO has grown. The WB and IMF—created within the framework of the Bretton Woods agreements—have begun to strongly affect the socio-economic development of most developing countries (and the countries with a transient economy) which depend on WB and IMF credits. This especially concerns countries which want to re-structure their debts. The programmes of re-structuring cover not only macro-economic but also social policy, including such spheres as privatization, financial policy, corporative legislation, and management. Coordination of the respective decisions with the problems of the debt of developing countries has become the main instrument to force privatization, de-regulation, and removal of the country's ability to solve the problems of economic and social development. Any disagreement entails serious sanctions in the sphere of foreign commerce.

Such a situation results in the fact that the national economic policy of developing countries is being discussed—not at the level of parliaments and governments—but mainly through the WTO. The northern countries try to increase the influence of the WTO on foreign investments, labor, and ecological standards. This can bring about a change in the Multilateral Agreement on Investments (MAI) made through the Organization for Economic Cooperation and Development (OECD). Though this agreement has not yet been finalized, it is essential that the MAI project foresees a liquidation of barriers for the functioning of foreign companies in almost all sectors of the economy and the accordance to foreign investors of, at least, no fewer rights than to national investors and companies.

Though the most substantial role in the development of globalization processes is played by such international institutions as the WB, IMF, WTO, and OECD, the role of the U.N. and their specialized organizations remains significant and has been manifested, in particular, by holding such large-scale international conferences as the World Conferences on the Environment and Development (UNCED, Rio de Janeiro, 3–14 June 1992), Population and Development (ICPD, 5–13 September 1994, Cairo, Egypt), Social Development (WSSD, 5–12 March 1995, Copenhagen, Denmark), Women's Problems (WCW4, 4–15 September 1995, Beijing, China), Food Problems (WFS, 13–17 November 1996, Rome, Italy), Disaster Reduction (WCRD, 18–22 January 2005, Kobe, Hyogo, Japan), Urban Problems (WUF3, 19–23 June 2006, Vancouver, Canada), and others. It is important to note that the U.N. approach to solving the problems of socio-economic development differs from that practiced by the WTO and Bretton Woods institutions (Furth, 1965). The latter much more strongly rely on the resolving role of market factors, minimization of State licensing, and rapid liberalization. In most specialist U.N. agencies, other approaches prevail, as a result of which the market itself cannot resolve existing problems. The role of governmental and public institutions should not be limited. However, the problem is that, as has been mentioned above, the role of Bretton Woods institutions like the WTO has recently increased, whereas the role of the U.N. has considerably decreased—which highlights the prevalence of the Bretton Woods–WTP approaches to resolving globalization problems (NSI, 2005; WB, 2006; Panarin, 2004; Pittock, 2005; Robertson and Kellow, 2001).

As for development of the globalization process, it is characterized, first of all, by a strong heterogeneity with a non-uniform distribution of advantages and losses among different countries. An imbalance has occurred between the few profiting countries and groups of people, and many countries and groups of people who suffer losses and even become marginalized. The processes of globalization, polarization, concentration of wealth, and marginalization have turned out to be closely interrelated. In the course of these processes, investment resources and progress in modern technology have been concentrated in a small number of countries in North America, Western Europe, Japan, and in a few east Asian countries, whereas most developing countries are either excluded from this process or participate little, which is to the prejudice of their interests: so, for instance, liberalization of import damages national producers, and financial liberalization leads to instability.

Thus, the liberalization process affects various countries and populations in different ways and, on the whole, it is characterized by the increasing role of a small number of the leading developed countries. The non-uniformity and heterogeneity of the globalization process result in a rapidly increasing social inequality and the gap between developed and developing countries.

According to the data of the 1992 report of the U.N. programme on the problems of socio-economic development (UNDP), 82.7% of total global income falls to the 20% of the global population living in developed countries, whereas for 20% of the poorest population this share constitutes 1.4%. In 1960 the average income of people in developed countries was 60 times higher than in poor developing countries. Since 1950 this ratio has doubled. It follows from the similar 1996 report that during the preceding three decades there are only 15 countries where considerable economic growth has been observed, whereas in 89 countries the economic situation has worsened during the last decade. In 70 developing countries the present level of income is lower than in the 1960s and 1970s. Thus, only a few countries have obtained economic advantages from globalization, and these advantages have been reached at the expense of many other countries. Despite 15 countries (mainly in Asia) reaching a higher rate of economic growth since 1980 than during industrialization in the West, most developing countries were characterized by a longer and deeper decrease than that taking place in the Great Depression in the 1930s. In many countries, people became poorer than they were 30 years ago.

The increasing inequality of countries and different populations is closely connected with the development of the processes of globalization and rapid liberalization. One of the unsettling consequences of these processes is an increase in the concentration of income in the hands of a few, not followed by a growth of investments and economy, resulting in increasing unemployment. Another important negative fact is that the factors causing an increase in inequality in the globalizing world hinder investments and slow down economic development. This situation can be illustrated by the fact that a rapid increase in financial liberalization has resulted in a gap between financial resources and, on the other hand, international trade and investment activity. A high level of loan interest determined by a limited monetary policy has led to an increase in expenditure on investments and prompted businessmen to concentrate their activity on buying–selling ready commodities instead of participation in the process of long-term investments in order to manufacture new products.

The main problem for most developing countries in the context of globalization processes consists in their inefficiency to use globalization advantages for a number of reasons. For instance, in the late 20th century, 11 developing countries turned out to be involved in the globalization process. In 1992, their share constituted 66% of the export of developing countries (compared with 30% in 1970–1980), 66% of annual foreign investment (1981–1991), and the greater part of portfolio investment. Since then, some of these 11 countries have suffered serious financial crises, increases in the size of external debt, and decreases in economic development, which means a lack of of progress in integration of the South into the world economy (this situation can be vividly exemplified by the events in Argentina that took place in early January 2002).

One of the reasons underlying the inefficiency of developing countries is, first of all, their economic and technological weakness (prevalence of raw materials export, undeveloped information and communication technologies, etc.). The weak position of the South in the negotiation processes—in view of its financial dependence on the North, decrease in the effect of U.N. institutions ready to support developing countries, insufficient organization of both individual countries and the South as a whole—is an important feature.

The widely manifested inability to use the processes of globalization and liberalization in the interests of national socio-economic development is a serious challenge for developing countries. There is no doubt that a selective approach to liberalization (instead of its rapid realization) is necessary, an approach that provides a reliable balance between opening up the national market and the needed protective measures. This approach should be based on consideration of the factors determining the negative consequences of globalization and liberalization, such as losses of political independence, financial openness, the risk of instability, and marginalization trends. On the whole, to survive under the conditions of the global economy, a number of particularly complicated problems should be resolved. Again, this might be favored by the growing authority and resources of the U.N. in contrast to the dominating and growing impact of Bretton Woods institutions. An important result of the way in which events developed in the second half of the 20th century indicates that for new participants of the globalization process to develop successfully they need to be supported by their governments. This process should develop in the spirit of sustainable development based on consideration of the interacting economic, social, and ecological factors. It is especially important that the 21st century be the period of changing economic policy in order to make the consumption/production ratio ecologically acceptable. Hence, it is necessary to work out models of development not only for southern countries but also (and to a greater degree) for northern countries in order to establish a new international economic order which would be capable of achieving the necessary social and ecological goals.

It is clear that fair globalization is the best way for civilization development. It is also evident that the present direction of civilization development has a long way to go to reach this level. Achievement of a fair globalization level needs fairness within countries. The ideologies of socialism and communism are close to the idea of fair globalization, but the world has turned back to capitalism and imperialism principles. The negative effects of this have shown themselves during the disintegration of the former Soviet Union when social inequality, unfair privatization, and violence cast many countries of the former Soviet Union back in their development and strengthened the expansion of the American interpretation of globalization. In this context, the U.N. Secretary-General, Kofi Annan, in his report to the U.N. General Assembly on 23 September 2003 succinctly warned the world body that it had "come to a fork in the road" (WCSDG, 2004).

## 1.3 SUSTAINABLE DEVELOPMENT

Though the problem of sustainable development has been widely discussed in the scientific literature (Agueman *et al.*, 2003; Altinbilek *et al.*, 2005; Andreev, 2003; Luken and Hesp, 2004; Springett, 2005), the respective information flow prompts us to return to the analysis of respective conceptual ideas. However, first it is necessary to remind ourselves of the ambiguity of the applied terminology. Let us note that translation of the words "sustainable development" (in Russian) is inaccurate. "Maintained stable development" would be more correct, but for the sake of convenience and taking into account that the term "sustainable development" has become commonplace, we shall use this term.

The concept of development sustainability in its present understanding first appeared in 1974 in the documents of the International Church Council (Grimstead, 2003) as a response to the appearance in developing countries of objections concerning fears about the environmental condition, with millions of people suffering from poverty and hunger. In 1980, the concept of sustainable development was put forward by the International Unit for Nature and National Resources Protection and became widely known in 1987 after publication of the report "Our Common Future" (Brundtland, 1987) prepared by the U.N. Commission on the Environment and Development headed by Gro Harlem Brundtland, who is now the head of the World Health Organization. According to the "Brundtland Report" (BR), sustainable development is defined as any development which meets the needs of the present generation without damaging future generations. In this connection, equality both within generations and between them is especially important.

A much more complete and concrete definition of the notions of sustainability (in the context of NSS dynamics) has been suggested by Kobayashi (2005) (several reviews of these problems can be found in Cropp *et al.*, 2005, and WB, 2003). According to this definition, SD means "actions accomplished by humans that show respect to a variety of living beings, which ensures the preservation of life, nature and culture for future generations within the ability of the environment, as well as establishing relationships in order to build a better society in the future and seeking a higher level of happy life for maximum population independent of time and space.'' Chapin *et al.* (2004) consider four categories of sustainability that refer to nature (A), economy (B), society (C), and wellbeing (D).

Based on analysis of data for Japan, the choice of 20 indicators of SD has been substantiated for these four categories of sustainability with reference to four areas (Kobayashi, 2005; Wijen *et al.*, 2005):

(A) *Nature*: (1) biodiversity, (2) global warming, (3) cycle of resources, (4) water, soil, and air, (5) ecological formation.
(B) *Economy*: (1) energy, (2) resources productivity, (3) food, (4) financial status, (5) international cooperation.
(C) *Society*: (1) security, (2) mobility, (3) sex characteristics, (4) traditions and culture, (5) financial fluxes.
(D) *Wellbeing*: (1) satisfaction with life, (2) science and education, (3) participation in social life, (4) health, (5) differences in life standards.

Analysis of estimates of the SD indicators enumerated above has shown that for Japan a conditional integral SD indicator in 2005 was 33.5 (with respect to 100 for a supposed stable society in 2050), whereas in 1990 this indicator constituted 41.3—that is, sustainability has decreased by 19%.

The slogan ''sustainable development'' has been picked up by governments and international organizations, becoming now widely used. At the U.N. Conference on the Environment and Development held in Rio de Janeiro in 1992 (Garaguso and Marchisio, 1993; Brown, 2005; Chistobayev and Solodovnikov, 2005; Daly, 2002; Dolak Ostrom, 2003), this slogan was supported by the leaders and governments of many countries, though such vigorous official support raised doubts among specialists in environmental problems about the SD concept as indefinably formulated and differently interpreted.

Dresner (2002) analyzed the historical development of the SD conception, the content of present discussions on these problems as well as obstacles in reaching SD, and prospects for bridging these obstacles.

As far back as 1627, F. Beckon (Subbotin, 1977) expressed an idea that science should be able to dominate nature and reach a better understanding of how nature develops: ''I now understand how to put nature with all its children at your disposal in order it served you and was your slave.'' However, soon after this, R. Decarte (1994) expressed an idea that understanding the processes of nature's laws is an endless task. Newton's laws of motion and the following rapid development of science in the 17th and 18th centuries agreed with these ideas.

The political theory (Belov, 1999) based on the deductive analysis of the rights to life, freedom, and property is important. This theory prompted T. Jefferson (Shuffelton, 1990, 1995) to develop the American Declaration of Independence inspired by philosophers such as Voltaire and Rousseau (Kuznetsov, 1989;

Rousseau, 1969; Voltaire, 1996). In Great Britain the idea of rationality has been widely acknowledged in the sphere of the economy. The works of the Scottish economist A. Smith (Weinstein, 2001) have been an intellectual support to capitalism and the market. The creation of the steam engine in the late 18th century marked the beginning of a new period which Friedrich Engels (1880) called the "industrial revolution" (Elster, 1986). In only two generations, human life dramatically changed. Technical progress in social organization made it possible for humans to reach a certain degree of control over nature on a massive scale.

As far back as 1798, when the industrial revolution had just begun, Malthus published his "Essay on principle of population" (Malthus, 1798, 1976) in which he expressed (now widely known) ideas on the growth of population size in geometrical progression, considerably surpassing the rate of food production. Objecting to the ideas of his opponents (Karl Marx and Friedrich Engels were the most radical opponents), Malthus (1798) wrote: "I've read some ideas about the perfection of humans and society with great pleasure. The respective appealing picture warms and delights me. I heartily wish to reach this perfection, but I see great and, to my mind, insoluble difficulties on this way." In this connection, it should be noted that the problem of natural resources turned out to be not as urgent as it seemed earlier (Lomborg, 2001, 2004), but the rapid increase in population (concentrated in poor developing countries) is in fact a serious problem for the modern world (Kondratyev, 1998; Kondratyev and Krapivin, 2005c; Kondratyev *et al.*, 2002, 2004). Even socialist Charles Fourier (1876), many years ago, expressed his concern about the physical limits of natural resources, and the writer Mary Shelley (Paley, 1993; Shelley, 1992) in her novel *The Last Man* (Shelley, 1826, 1994) was the first to warn about the threatening global ecological disaster. John Stuart Mill (1909) in his famous book *Principles of Political Ecology* touched on the same theme.

At present this concern is better formulated in terms of the broken closedness of global biogeochemical cycles and broken mechanisms of biotic regulation of the environment. In this connection, one can recollect the words of Friedrich Engels (1940): "Thus at each step we are reminded that we do not control nature like conquerors, being outside nature. In fact, we, with our body, blood and mind, belong to nature and exist within it. Our regulation of nature consists only in that we have an advantage over all other living beings, with our ability to know and correctly apply the laws of nature."

This concern was first formulated at the international level at the U.N. Conference on the Human Environment held in Stockholm in 1972 (Ward and Dubos, 1972) when the slogans "Only One Earth" and "Space Vehicle Earth" were used. These metaphors were aimed at emphasizing two principal circumstances:

(1) limits to human activity; and
(2) needed human control of environmental dynamics.

It was time for the re-appearance of Malthusianism (Malthus, 1798) and comprehension of global ecological limits. A series of publications dedicated to these problems was stimulated by the Roman Club (this mainly refers to the book by Meadows *et al.*, 1972; Moiseyev, 2003; Nikanorov and Khoruzhaya, 2000; Pollack, 2003). An important stage was the preparation of the report initiated by U.S. President Carter (1977–1981) aimed at analysis of the global ecological situation and forecast of its development up to the year 2000. In particular, it was mentioned in the report: "If the present trends continue, the world in 2000 will be more populated, more contaminated, and more subjected to destruction. Serious stresses are coming connected with the problems of population resources and the environment. Despite an increase in material production, people in the world will become in many respects poorer than they are now."

### 1.3.1 Formation of the sustainable development conception

A fundamental role in understanding the fact that environmental problems constitute a critically important aspect of socio-economic development was played by the U.N. Conference on the Environmental Problems held in 1972 at which a declaration on environmental problems was accepted, and then UNEP—the U.N. organization on environmental problems—was created. In 1974, the conception of sustainable society was first put forward at the Ecumenical Conference on the problems of science and technologies concerning the socio-economic development convened by the World Church Council (WCC, 2005).

According to this conception (WCSDG, 2004): "First, the social stability cannot be reached without a uniform distribution of deficit or without possibility of combined participation in resolving the social problems. Second, the global society cannot be stable if food needs at any time are not much below the global level of their provision, and the level of pollutants emissions is below the ability of their accumulation by ecosystems. Third, a new social organization will be stable only under conditions that the scale of the use of non-retrievable natural resources does not exceed the growth of resources due to technological investment. Finally, a stable functioning of society needs a level of societal activity which can never be affected negatively by the unceasing strong and frequent natural global climate changes."

An important feature of the conception of sustainable society consists in that it was based, first of all, on the principle of the uniform distribution of resources, which then became the cornerstone of the "Brundtland Report" (BR) as well as on the idea of "democratic participation" whose importance was emphasized at the Second U.N. Conference on the Environment and Development in Rio de Janeiro (Kondratyev, 1998, 1999a; Kondratyev and Krapivin, 2003; Kondratyev *et al.*, 2003a, 2004).

The term "sustainable development" first appeared in the document "The World Conservation Strategy" published in 1980 by the International Union for the Conservation of Nature and Natural Resources (Workman, 2005), and the notion itself has been defined as an "integration of preservation (nature protection) and development to provide a planetary change which can ensure safe survival and welfare of all people." "Development" is defined as "a modification of the biosphere with the use of human, financial, living, and non-living resources to satisfy the humans' needs and improve life standard." Development should be combined with preservation (nature and resources protection), defined as human control of the biosphere to ensure maximum usefulness for the present generation with maintained biospheric potential that would meet the needs and aspirations of future generations.

In 1982, the U.N. General Assembly organized the World Commission on Environment and Development (WCED) (Ewing, 2005). It was headed by Gro Harlem Brundtland (later this commission was called the Brundtland Commission, or BC). The key thesis of the BC report "Our common future" published in 1987 was as follows (Ebel *et al.*, 2004):

"Many contemporary trends of development lead to the growing number of the poor and are subjected to various negative effects, and at the same time, the environment is degrading. How can this development serve the world during the next century in the same environment but with a doubled population size? Understanding this situation has broadened our views of the future. We should perceive a necessity of such a new way of development which would provide a stable progress in humans' life not in several places during several years but over the whole planet till the far-away future." This "new way" was called "sustainable development (SD) capable to meet the present needs without any damage to future generations." It should be emphasized that this simple and vague definition has been further interpreted in various and contradictory ways, though the BC report contains a detailed comment on the SD conception. In particular, the following opinions attract attention:

"The SD conception states the presence of limits—not absolute limits but limitations imposed by the present level of technologies and social organization development on the environmental resources and the ability of the biosphere to 'assimilate' the consequences of humans' activity. However, new technologies and social organization can be controlled and improved to establish the ways of development of a new era of economic growth. The commission believes that the widely spread poverty is no more inevitable. Poverty is not only evil in itself. It is important that SD requires satisfaction of the basic needs of all people and giving everybody a possibility to satisfy their aspirations for a better life. The poor world will be always subjected to ecological and other disasters." And then:

"In the past, we were concerned about the impacts of the economic growth on the environment. Now we should show our concern with respect to the effects of ecological stresses—degradation of soils, water resources, atmosphere and forests—on the prospects of economic development. During the recent past, we had to face an intensive economic interdependency of the countries. At present, we are to get accustomed to their accelerating ecological interdependency. Ecology and economy have become more and more interrelated—on local, regional, national and global scales—under the influence of numerous cause-and-effect feedbacks."

The BC report was an important stage of preparation of UNCED held in Rio de Janeiro in 1992. The most important documents adopted by the conference included: the Framework Convention on the problem of climate change; Convention on biodiversity; Statement on forest problems; Rio Declaration; Agenda for the 21st century. As Dresner (2002) has justly noted, the major value of UNCED consists in the perception of global environmental problems as one of the key aspects of international politics. One of the important indicators of the changed situation was the participation in UNCED of more than 100 leaders of various countries, whereas at the Stockholm Conference (5–16 June 1972) only the prime ministers of Sweden and India took part.

Since the respective problems have been widely discussed in the scientific literature (Kondratyev and Krapivin, 2003; Kondratyev and Losev, 2002), we will not dwell on them here or on further events connected with the 19th Special Session of the U.N. General Assembly (23–27 June 1997, New York) and with the World Summit on Sustainable Development held in Johannesburg in 2002 (Kondratyev, 2001, Kondratyev *et al.*, 2004; Kondratyev and Krapivin, 2005a). But, analysis of the situation concerning sustainable development has recently become more urgent.

Discussions at the U.N. Special Assembly have demonstrated the gigantic gap between declarations and reality. The creation of the World Trade Organization (WTO) has caused a serious concern for ecologists who fear the trend of decreasing ecological standards in the short-term interests of increasing the competitive ability of goods. Also important is the fact that—by favoring free trade expansion—the WTO involves economically weak countries in the system of international competition, which they cannot endure, and runs the risk of giving them the status of economically under-developed countries without the necessary resources—in particular, to resolve the problems of ecological security.

A convincing illustration of the inadequacy of numerous declarations adopted at world forums is the history of the Kyoto Protocol (KP) and its coming into effect on 16 February 2005. Since these problems have been discussed earlier (Kondratyev, 2004a–c), we shall only note the paradoxical absurdity of the present situation: it has been widely acknowledged that even a complete realization of KP recommendations on reduction of GHG emissions to the atmosphere will not affect the global climate, but—despite great expenditure on emissions reduction—the real situation is that on the whole they go on growing. Note, by the way, that global change has been recently characterized mainly by negative trends: deforestation and biodiversity decrease has continued, problems of freshwater resources, and impact on the coastal and marine environment have all become more acute.

One of the possible reasons for the gap between declarations and reality is connected with the vague definition of the notion of sustainable development, which, on the one hand, concentrates attention on stability problems, while, on the other hand, emphasizes the importance of development—in this connection, it is apparent that the notions of "stability" and "sustainable development" cannot be considered identical and therefore their interchangeable use in *Agenda for the 21st Century* (Kidder, 1997) is, of course, incorrect. Provision of a balance between the two aspects of this notion is a difficult problem, the more so because manifestations of "extremism" exist on both sides. The situation becomes even more complicated by the ambiguous notion of "development" (of course, this concerns socio-economic development and hence the human dimension should occupy the key position).

Since the concentration of effort on provision of economic growth has led to an under-estimation of other aspects of social progress, the concept of meeting the "base needs", widely used in the mid-1970s, should foresee both material (food, home, health, education, etc.) and non-material needs (basic human rights, participation in public activity, etc.). In the early 1980s, the problem of debt, however, seriously intensified, which forced many governments of developing countries to reduce the level of expenditure on reaching goals concerning "human dimensions". This change has also been favored by the requirement to cut social expenditure and structural change as a condition for further loans, brought forward by the WB and IMF. As for structural change aimed at accelerated movement towards a free market, this was connected with such measures as a weaker role of government in the economy, liquidation of subsidizing some spheres of the economy, liberalization of prices, expansion of privatization, and an unlimited openness of national economies to international trade and finance. In this connection, the announcement in the Brundtland Report—that meeting base needs is of first priority—is important. However, the report emphasized the importance of ensuring economic growth, which should be favored by structural changes.

In this context, of key importance was an alternative conception of human development put forward by the Indian economist Amartya Sen, the 1998 Nobel Prize winner (Amartya Sen and Gates, 2006; Dresner, 2002). According to this conception, the standard of living should not only ensure an average level of income, but also satisfy the basic vital values of society. The Human Development Index (HDI)—worked out within the framework of the U.N. Development Programme (UNDP)—can serve as a criterion of the level of human development and involves such indicators as length of life (life time), education, and material wellbeing. With HDI taken into account, one could rank the countries, which gave rather interesting results.

For instance, it turns out that poor countries—like Costa Rica, Cuba, and Sri Lanka—can compete with developed Western countries with their high levels of life time and literacy. Despite a low level of income in Kerala (India) determined by per capita GDP—it is the poorest state in India—life time averages 73 years compared with 61 years in the rest of India and with 76 years in Great Britain and the U.S.A. The literacy of grown-ups in Kerala is 91% compared with 65% in the rest of India and 99% in Western countries. The birth rate averages 1.7 children per family (as in Great Britain) but for the rest of India the respective index averages 3.1. The achievements by Kerala despite limited financial resources are a result of a considerable part of them being spent on the health service and education. An opposite example is Brazil, where economic growth has been reached at the expense of a low level of human development. Some of the west Asian economic "tigers" have demonstrated societal development by giving priority to human development and slow economic growth and intensifying the progress of both aspects over the coming decades.

Though, according to WB estimates, the progress of east Asian countries is determined by successive application of the principles of the market economy, in fact, these countries are characterized by the intensive intrusion of their governments in economic development, the use of subsidies (especially in agriculture), and taking protective measures in trade. Of importance was the investment in education and the health service. The economic development of east Asian countries has been reached, however, at the expense of serious ecological losses (deforestation, environmental pollution, biodiversity decrease, etc.). In this connection, there is no doubt that the development of east Asian countries (as happened in the West, on the whole) has been ecologically unstable.

A certain role in the evolution of the idea of SD was played by the notion of "natural capital" defined (in contrast to the notions of financial or "human" capital) as an expression of the value of nature to humans. This approach is similar to the conception of biotic regulation of the environment proposed later (but is more constructive).

Having analyzed various ideas of SD, Dresner (2002) came to the conclusion that in this context special attention deserves to be given to three principles of SD put forward by H. Daly (2002), which determine the necessity:

(1) To limit the scale of population size growth, taking into account the "carrying capacity" of the Earth.
(2) To ensure that technological progress guarantees the growth of efficiency but not production levels.
(3) To balance the relationship between the use of renewable natural resources and their supply, the level of emissions not to exceed the assimilating ability of the environment.

The contradictory evolution of developments connected with the SD notion has stimulated further discussions. In particular, a critical analysis of this evolution has been undertaken in a series of papers published in the journal *Sustainable Development* (Bagheri and Hjorth, 2005; Koroneos and Xydis, 2005; Mont and Dalhammar, 2005). So, for instance, Redclifte (2005) carried out an overview of the conceptual history of the notion of sustainable development from the time of the BC report till now, emphasizing the necessity of broadening this notion with "human dimensions" (social aspects, ecological fairness) taken into account. Such aspects are sometimes ignored in the market (liberal) conception of development. The views expressed by Redclifte (2005) closely correspond to the ideas of the report entitled *Fair Globalization* (ILO, 2004).

According to Luke (2005), the notion of sustainable development has become mainly the subject of rhetorical discussions as an ideological "construction" used in the context of the problems of the functioning of present-day global society. The term is used more and more as a "label" denoting the regimes of existence which are neither stable nor reflect socio-economic development. Discussing the sustainable development of economics in the context of ecodynamics of natural systems, Hudson (2005) emphasized the priority of two aspects of these problems: (1) natural resource consumption; (2) emission of pollutants and waste to the environment.

Orsato and Clegg (2005) have analyzed the possibilities of radical reforms (with the data of ecological sociology and organization theory taken into account), which might be able to ensure sustainable industrial development.

Two basic related problems are particularly important in the context of sustainable development: globalization and how the present consumption of society develops (Kondratyev, 1998; Kondratyev *et al.*, 2003a, d, 2006b; Kondratyev and Krapivin, 2003, 2005a–c).

### 1.3.2 Sustainable development and public health

Different workshops and other international initiatives that have taken place in recent years have tried to answer the main question: Do we know enough about NSS dynamics or is more research needed, and how can we achieve sustainability? The sustainability topics that were discussed during recent years did not consider the problem of human health in the context of globalization and sustainable development. Environmental changes—both at global and local levels—have an increasing effect on health, particularly that of poor and vulnerable populations. A healthy environment is one in which people have access to safe food and water, have adequate sanitation, and are protected from the risks associated with disasters and environmental pollution and degradation. Environmental health includes different aspects that can improve human health to attain sustainable development of the NSS. Kogan and Kruglova (2005) established that globalization processes are connected with the global biofield.

It is known that one condition for people to exist safely is having the possibility to adapt to the changing environment. In the 21st century people must adapt to numerous changing factors of social, political, economic, informational, and religious natures. In globalization conditions, when people interact in human society they can develop their behaviour depending on these aspects. Kogan (2006) proposed a model of global biofields. This model parametrizes the people biofield as an element of human society. A biofield is defined as a system of fields having different natures and arising during the living process. The term "biofield" is debatable. Kogan (2004) considered it as some display form of vital activity for people and human society. A holism of this term gives the possibility of combining the physical and spiritual aspects of globalization and sustainable development. Anderson *et al.* (2006) explained that global public health problems can be solved through the emergency medicine approach as a global discipline. The core concepts and strategies of emergency medicine care require focused medical decision-making and action with the goal of preventing needless death or disability from time-sensitive disease processes. It assumes the conditions that must be treated within a certain time period to prevent or minimize mortality or morbidity. It says that an interaction of medicine with the environment is therefore essential for assessment of the situation with a view to undertake preventive and curative measures.

One of the characteristics that show the interactive harmony between the sustainable environment and public health is the Environmental Public Health Indicators (EPHIs). EPHIs were first developed by CSTE in 2000 (CDCP, 2006; Goldman and Coussens, 2004; Johnson, 2005; Levesque and King, 2003; Tolba, 2001). These indicators were identified in part to provide a means of placing non-infectious diseases and conditions under surveillance towards building a comprehensive

National Public Health Surveillance System. The purpose and goals of EPHIs include:

- to describe the elements of environmental sources, hazards, exposures, health effects, and intervention and prevention activities that may stand alone or be combined to describe their interaction;
- to assess the positive and negative environmental determinants of health including measures of the man-made environment;
- to serve as communication tools for making environmental health information available to stakeholders; and
- to identify areas for intervention and prevention and evaluate the outcomes of specific policies or programs aimed at improving environmental public health.

The measures and sources for EPHIs include numerous NSS characteristics and different databases that can be used as input information for the Global Model of the NSS (GMNSS) (Kondratyev *et al.*, 2002). Hazardous or toxic substances in different environments can be taken into account under simulation experiments by means of the GMNSS to assess the global pattern of public health distribution. Knowledge of the health effect dependence on the environment is a basic element for the establishment of NSS stability.

### 1.3.3 Sustainable development and spiritual aspects of life

Civilization development depends on the philosophy of life, spiritual traditions, religious convictions, and other aspects of population culture. The spiritual dimension of sustainable development was discussed at the World Summit on Sustainable Development (30 April–2 May 2001, New York) where a focus on the spiritual dimension of human reality was made. It was concluded that development policies and programs can truly reflect the experiences, conditions, and aspirations of the planet's inhabitants and elicit their heartfelt support and active participation. Spiritual principles are to be taken into account when governments form their strategies within the framework of global NSS dynamics. In seeking to bring spiritual principles into their deliberations, it is necessary to focus on religious and spiritual values, particularly as they relate to and impact on development processes. The importance of spiritual values for development was noted by the Resolution of the 55th Session of the U.N. General Assembly (5 February 2001, New York).

The development of environmental ethics and philosophy by many authors during recent years has helped to understand the importance of knowledge about the larger relationship with nature (Heynemann, 2005). Without discussion on these subjects it is impossible to formulate cardinal principles for the constructive consideration of spiritual values in the GMNSS. This is very important for all countries throughout the world, and for Russia, in particular, where a new political philosophy has only just been formed and environmental ethics are not observed. In this context, both Russia and other countries need to develop a long-term environmental strategy that could become the basis of practical policy-making. This will help governments to identify their new place and role in world dynamics.

Religion has an important role in sustainable development (Bates, 2005; Blancarte, 2006; Jennings, 2006). Many scientific centers study this role. For instance, the Earth Institute's Center for the Study of Science and Religion (CSSR) was founded in 1999 as a forum for the examination of issues that lie at the boundary of two complementary ways of comprehending the world and our place in it. By examining the boundaries and intersections of science and religion, the CSSR is to stimulate dialogue and promote understanding. Another center, the Chicago Center for Religion and Science (CCRS), is dedicated to relating religious traditions and scientific knowledge in order to gain insight into the origins, nature, and future of humans and their environment, and to realize the common goal of a world in which love, justice, and ecologically responsible styles of living prevail. The purpose is to provide a place of research and discussion between scientists, theologians, and other scholars on the most basic issues pertaining to:

- how we understand the world in which we live and our place in that world;
- how traditional concerns and beliefs of religion can be related to scientific understandings; and
- how the joint reflection of scientists, theologians, and other scholars can contribute to the welfare of the human community.

Conflicts and the occasional agreement between science and religion are key mechanisms for the development of a constructive approach to global environmental science. The relationship between science and religion is mainly determined by disagreements in two main areas:

- There have been hundreds of disputes since the end of the 16th century in which scientists and theologians have taught opposing beliefs. At any given time, in recent centuries, there has been at least one active, major battle. Dozens are going on at the present time.
- Science evaluating religion involves the use of the scientific method to evaluate the validity of a religious belief.

A place for religion in science is defined by Pollack (2006): "Science and Religion both deal with explanations and understanding mechanisms, but if one can say 'so what?' to either of the first two pairs, why not say it about this third? Are the shared attributes of science and religion rich enough to outweigh their manifold intellectual, emotional and intentional differences? I say yes." And further: "Globally, all that makes us human in a biological sense is that despite these differences the six billion different human genomes are all in principle capable of coming together with each other through sperm and egg to make another generation of people. The biology of us makes us truly all equal." Chernavsky *et al.* (2005) expanded the biological sense of the present person and gave a new view on human intellect and their constructive activity. These characteristics will play a principal role in the near future when the spirit of contradiction between nature and society reaches a critical level and when religion cannot help to overcome this level. In this sense, the perspectives of global model development again return to the Club of Rome model (Forrester, 1971; Meadows *et al.*, 1972).

Having considered approaches to an assessment of NSS dynamics—from Forrester and Meadows to recent publications (Krapivin *et al.*, 1982; Krapivin and Kondratyev, 2002; Kondratyev *et al.*, 2003c,d)—one can draw the conclusion that appreciable progress in the search for ways to reach global sustainable development can be made with a systematic approach to multi-functional monitoring of the NSS. Thanks to the Club of Rome authors, more than 30 years ago the contradiction was first emphasized between the growth of population size and limited natural resources, and for the first time since the works by Vernadsky (1944, 2002) an attempt has been made to use numerical modeling to study NSS evolution. Of course, the CR model oversimplifies the real internal connections in the NSS, describing interactions between its elements by averaged indirect relationships, without direct account of economic, ecological, social, and political laws. The possibility of such an account appeared later in connection with studies in the field of simulation and evolutionary modeling and the theory of optimization of interaction for complex systems resulting in creation of methods and algorithms for prognostic estimation of dynamic processes in conditions of *a priori* uncertainty. However, the problem of creating a global model adequate to the real world still cannot be solved even in present conditions. On the one hand, a complete consideration of all the NSS parameters leads to insurmountable multi-variance and information uncertainty with irremovable problems. Moreover, in such spheres as the physics of the ocean, geophysics, ecology, medicine, sociology, etc., an adequate parametrization of real processes will always be problematic because it is impossible to have a complete database. Nevertheless, a search of new, efficient ways of synthesizing the global system of NSS control based on adaptive principles of the use of the global model and renewed databases seems to be a prospect and raises hopes of getting reliable forecasts on NSS dynamics. Preliminary calculations using the GMNSS have shown that the role of biotic regulation in the NSS has been under-estimated, and forecasts, for instance, of the levels of the greenhouse effect have been overestimated (Streets *et al.*, 2001). Therefore, in this book an attempt has been made to synthesize the GMNSS taking into account the earlier experience and accumulated databases as well as knowledge about the environment and human society.

Further improvement of the GMNSS will be connected with the balanced development of studies both in the sphere of parametrizations in the NSS and in modernization of Earth observation systems, covering the whole thematic space of the NSS (Kramer, 1995; Chukhlantsev, 2006; Crossland *et al.*, 2005; Foster and Kreitzman, 2004; Gibson *et al.*, 2005; Golubeva, 2002):

- Sun–land interactions (physical mechanisms of the transport of mass, momentum, and energy in the geosphere).
- Atmospheric dynamics (atmospheric chemistry, atmospheric physics, meteorology, hydrology, etc.).

- Dynamics of the World Ocean and coastal zones (winds, circulation, sea surface roughness, colour, photosynthesis, trophic pyramids, pollution, fishery).
- Lithosphere (geodynamics, fossil fuel and other natural resources, topography, soil moisture, glaciers).
- Biosphere (biomass, soil–plant formations, snow cover, agriculture, interactions at interfaces, river run-off, sediments, erosion, biodiversity, biocomplexity).
- Climatic system (climate parameters, climate-forming processes, radiation balance, global energy balance, greenhouse effect, long-range climate forcings, delay of climate effects).
- Socio-political system (demography, geopolicy, culture, education, population migration, military doctrines, religion, etc.).

## 1.4 GLOBAL CHANGE: PRIORITIES

Global ecodynamics is a unifying element for globalization and SD problems. Problems of global ecodynamics (changes in the NSS on a global scale) attract growing attention (Kondratyev and Krapivin, 2004; Kondratyev and Losev, 2002). Respective developments are aimed at assessing changes in the NSS taking place in the past, those observed now, and possibly those in the future. Even estimates of the present ecological situation are more than contradictory, varying from substantiation of wellbeing (as has been done, for instance, in the much-talked-of monograph by Lomborg, 2001) to conclusions about a looming global disaster (such statements are most popular in the mass media). Illustrating apocalyptic prognoses, Lomborg (2001, 2004) selected a series of ecological "litanies", such as, for instance, "for more than 40 years the Earth has been sending out distress signals" (Lomborg, 2001). Perhaps, the most vivid case of politically motivated "ecological extremism" is the statement of Sir David King, the Science Advisor to the Prime Minister of Great Britain, according to which climate change is a more threatening global danger than terrorism. Specifically, he said in one lecture on global warming (King, 2002; Kreisler, 2006):

"The earth's energy budget is a key to understanding this process and I'm very keen to explain to people it wasn't discovered yesterday, this goes back to the great French mathematician Fourier who first understood the greenhouse effect and simply formulated it by saying the sun's energy comes in through the atmosphere, warms up the earth and the earth being warm then generates heat back into space and the atmosphere around the earth contains some of that heat which means that daytime and nighttime temperatures are not that different as they would be, for example, on the moon where there is no atmosphere. That's the greenhouse effect and rather like going to bed with a duvet, if you're not warm enough you put another blanket on and you'll warm up. If you add to the greenhouse gases then something has to warm up and that something is, of course, the earth. So what we have here is a fairly complex picture. We've got solar energy coming in and you've got the energy penetrating the atmosphere which is essentially the high energy part of the solar spectrum, the visible and ultraviolet light that gets through very effectively warms up the earth and the lower atmosphere and then we get what we call black body radiation, you get infrared radiation back out. So this is at the other end of the spectrum and what this means is that this can be absorbed, this energy coming back from the earth, by the atmosphere."

A detailed realistic analysis of the prospects of civilization development is contained in several monographs (Kondratyev *et al.*, 2002, 2003a, c, 2004).

The main priorities of studies into NSS dynamics include:

- conceptual and analytical approaches to studies of interactions in the NSS;
- discovery of connections of the environment and socio-economic development with ecological policy;
- establishing connections between environmental changes taking place now and possibly in the future;
- substantiation of the means that can be used to study these connections;
- observance of the laws of the international ecological rights in national environmental programmes; and
- revealing gaps in scientific developments and estimates.

These basic principles should be realized by taking the following four general requirements into account:

(1) Coordination of scientific development, ecological policy, and their accomplishment.
(2) Consideration of the interaction of ecodynamics with socio-economic development.
(3) Examination of the special features of processes on different spatial scales.
(4) Reflection of the special features of processes on different spatial scales.

Under this, it is necessary to take into account an interaction of the key problems of ecodynamics. This interaction is defined by the presence of numerous feedbacks and non-linear correlations in the NSS that can lead to the "threshold effects" (Frankham, 1995). This is essentially important in the study of global climate change.

Adequate analysis of the role of feedbacks and NSS non-linearity is seriously complicated by available fragmentary information. It is disappointing that the conception of biotic regulation of the environment developed in Russia, which could have served a conceptual basis for resolving the problems of global ecodynamics, has still not been recognized (Gorshkov, 1995; Gorshkov *et al.*, 1998, 2000, 2004). Unfortunately, the biotic regulation conception remains "unnoticed" in the West, which has been illustrated, in particular, by recent polemics with respect to the Gaya conception (Ebel *et al.*, 2004; Kerr, 2005).

The biotic regulation conception is based on the following conclusions:

(1) The Earth is unique in the Solar System, since on this planet life exists in the form of biota—the totality of all living organisms, humans included. The important properties of life include the biological stability of species and their communities

as well as a very rigid distribution of energy fluxes absorbed by biota among organisms of different sizes. Biota is responsible for the formation of environmental properties and their stability according to biotic needs. This is the reason making the long existence of biota on the Earth based on the principle of biotic regulation possible. Maintenance of environmental stability is one of the major goals of all living organisms.

(2) Like many other species, *Homo sapiens* is a biotic species and therefore the problem faced consists likewise in maintaining global biospheric stability. Otherwise, sustainable development would be impossible. Long ago, humans left their natural ecological niche and began to consume much more biospheric resources than permitted by the requirements of ecological balance. Since the industrial revolution, this process of breaking the natural balance has been accelerating under conditions of a rapid growth of population size.

(3) Approximate estimates have shown that to provide a stable state of the biosphere, one can use no more than 1% of its resources. At present, this share is close to 10%. A similar situation exists with respect to the evolution of global biogeochemical cycles of matter. For instance, the completeness of the global carbon cycle before the beginning of the industrial revolution had been close to 0.01% (biodiversity had played an important role in establishing biospheric stability). But now, the completeness has decreased to 0.1% and a threat of a global ecological disaster becomes more and more pronounced. So, the biosphere should be considered not as a resource but a fundamental condition for life on Earth. At present, a major goal should be the restoration of the biosphere, already subjected to substantial disturbing forcings, and its maintenance in a state that would ensure sustainable development. However, a difficulty lies in the presence of many insufficiently studied aspects of biospheric dynamics. In this context, the creation of adequate observational systems and further improvement of numerical modeling methods play decisive roles. The first of these problems is especially urgent, since up to the present an adequate system for global change monitoring not only does not exist but has not even been substantiated conceptually.

(4) A critically important feature of current civilization development consists in the inequality between developed and developing countries, which is mainly manifested through the unequal use of existing resources. The per capita consumption of the "golden billion" (this term is popular in the Russian-speaking world, referring to the relatively wealthy people in industrially developed nations, or the West) is incomparably higher than for the rest of the world. Therefore, it is necessary to reach an agreement on a new social order based on cooperation and partnership. The accomplishment of adequate systems for life support can only be possible on the basis of respective international agreements.

(5) It is becoming more and more clear that the present system of market economy does not provide for transition from the current unstable to a stable trajectory of development. The respective effort undertaken in such spheres as education, legislation, and management cannot be considered sufficient. Non-governmental and religious organizations must play a more constructive role.

(6) To stimulate the transition to sustainable development, accomplishment of a series of measures is needed based on understanding the complexity of the present situation.
   6.1 It is extremely important to realize the fact that further development of the consumption society will lead to a global ecological disaster and civilization collapse. The only resolution of the problem consists in rejection of the traditional paradigm of the consumption society and in radical change in the way of life.
   6.2 To overcome North–South social contrasts, immediate measures are needed to render help to developing countries based on the realization of accepted U.N. recommendations.

It is the biotic regulation conception that should serve as the basis for the first priority of the UNEP Program (UNEP, 2002, 2004), which is defined as the interaction between man and the environment: a conceptual approach to the analysis of interactions. This problem can be resolved using the UNEP DPIR (Drivers, Pressures, Impact, Response) model, which describes the cause-and-effect connections between ecological and socio-economic NSS components.

The concrete directions of ecological development include:

- *Natural and anthropogenic environmental changes.*
  Of key importance here is a system classification and priorities of interaction between natural and anthropogenic changes in the environment (it should be emphasized that this unresolved problem is one of the major reasons for the extremely contradictory character of current conclusions about the nature of the present global climate change (Kondratyev, 2004a–c). The UNEP Program document has justly emphasized the priority of the problems of biogeochemical interactions and cycles—even in the case of the carbon cycle we are still far from an adequate understanding of the most substantial mechanisms behind its formation (Kondratyev and Krapivin, 2004; Diamond, 2005; EPA, 2001, 2005, 2006; Howe, 2004).
- *Ecological factors and human wellbeing.*
  Both positive and negative ecological factors demand attention. Positive factors include ecosystems that perform the functions of life support (water, food, recreation, among many others). Various factors of ecological stress (disease, agricultural pests, natural disasters, etc.) are of course negative.
- *Anthropogenic impacts on the environment.*
  Of special interest here is an analysis of demographic dynamics and conditions of socio-economic development.
- *Ecological policy and its connections with ecodynamics.*
  In particular, the use of technologies, institutional measures, and risk control need to be borne in mind. Quite separate are the problems of the interaction between science and ecological policy. As for the concrete factors of ecodynamics, the UNEP Program (2004) has justly emphasized the first priority of analysis and assessment of the role of the interaction between such problems as

biodiversity, climate change, degradation of soil fertility, freshwater, coastal and marine environment, local and regional air quality, ozone layer depletion, persistent organic pollutants (POPs), and heavy metals.

Questions to be answered are as follows:

- What are the key interactions between various environmental changes and what factors determine them?
- How are interactions between various anthropogenic impacts realized and how can they be "separated" if needed?
- How do interactions between products provided by ecosystems and ecosystem functioning manifest themselves?
- What are the present interactions between impacts and responses to them and how can they be re-grouped or removed in case the need arises to change interactions in the NSS?

As for the goals of socio-economic development up to 2015 (compared with 1999 conditions), the main problems include (Dodds and Pippard, 2005; Hawken *et al.*, 2005; Helming and Wiggering, 2003; Islam and Clarke, 2005):

(1) Liquidation of extreme poverty and starvation (halving the number of people with an income less than $1 per day).
(2) Realization of universal elementary education.
(3) Provision of sexual equality and a raising of the role of women.
(4) Reduction of infant mortality (children under five years) by two-thirds by the year 2015.
(5) Improvement of mothers' health (reduction of mortality in childbirth by three-quarters).
(6) The struggle against HIV/AIDS, malaria, and other diseases.
(7) Provision of ecodynamic stability (liquidation of decreasing trends in environmental quality, halving the number of people without access to good-quality drinking water, improvement of living conditions for, at least, 100 million people living in slums).
(8) Development of cooperation in the interests of socio-economic development (with an emphasis on the problem of developing countries' debt).

Serious efforts to substantiate the priorities of environmental studies in the 21st century have been undertaken by NOAA scientists who developed the Strategic Plan (SP) for 2003–2007 and later (Lautenbacher, 2005; NOAA, 2003). The SP determines the priority of the four directions of developments for NOAA:

(1) Protection, reconstruction, and control of the use of coastal and ocean resources based on understanding the laws of ecosystem dynamics. The urgency of these problems is explained by the fact that about half the U.S. population is located in the coastal zone, whose extent along the coastline constitutes about one-fifth of its total length. Coastal regions are developed faster than others. Here, in particular, population size grows by 3,600 people per day, there are more than

28 million workplaces, and the gross production of goods and services exceeds \$54 billion.

(2) Understanding the laws of climate change in order to increase the ability of society to respond to them (note that the document uses two synonymous notions: variability and change with respect to climate, without their clear definition). In this context, it should be noted that provision of about one-third of the GDP in the U.S.A. depends directly or indirectly on weather conditions and climate. The annual and inter-annual climate change—like that determined by El Niño—resulted in 1997–1998 in economic damage of about \$25 billion with losses of property and yield estimated at \$2.5 billion and \$2 billion, respectively. Under U.S. weather conditions, the provision of reliable weather forecasts for the agricultural sector alone makes it possible to avoid losses of about \$700 million.
(3) Provision of information on climate and water resources. On average, the annual economic damage caused in the U.S.A. by hurricanes, tornadoes, tsunamis, and other disasters constitutes about \$11 billion. About one-third of the national economy (~\$3 trillion) is "weather-sensitive".
(4) Maintenance of national commercial interests by provision of ecologically reliable information to ensure safe and efficient transportation. Of course, safe and efficient systems of transportation are one of the key components of the economy. This concerns all kinds of transportation: land, sea, and air.

Six directions for development are considered most substantial:

(1) Creation of a system of integral (complex) observations of the global environment, processing and analysis of observational data.
(2) Efficient resolution of the problem of propagation of knowledge about the environment.
(3) Accomplishment of the required scientific studies.
(4) Provision of an efficient scientific cooperation.
(5) Resolution of national security problems.
(6) Adequate resolution of organizational problems.

The NOAA strategic plan foresees achievement of the following strategic goals for the first of the four directions for developments mentioned above:

- Increasing the number of coastal and marine ecosystems that are not subjected to anthropogenic impact and are in a stable state.
- Raising the socio-economic importance of the marine environment and respective resources (sea food, recreation, tourism, etc.).
- Improving the living conditions for coastal and marine flora and fauna.
- Increasing the number of species protected from unfavorable impacts.
- Increasing the number of cultivated species to an optimal level.
- Improving environmental quality in coastal and marine regions.

These goals should be achieved by using the ecosystem approach and in this connection by extending the scale of studies of various ecosystems dynamics within the framework of three strategic directions:

(a) Protection, reconstruction, and control of the use of resources of coastal and marine regions as well as the Great Lakes.
(b) Reproduction of species that are to be protected.
(c) Expansion and maintenance of stable fisheries.

To achieve the goals of the NOAA Strategic Plan, of key importance is to create a complex system for environmental monitoring which would provide for obtaining, processing, and analysis of data on the ocean–atmosphere–land interactive system from local to global scales.

Assessing the NOAA SP on the whole, it needs to be said that, unfortunately, it is too strongly directed towards meeting national interests. Even in this context it is impossible to adequately resolve the problems of ecodynamics without taking into account the interactivity of processes of local, regional, and global scales. The section of the plan concerning climate problems is a source of serious bewilderment. With the extreme urgency of these problems (mainly due to their strongly politicized character), of extreme importance is concrete understanding of the limited means of numerical modeling and systems of climate observations as well as planning for efforts to overcome this limitation.

Without dwelling on the concrete characteristics of present global ecodynamics (see the detailed overview in Kondratyev and Romaniuk, 1996), we only emphasize the absence of long-term perspectives of development of the modern consumption society. Therefore, at the World Summit on Sustainable Development held in Johannesburg in 2002, the necessity was emphasized to accomplish 10-year programs on reaching stable production and consumption including the following recommendations (Southerton *et al.*, 2004; Ocampo, 2005; Porritt, 2005; Princen *et al.*, 2002):

- Giving developed countries the leading role in provision of stable production and consumption.
- Achieving these goals based on general responsibility.
- Making a priority of the problem of stable production and consumption.
- Laying emphasis on involving the younger generation in resolution of the problems of sustainable development.
- Using the "polluter pays" principle.
- Controling the cycle of product evolution from production to consumption and waste to increase productive efficiency.
- Backing a policy that favors both the output of ecologically acceptable products and ecologically adequate services.
- Developing more ecological and efficient methods of energy supply.
- Supporting the free-will initiatives of industry in order to raise its social and ecological responsibility.
- Studying and applying ecologically pure production, especially in developing countries as well as in small- and medium-sized businesses.

Though these recommendations are rather declarative, they still demonstrate a necessity to change the paradigm of socio-economic development (this mainly refers to developed countries) from a consumption society to priorities of social and spiritual values. A concrete analysis of the means of bringing about this development requires the participation of many specialists in the field of social sciences.

As mentioned above, the main aspect of global ecodynamics (in the context of biotic regulation of the environment) is the state of ecosystems on the planet. Using data from the recent fundamental UNEP Report (Reid, 2005), let us consider the basic related problems.

## 1.5 GLOBAL DYNAMICS OF ECOSYSTEMS: NATURAL AND ANTHROPOGENIC IMPACTS

### 1.5.1 Introduction to the global ecodynamics theory

The report of the U.N. General Assembly presented by U.N. Secretary-General Kofi Annan (U.N., 2000, *We the Peoples: The Role of the United Nations in the 21st century*) contained, in particular, an appeal to prepare for discussions an analytical report on the problems of global ecosystem dynamics "The Millennium Ecosystem Assessment" (MEA). In response to this appeal, the respective document was prepared in 2001 aimed mainly at assessing the consequences of global change in ecosystems for human wellbeing, scientific substantiation of measures needed to ensure and maintain the sustainable development of ecosystems, and analysis of the importance of ecosystem dynamics for the sustainable development of society (Reid, 2005).

The main content of the MEA (2005b) report prepared by the four working groups ("Conditions and Trends", "Scenarios", "Responses to Forcings", "Sub-global Assessments") with the participation of more than 2,000 experts and reviewers was to answer the following questions:

(1) What changes have taken place in ecosystems?
(2) How have "ecosystem services" (functioning of ecosystems as life support systems) and their application changed?
(3) How have changes in ecosystems affected human wellbeing and reduction in the scale of poverty?
(4) What are the most substantial factors that cause changes in ecosystems?
(5) What are the possible scenarios of future changes in ecosystems and associated systems of life support ("ecosystem services")?
(6) What can be learned about the consequences of ecosystem changes for human wellbeing by analyzing sub-global processes?
(7) What is now known about the time scales, risk, and non-linear dynamics of ecosystems?
(8) What choices are there to ensure ecosystem stability?
(9) What are the most substantial uncertainties that hinder the making of adequate decisions to ensure the stable functioning of ecosystems?

Of great importance for answers to these questions were the results of developments accomplished within the programs of four international conventions: the Convention on Biological Diversity (UNEP, 2006); the U.N. Convention to Combat Desertification (Lean, 1995), the Ramsar Convention on Wetlands (Ramsar COP8, 2002; Ramsar COP9, 2005), and the Convention on Migratory Species (UNEP, 2005). For example, the Convention on Wetlands (signed in Ramsar, Iran, in 1971) is an intergovernmental treaty which provides a framework for national action and international cooperation for the conservation and wise use of wetlands and their resources. There are presently 153 Contracting Parties to the Convention, with 1,629 wetland sites, totaling 145.6 million hectares, designated for inclusion in the Ramsar List of Wetlands of International Importance. "The Convention's mission is the conservation and wise use of all wetlands through local, regional, and national actions and international cooperation, as a contribution towards achieving sustainable development throughout the world" (Ramsar COP8, 2002; Ramsar COP9, 2005). Naturally, the MEA report is based on the use of results published earlier.

Emphasis in the report (Reid, 2005) was placed on analysis of the interconnections between ecosystems and human wellbeing—in particular, in the context of "ecosystem services". In this case the ecosystem is defined as a single interactive dynamic complex of communities of plants, animals, and microorganisms together with the inert environment. The MEA report contains a consideration of practically all diversity of ecosystems ranging from relatively weakly changed (as, e.g., natural forests) to landscapes with anthropogenically changed structures and agricultural lands as well as urban agglomerations, where ecosystems have been changed radically by man.

As for "ecosystem services" (ES), they are defined as performing the functions of life support by using the respective potential of ecosystems. Such "services" can be classified as *provisioning* with food, water, wood, fiber; *regulating*, connected with impacts on climate, water regime, diseases, waste, and water quality; *cultural*, on which depend conditions of recreation, esthetic and spiritual aspects; and *supporting*, which include soil formation, photosynthesis, and biogeochemical cycles. In this connection, it is important to remember that *Homo sapiens*, while being protected from ecological impacts due to achievements in technologies and culture, nevertheless depends critically on the functioning of "ecosystem services".

The fundamental goal of the MEA report is to analyze the ES impact on human wellbeing as determined by the totality of such aspects as: basic materials for a good life, including secure and adequate means of living, sufficient amount of food (at any time), dwelling, clothes and various goods; health, including good state of health and favorable environment (pure air and water); good social relations which ensure social prosperity, mutual respect, ability and readiness to render help (especially to children); security that foresees access to natural and other resources, personal security, as well as security against natural and anthropogenic disasters; and freedom of choice and action that guarantees achievement of goals with individual qualities taken into account. Freedom of choice and action is also determined by other components of prosperity (as well as some other factors, especially education), and it is a condition for achieving its other components, especially equality and justice.

An important conceptual side of the MEA report consists in the consideration of humans as an integral part of ecosystems and in taking into account their interaction with these ecosystems which (as human living conditions change) leads (directly or indirectly) to changes in ecosystems, causing thereby changes in human wellbeing. On the other hand, the impact of social, economic, and cultural factors not related directly to ecosystems causes changes in human living conditions, and many natural factors affect ecosystem dynamics. Though the MEA report pays special attention to the connection of ecosystem dynamics with human wellbeing, it emphasizes that human impacts on ecosystems have not only consequences for the care of humans, but also the result of recognition of the value of individual species and the whole ecosystem. Such immanent values are qualities that are independent of their use by anybody else.

The irrational use and over-exploitation of natural resources head the list of unresolved problems for ecosystem stability. Though some ecosystems ensure the growing volume of production in the form of fish, cattle, and various agricultural commodities, the integrity of such ecosystems has been breached and in many cases the efficiency of ecosystem functioning has decreased. As Stokstad (2005) noted, of 24 kinds of ecosystem services considered in the MEA report about 60% have degraded. Therefore, the Co-Chair of the Board of Directors of the Millennium Ecosystem Assessment Robert T. Watson (2002) emphasized that inceases in population undermine the ecological capital in the world. One of the main dangers of this trend is that it enhances the risk of sudden and severe changes in ecosystems (e.g., collapse of fisheries or the growth of diseases such as HIV/AIDS).

An important aspect is that the negative consequences of the impact on ecosystems mainly falls on the poorest—this refers, in particular, to the arid regions of sub-Saharan Africa, Central Asia, and Latin America, where about one-third of the global population is concentrated. Provision of the stable functioning of ecosystems is of key importance for overcoming poverty and achieving the U.N.-formulated goals of socio-economic development. The Co-Chairman of the WG on business and industry J. Lubchenko (Lubchenko *et al.*, 1991) has justly noted that only by protecting and restoring the ecological functions of ecosystems can we adequately resolve the problems of starvation and poverty.

In the context of the problem of the stable functioning of ecosystems special attention is paid to anthropogenic impacts on the global biogeochemical cycles of nutrient salts. For instance, the wide use of nitrogen fertilizers has resulted in dangerous intensification of the processes of eutrophication, algal blooms, large-scale deaths of mammals, and groundwater contamination.

The MEA report contains an informative analysis of the aspects of the global dynamics of ecosystems, but, unfortunately, one key aspect of the problem has been ignored. This concerns the systems' ability to self-regulate, and in this connection the conception of the biotic regulation of the environment. Among the many participants to the MEA preparation, there were no specialists familiar with the numerous publications on these problems (Gorshkov, 1995; Gorshkov *et al.*, 1998, 2000, 2004; Kondratyev *et al.*, 2004; Kondratyev and Krapivin, 2005a; Losev, 2001, 2003). The biotic regulation conception is of fundamental importance both for understanding the laws of present ecodynamics and, in particular, for prognostic estimates.

To stimulate further discussions on the problems of ecosystem dynamics, mainly bearing in mind their key ethical aspects concerning the forecasts of civilization development, Ehrlich and Kennedy (2005) proposed carrying out a "Millennium Assessment of Human Behaviour" (MAHB) on the basis of MEA and IPCC experience in order to understand what actions are needed to resolve ecological problems from the viewpoint of changes needed to bring about an ecologically stable society that is peaceful and provides equal rights for everybody. It is important, in particular, to ask the people of rich countries whether they are ready to agree with the existing opinion about the fact that their lifestyle can not be the subject of discussions (Kondratyev, 1996, 2002; Kondratyev and Krapivin, 2005f; Kondratyev *et al.*, 2006b; Kuznetsov *et al.*, 2000). It is important to analyze which (from the ethical point of view) is most adequate, considering the needs and possibilities on a global scale.

Although mainly confined to the discussion of social aspects, the MAHB initiative should be connected with holding open and transparent forums dedicated to discussions about ethical interactions between people and stated as well as approaches to exploitation of natural systems of life support. For this purpose, working groups should be formed to discuss, in particular, the following problems:

(1) What is currently known about the evolution of cultures and, in this connection, what democratic changes are possible?
(2) How deficient and non-uniformly distributed non-renewable natural resources are used and what are the ethical relationships between their distribution, economic capabilities, and accessibility.
(3) Ethical aspects concerning the global commercial system.
(4) Conflicts between the needs for resources and ecological consequences of their usage.
(5) Inequality of economic conditions, races, and sex as factors of environmental degradation.

Other possible directions for discussions include problems of war and peace, international management, health maintenance, and others.

The main goal of MAHB should be to integrate the efforts of different WGs on a continuing basis and to substantiate respective recommendations. The MAHB could, for instance, coordinate the efforts of public organizations to limit (under certain conditions) the monopoly of some corporate actions. An example of the fruitfulness of such efforts can be preparation and further signing of the Montreal Protocol on reduction of anthropogenic impacts on the global ozone layer, which has demonstrated the successful interaction of governmental and public organizations.

The MAHB is supposed to concentrate on analyzing the mechanisms of decision-making concerning the use of natural resources and respective risks. The problem of relationships between the interests of individual people and society as a whole demands serious attention, as well as finding institutional structures to resolve this problem, which is especially important from the viewpoint of coordinating the interests of groups of people with different traditions and little contact. Much more serious efforts should be applied to stimulate interdisciplinary developments.

As for the problems of globalization and sustainable development, especially taking into account that present losses of biodiversity and damage to natural ecosystems are sometimes considered as having a positive value for human wellbeing from the viewpoint of short-range perspectives, the preservation of biodiversity and stable functioning of "ecological services" are most important. In this connection, Balmford *et al.* (2005) analyzed the prospects of achieving by 2010 the goals formulated in the Convention of Biodiversity at the World Conference on Sustainable Development in 2002 as "... an achievement by 2010 of substantial reduction of the present rate of biodiversity losses on global, regional, and national scales." However, the problem is that the actual data on biodiversity are rather incomplete, and the present understanding of the role of interactive biological, geographical, and geochemical processes as factors of human wellbeing is only rudimentary. Therefore, concrete substantiation of ways of achieving the goals of the Convention on Biological Diversity (CBD) by 2010 is a complex problem. Of particular importance here is the problem of adequately choosing quantitative criteria (indicators) of biodiversity and ecosystem functioning ("ecological services").

In early 2004, the participants of the CBD formulated preliminary framework conditions which determine an approach to the choice of such indicators (Table 1.4). Of course, the information contained in Table 1.4 needs further serious discussions and optimization. An Advanced Workshop was organized by the Global Biodiversity Information Facility in this connection in July 2004 in Great Britain (Heywood, 2004). It reached a conclusion about the expedience of selecting 18 indicators which ensure a complex quantitative estimate of biodiversity losses, though there are many uncertainties with respect to the causes of these losses (MMSD, 2001). Therefore, it is possible that later on there may appear a need for additional indicators. An analysis of ecosystem dynamics in the U.S.A. has led, for instance, to the conclusion that 102 indicators should be used. On the other hand, there is no doubt that solution of the problem cannot be achieved by an assessment of indicators. Of course, it is necessary to develop respective simulation models of the NSS—interactive biological, chemical, and physical processes with human dimensions taken into account (Kondratyev *et al.*, 2003a; Kondratyev and Krapivin, 2005a–c; Kondratyev *et al.*, 2004, 2006b), which is also important for substantiation of the need for information about ecosystem dynamics obtained, in particular, by using remote sensing. Up until recently it has been mainly biologists who have dealt with the biodiversity problem, but the participation of specialists in the field of socio-economic sciences and geography as well as experts in geophysical and geochemical processes is important too.

### 1.5.2 Present state of global ecosystems

During the last 50 years global ecosystems have been subjected to anthropogenic changes that they had not suffered during their evolution in the Holocene. Such changes have been connected with growing population size and an increase in the

**Table 1.4.** Framework considerations which can serve as the basis for choosing biodiversity indicators and "ecological services".

| Possible indicators | Proposed indicators |
|---|---|
| *Biodiversity components*<br>• area of forested regions<br>• trends in the amount and distribution of individual species<br>• area of protected territories<br>• change of status of threatened species<br>• trends of genetic diversity of domestic plants and animals<br>• extent and localization of rhizophora (mangrove forests) and sea algae<br>• control of protected territory efficiency<br>• financial investment to protect territories | • conditions of forests<br>• extent and conditions of shrubs, grasslands, and deserts<br>• extent of wetlands and large water basins<br>• conditions of watershed basins and extent of coastal vegetation<br>• relative extent of coral reefs<br>• extent and conditions of estuaries |
| *Stable use*<br>• areas of forests, agricultural lands, and fisheries under conditions of stable management<br>• share of products obtained from stable sources | |
| *Threat to biodiversity*<br>• nitrogen deposition<br>• number of parasite species and expenditure on prevention of them becoming too intrusive | • scale of sea fishery<br>• road-free areas<br>• outbreaks of epidemics among wild species |
| *Integrity of ecosystems, produced goods, and services*<br>• marine trophic index<br>• water quality in inland water bodies<br>• trophic index of freshwater<br>• relationship and fragmentation of ecosystems<br>• cases of anthropogenic destruction of ecosystems<br>• human health and wellbeing under conditions that depend on biodiversity<br>• use of biodiversity in food and medicine<br>• fish catches per unit effort<br>• wood and fuel production per unit effort | • number of dams<br>• bottom sediments in rivers<br>• share of people without good-quality drinking water<br>• carbon supplies in ecosystems<br>• the market share in ecotourism<br>• interpretation of biodiversity problems in web networks<br>• use of pesticides per unit agricultural output<br>• level of harvest per unit effort |
| *Traditional knowledge, innovations, and practice*<br>• status and trends of linguistic diversity and the size of population speaking native languages | |
| *Transfer of resources*<br>• official measures to render help to support CBD | |

need for food, freshwater, wood, fiber, and fuel. Naturally, these changes have favored the progress of socio-economic development and people's wellbeing, but so selectively (in different global regions and for different groups of population) that in some cases the consequences of the use of biospheric resources for people have turned out to be negative. It is also important to remember that the actual consequences of the impact on the global biosphere have only been understood during recent years. For some people these consequences have been negative due to the three problems considered below. Solution of these problems will make possible the long-term stable use of biospheric resources.

First, as the estimates of the state of ecosystems at the turn of the century (Reid, 2005) have shown, about 60% of ecosystems (15 of 24 ecosystems studied in MEA) have suffered either degradation or unstable use, including such resources as freshwater, fish catches, local and regional climate, as well as the impact of natural disasters and pests. Though material losses are difficult to assess, there is no doubt that they are significant and characterized by increasing trends. Many ecosystems have been damaged due to impacts on other "ecosystem services"—for instance, those connected with food production.

The following facts refer to concrete manifestations of impacts on ecosystems:

- During the 30 years after 1950 a much larger amount of land was transformed into agricultural lands than during the preceding 150 years (1700–1850). Nowadays, cultivated lands (regions where $<30\%$ of land are agricultural) cover about one-quarter of the land surface.
- During the last few decades of the 20th century about 20% of global coral reefs have been lost, and about 20% have suffered considerable degradation. During the same period 35% of mangrove forests were lost (about half of these forests are in regions where documentation is good).
- After 1960, the volume of water in reservoirs has grown fourfold. Artificial reservoirs contain four to six times more water than rivers. The volume of water from rivers and lakes has doubled since 1960, and the input of phosphorus to them has tripled. On a global scale, about 70% of water is used in agriculture.
- Since 1960, the input of chemically active (biologically accessible) nitrogen to land ecosystems has doubled, and the phosphorus flux has tripled. More than half of all synthetic nitrogen fertilizers, first produced in 1913, were used after 1985.
- Since 1750 the $CO_2$ concentration in the atmosphere has increased by about 32% (from 280 to 376 ppm in 2003), which was mainly caused by fossil fuel burning and change of land use; about 60% of this growth occurred post-1959.

The decrease in biodiversity has become apparent and threatening:

(1) by 1990 more than two-thirds of the area of the largest 14 biomes of land and more than one-half of the area of 4 other biomes had been transformed mainly into agricultural lands;

(2) within the natural habitats of various taxonomic groups there was a decrease in either the number or the range of diversity of most species;
(3) the global distribution of species has become more homogeneous: the totality of species observed in one region has become very similar to that in other regions, which is mainly the result of intentional or non-intentional introduction of species together with the intensive long-range travel and transportation;
(4) there was a reduction in the total number of species on the planet. During the last several hundred years the rate of anthropogenic extinction of species has increased more than 1,000 times compared with the background level characteristic of the planet's history. About 10–30% of species of mammals, birds, and amphibians is under the threat of extinction. The largest share of threatened species is characteristic of freshwater ecosystems; and
(5) there has been a global decrease in genetic diversity, especially in the cases of cultivated species.

Strong changes in ecosystems have been determined, as a rule, by a necessity to satisfy the growing needs of food, water, wood, fiber, and fuel. However, intentional damage to some ecosystems has also been determined by the construction of roads, sea ports, urban development and there has been an increasing level of environmental pollution.

The global population size exceeded 6 billion people by the end of the second millennium, and more than a sixfold increase in economies have brought about a radical increase in requirements with respect to "ecological services". This situation has resulted in a 2.5-fold increase of food production between 1980 and 2000, more than doubled freshwater consumption, tripled the use of wood for paper production, and doubled the power of hydroelectric stations. The related needs of ES were provided through ES intensification (e.g., through wider use of natural waters to irrigate or to raise fish catches as well as through increasing the productivity of agricultural crops and domestic animals). The growth in productivity was achieved due to the use of new technologies (new sorts of agricultural crops, fertilizers, and irrigation) as well as by broadening the area of agricultural lands (Camillo *et al.*, 1986; Guhathakurta, 2003; Hillel, 2003; Jones *et al.*, 2005).

Second, of importance is the fact (though not completely specified) that the observed changes in ecosystems can raise the probability of the non-linear dynamics of ecosystems (including accelerating sudden and potentially irreversible changes) being able to negatively affect the stable functioning of ES as systems of life support. Outbreaks of diseases, drastic changes in drinking water quality, formation of "dead zones" in coastal waters, collapse of fishery, regional climate change, etc., can exemplify such a situation.

The third problem connected with ES degradation, consisting of a persistent decrease in the ability of ecosystems to function properly, is determined by the socio-economic conditions that lead to poverty, further favors an increase in social inequality and is a potential source of social conflict. This situation is especially characteristic of sub-Saharan Africa, where the living conditions and the rational control of ES play a very important role.

The increasing (though slower than before) global population size and growing needs for ES (it is supposed that by 2050 the global GDP will increase three- to sixfold) seriously threaten sustainable development, and these problems do not have any simple solutions in view of the multi-component and interactive character of related processes. Three of the four scenarios developed within the MEA show that ecosystem degradation can be partially stopped by serious reforms within the spheres of politics, institutional aspects, and practical measures. These reforms, however, need to be radical, but they have not been started as yet.

The summarizing Table 1.5 compiled by the MEA authors characterizes the general features of present ES functioning (Reid, 2005). The notion of degradation is defined in Table 1.6 as conditions when the present level of the use of any resource turns out to be below the level of stability, and in the case of the regulating ES as a reduction in their usefulness (e.g., mangrove forest degradation on coastlines enhances the negative consequences of storms for ecosystems, and their reduced ability to maintain the quality of natural waters results from growing pollution).

The negative economic consequences of ES degradation cannot be denied: while, on the one hand, the growth of production is observed during degradation processes, on the other hand, this degradation turns out to be, as a rule, unacceptable from the viewpoint of the common interests of society. In many cases the physical, economic, or social consequences of ecosystem degradation become transboundary. The non-market-specific character of ES is extremely important (Reid, 2005).

Despite progress in the use of ES, the level of poverty remains high and social inequality grows on the whole: so far, many people cannot use ES adequately or do not even have access to them. In this connection, the following facts should be mentioned:

- In 2001 about 1.1 billion people had a daily income of <US$1, about 70% of these people lived in agricultural regions.
- During the last decade, the income and other indicators of wellbeing became more uneven. In sub-Saharan Africa, the death rate of children under five years of age exceeded 20 times that in developed countries, and this ratio is higher than 10 years ago. During the 1990s, in 21 countries there was a decrease in the index of socio-economic development. The Human Development Index (HDI) is characterized by a combination of factors about socio-economic welfare, health, and education. Fourteen of these countries are located in sub-Saharan Africa.
- Despite the increase in per capita food production observed during the last four decades, 852 million people were under-nourished in 2000–2002 (compared with 1997 this figure has increased by 37 million people).
- Presently, about 1.1 billion people have no access to good-quality drinking water, and 2.6 billion people live under anti-sanitary conditions. More than half the urban population of Africa, Asia, Latin America, and the Caribbean suffer from diseases caused by inadequate water quality and sanitary conditions.

**Table 1.5.** The global status of ecological service (ES) functioning.

| Ecological services | ES product | ES change |
|---|---|---|
| *ES that provides resources* | | |
| Food | Agricultural crops | Substantial increase of production |
| | Stock-breeding | Substantial increase of production |
| | Fish | Decrease in production due to over-catching |
| | Aquatic species | Substantial increase of production |
| | "Wild food" | Decrease in production |
| Fiber | Wood | Loss of forests in some regions but an increase in other regions |
| | Cotton, hemp, silk | Decrease in production of some fibers and an increase in others |
| | Wood fuel | Decrease in production |
| Genetic resources | | Losses due to extinction and decrease of genetic resources |
| Biochemical components, natural medicine, pharmaceutical preparations | | Losses due to extinction and over-exploitation |
| Fresh water | | Unstable use of drinking water and water for industry and irrigation |
| *ES that regulates environmental properties* | | |
| Air quality regulation | Global | Reduction of atmospheric ability to self-regulate |
| Climate regulation | | Intensified carbon supply by the mid-century |
| Water resource regulation | | Unsteady depending on ecosystem change and their geographic location |
| Soil erosion regulation | | Enhanced degradation of soils |
| Water purification and waste-processing | | Deterioration of water quality |
| Regulation of diseases | | Unsteady depending on the change of ecosystems |
| Regulation of parasites | | Degradation of natural control due to the use of pesticides |
| Pollination | | Global decrease in the number of pollinators |
| Regulation of natural disasters | | Losses of natural buffers such as wetlands and mangrove forests |
| *Impacts of ES on cultural conditions* | | |
| Spiritual and religious values | | Rapid decrease of diversity of plants and species |
| Esthetic values | | Decrease in the amount and quality of natural land |
| Recreation and eco-tourism | | Increased accessibility to various regions but degradation of many of them |

**Table 1.6.** Shares of global consumption and population (%) in different regions.

| Region | Consumption | Population |
|---|---|---|
| U.S.A. and Canada | 31.7 | 5.3 |
| Western Europe | 28.9 | 6.5 |
| Eastern Asia and Pacific Ocean region | 21.6 | 32.9 |
| Latin America and Caribbean region | 6.9 | 8.7 |
| Eastern Europe and Central Asia | 3.7 | 8.2 |
| South Asia | 2.2 | 22.5 |
| Australia and New Zealand | 1.8 | 0.6 |
| Near East and North Africa | 1.7 | 4.4 |
| Sub-Saharan Africa | 1.5 | 10.9 |

### 1.5.3 Scenarios for possible evolution of global ecosystems

Within the framework of the MEA project (Reid, 2005), four scenarios of the possible long-term dynamics of global ecosystems have been worked out based on two major initial suppositions:

(1) accelerating globalization takes place in the world; and
(2) the world is being intensively regionalized.

Two different approaches have also been considered in order to provide the long-term stable functioning of ES. The four scenarios foresee a continuous enhancement of the pressure on ecosystems during the first half of the century. The most important factors of the impact on ecosystems include a change of habitats (change of land use, diversion of river waters), over-exploitation, intrusion of parasite species, pollution, and climate change. It is important that the impacts of these factors are, as a rule, interactive.

The most substantial results of scenario estimates can be characterized as follows:

- *Transformation of habitats, especially as a result of the growth of the area of agricultural lands.* According to the above-mentioned scenarios, in 2000–2050 an additional 10–20% transformation of grass and forests mainly into agricultural lands should take place. This will happen mainly in arid regions and in the countries with low incomes. In industrial countries the extent of forested territories should grow.
- *Over-exploitation of natural resources, especially fish over-catching.* Over most of the World Ocean, the industrial fish biomass will decrease by 90% compared with that before the beginning of industrial fishery.
- *Intrusion of parasite species continues to grow as well as related diseases.* Such intrusions will have a negative impact on local species and hinder normal ES functioning.

- *The level of environmental (especially biogenic) pollution grows*. The content of anthropogenic nitrogen in continents has doubled and will increase by two-thirds by 2050. According to three of the four scenarios under consideration, the global flux of nitrogen to coastal ecosystems should increase by 10–20% by 2030, mainly in developing countries. Excess nitrogen fluxes cause an enhancement of eutrophication of freshwater entering coastal marine ecosystems and souring in land ecosystems, which leads to biodiversity reduction. The growth in nitrogen inflow will also have an effect on the increase in surface ozone concentration (this will lead to a decrease in bioproductivity of agricultural crops and forests) and on stratospheric ozone depletion—resulting in the growth of biologically dangerous UV-B (290–320 nm) solar radiation (Diffey, 1991) at the surface level causing skin cancer, respiratory, and other diseases.
- *Anthropogenic climate change*. The climate warming observed in many regions has already affected biodiversity and ecosystems, including distributions of species, size of populations, time of reproduction, migration dynamics, increases in the frequency of outbreaks of diseases, and the numbers of parasites. There will be considerable but, apparently, partially reversible damage to coral reefs and other consequences.

An analysis of scenario estimates has led to the conclusion that by the end of the 21st century the impacts of climate change will be the dominating factors that determine the global dynamics of biodiversity and ecosystems. However, in some regions at first one may observe the positive impact on biodiversity of such climate change as temperature increase and precipitation enhancement, but in the long-range perspective dangerous effects on ecosystems should prevail compared with positive consequences—this can happen if an increase in global mean SAT compared with the pre-industrial level exceeds 2°C or the rate of SAT increase exceeds 0.2°C over 10 years. Such projections are, however, highly uncertain (Kondratyev and Krapivin, 2005a).

According to the four scenarios under discussion, a considerable increase in the scale of ES usage, continuous losses of biodiversity, and further ES degradation should take place. It is supposed that during the next 50 years the need for edible agricultural crops will grow by 70–85% and for water by 30–85%. The irretrievable use of water should increase in developing countries but decrease in developed ones:

- *Food security* cannot be reached by 2050 within the scenarios considered, and in some regions the shortage of children's food will remain.
- *Decrease in ES quality connected with freshwater resources* is reflected in every scenario (especially when the dependence on environmental conditions is clearly pronounced).

According to prognostic estimates, the loss of habitat and other changes in ecosystems will lead to a decrease in the biodiversity of local species. On a global scale, by

2050 the total number of plant species may decrease by 10–15% due to the loss of habitats and other factors.

Naturally, the degradation of "ecosystem services" creates serious barriers which hinder achievement of the goals of socio-economic development planned by the U.N. in 2000 to provide an improvement in people's living conditions by reducing the levels of poverty, starvation, infant and maternal mortality, diseases, liquidation of sexual discrimination, achievement of ecologically sustainable development, and intensification of global scale cooperation. The specific feature of the global situation is that these goals are most difficult to achieve in regions where the degradation of ecosystems is most intensive.

Though an adequate socio-economic policy should play the main role in achieving MDC goals, many of the problems can be resolved only with considerably improved methods of ecosystem control. This refers, for instance, to the problems of poverty and drinking water quality, which cannot be resolved without reconstruction of degrading ecosystems:

- *The problem of starving people* will be resolved slower (according to all the four scenarios) than required by the international agreement to halve the number of starving people in the period 1990–2015. Moreover, this requirement will be met most slowly in regions where this problem is most urgent (sub-Saharan Africa and South-East Asia). The effect of the state of ecosystems on resolution of the starvation problem is manifested through the impacts of climate conditions, ploughed land degradation, and availability of water resources for agricultural harvests and accessibility of "wild" sources of food.

According to three of the four scenarios, by 2050 under-nourishment that causes infant mortality should decrease by 10–60%, but according to one of the scenarios, by the year 2050 there will be a 10% increase in the shortage of children's food. The dynamics of the rate of infant mortality depends strongly on the level of diseases and water quality (one of the medical causes of infant mortality dominating in the world is diarrhea, and in sub-Saharan Africa malaria too).

The struggle against diseases is possible according to one of the (most optimistic) scenarios, but it follows from another scenario that in many poor regions this is unachievable. Unfavorable changes to ecosystems will enhance the danger of pathogens such as malaria and cholera, as well as the appearance of new diseases (Pelling, 2003; Rees, 2003; Rosenzweig, 2003; Wallström *et al.*, 2004).

### 1.5.4 Preliminary assessments and conclusions

Analysis of the four forecasts of global ecodynamics shows that according to three of them, an important role in withstanding many of the negative consequences could be played by an adequate ecological policy, institutional improvement, and various practical measures for nature protection. It is important, however, that these measures should be large-scale, but putting them into effect has not as yet started. Therefore, the predicted dynamics of cultivated ecosystems up to 2050 turns out to be negative. Only one scenario is excepted and—as for other three scenarios—only one of the three ES will be in a better state by 2050 than they were in 2000. The scale of interference needed to achieve even this result is large, including an improvement in technologies, introduction of adaptive control, and forestalling actions to resolve environmental problems as they appear, radical growth in financial support for education and the health service, and measures on liquidation of socio-economic inequality (especially on poverty liquidation). It is important to point out, however, that—despite the improving state of one or several ES—biodiversity losses will continue, and thus the long-term perspectives of sustainable development will remain under threat. Measures to overcome ecosystem degradation taken in the past that have been markedly useful have always lagged behind the growing load on ecosystems.

It is still critically important to obtain more reliable information for analysis of the present dynamics of ecosystems and for respective prognostic estimates, which will serve as the basis for substantiation of adequate measures to ensure stable ES functioning. In particular, the current global observational system needs to be improved for thorough monitoring of ecosystem dynamics. Further development is needed for numerical modeling of the NSS, especially when non-linear changes in ecosystems and forecasts of sudden change are considered.

The most important conclusion drawn from the Report of the Millennium Ecosystem Assessment (MEA, 2005a,b; Reid, 2005) is that—as mentioned above (Kondratyev *et al.*, 2003a, 2004; Kondratyev and Krapivin, 2005b)—the present consumption society cannot be sustained. Therefore, an immediate problem is to substantiate and realize a new paradigm for socio-economic development. The report (MEA, 2005; Reid, 2005) contains many rhetorical propositions to ensure the preservation of the present path of socio-economic development, but they are, as a rule, impracticable.

Let us now consider some concrete data that characterize present global ecodynamics, which illustrate the grounds for the recommendations enumerated above (mainly in Section 1.5.3).

## 1.6 PRESENT GLOBAL ECODYNAMICS

The most substantial feature of current global ecodynamics (i.e., the 20th and the early 21st centuries) is the rapid increase in population size (mainly in developing countries), an increase in the urban population (a considerable growth in the number of megapolises, which also refers to the Third World), and expansion of the scale of such dangerous diseases as HIV/AIDS, hepatitis, tuberculosis, among others (and once again it is the developing countries who bear the brunt). Since demographic problems have been discussed in detail earlier (Kondratyev, 1998; Kondratyev *et al.*, 2003c, d, 2004; Kondratyev and Krapivin, 2003; Obaid, 2003; Rimashevskaya, 2002), we only note that—according to U.N. estimates—by 2050 the global population will increase to 8.9 billion people. This means, in particular, that any decrease in possible per capita consumption (due to increased efficiency of technologies) with the view to contributing to cumulative consumption will be overtaken by the impact of the growth in population. If, for instance, by 2050 the consumption of meat by an average American decreases by 20% (compared with 2000), due to the growth of the U.S. population the total consumption of meat will increase by 5 million tons. An important common fact is that despite the overwhelming population growth in developing countries their contribution to the impact on the environment will not necessarily exceed that in developed countries. For instance, the U.S. population is growing now by about 3 million people per year, and in India by 16 million people per year, but the impact of the U.S.A. on the environment (due to high per capita consumption) will be much greater. This is illustrated, in particular, by estimates of the growth of $CO_2$ emissions to the atmosphere: 15.7 and 4.9 million tons of carbon in the U.S.A. and in India, respectively.

We begin by considering the problem of key importance for ensuring sustainable development—the production/consumption relationship. At the World Summit on Sustainable Development in Johannesburg (26 August–4 September 2002) it was emphasized: "To provide the global sustainable development, fundamental changes are needed in production/consumption ratio." Mr. Børge Brende, the Norwegian Minister for Environmental Problems, Chairman of the U.N. Commission on Sustainable Development, emphasized (Starke, 2004) that, to save billions of people from poverty, a new structure of consumption is needed, with the major responsibility for resolving this problem in the present consumption society resting with the developed countries, though, of course, joint efforts are also important. It is necessary to take into consideration that the present structure of personal consumption expenditure includes the consumption of durable goods, non-durable goods, and services. All this expenditure often depends on seasonal circumstances and other social or religious parameters. In other words, the structure and dynamics of personal consumption expenditure are functions of numerous parameters of the NSS—the parametrization of which in the GMNSS demands their detailed analysis and study.

### 1.6.1 Global demography

The notion of globalization and sustainable development relates of course to human society. Future development of human society depends on the numerous uncertainties connected with the understanding that Earth's resources are restricted. One possible approach to help overcome these uncertainties is a new science called global ecoinformatics. This has been intensively developed in recent years. It is within its framework that information technologies have been created, which can facilitate the combined use of various data on the past and present state of the NSS. The creation of a model of NSS functioning based on knowledge and available data that can be combined with an adaptive–evolutionary concept for geoinformation monitoring—which enables one to realize the interconnection between the NSS model and the regime of global data collection—can be considered an important step in global ecoinformatics. As a result, the NSS structure can be optimized to achieve a sustainable interaction between nature and human society and to create an international strategy for the coordinated use of natural ecosystems. But, for this to happen it is necessary to have efficient mechanisms for the parametrization of demographic processes, the description of which in a global model will increase the validity of global ecodynamic forecasting.

The demographic situation determines the dynamics of anthropogenic processes. Therefore, attempts to develop models of demographic processes are justified. The existing predictions of changes in population size and variations in its spatial distribution enable one to synthesize the scenarios to be used in a global model as well as to try to solve the problem of verifying it. An adequate model of the global demographic process requires an extensive database covering the characteristics of changes in the standards of demographic behavior, detailed information on intensities of demographic processes in various regions of the planet, assessments and criteria of demographic policy, and, which is especially important, the structural indicators of the human society (Demirchian and Kondratyev, 1999).

Many authors have considered the problem of parametrization of demographic processes. Demographic processes are considered part of the planetary biospheric processes in which *Homo sapiens* plays the role of a user of biocenoses products and a regulator of energy and matter fluxes between them.

#### *1.6.1.1 Matrix model of population size dynamics*

In accordance with the possibility and needs of the global model, the unit of population dynamics includes the impact of the following factors (Krapivin and Marenkin, 1999; Logofet, 2002; Kondratyev, 1996; Leslie, 1945; Meadows *et al.*, 2004):

- per capita food provision $F$ (calculated as the sum of several shares of vegetation and animal populations of the region as well as fish catch);
- the share $A$ of animal protein in the human diet (determined from the contribution to $F$ from animals and fish);
- the level of the public medical service $M$ (in this model version it is in proportion to per capita financing); and
- genetic load for human population $G$ (grows slowly with population development and depends on the level of environmental pollution).

The sex structure of a population and the processes of population migration between regions are ignored. The age structure contains three groups (0–14 years, 15–64 years, 65 years and older). There is also a fourth group of population (disabled people within these three age groups). For terminology convenience the fourth group is considered as not affecting the age structure.

Let $S_t$ be the age structure at a time moment $t$, then the population size dynamics can be described by the following matrix equation:

$$S_{t+1} = D \times S_t$$

where $D$ is the demographic matrix $4 \times 4$ including the effect of the factors $F$, $A$, $M$, and $G$:

$$D = \left\| \begin{matrix} d_{11} & d_{12} & 0 & 0 \\ d_{21} & d_{22} & 0 & 0 \\ 0 & d_{32} & d_{33} & 0 \\ d_{41} & d_{42} & d_{43} & d_{44} \end{matrix} \right\|$$

Matrix $D$ differs from Leslie's traditional matrix for the model of population in its age structure: first, by the inequality to zero of its diagonal elements, which is explained by the overlapping of the next generations; and, second, by the last line that reflects the non-age character of the fourth group. The diagonal elements of matrix $D$ are determined from the apparent balance relationships:

$$d_{ii} = 1 - \mu_i - \sum_{j>i}^{4} d_{ji} \quad (i = 1, \ldots, 4) \tag{1.1}$$

which include the coefficients of mortality $\mu_i$ of the $i$th group, descending functions of per capita food provision and medical service level $\mu_i = \mu_i(F, M)$ with:

$$\lim_{F,M\to\infty} \mu_i(F, M) = \mu_{i,\min} > 0$$

where $\mu_{i,\min}$ characterizes a minimum physiological mortality with optimal food provision and medical service.

It is supposed that the reproductive potential of the human population is entirely concentrated in the second age group. The $d_{12}$ coefficient value is considered to be a regional constant. Note that the birth rate coefficient is a complex and poorly studied function of many variables, ethnic traditions being one of the important variables. The birth rate is affected by religion; for instance, the Muslims consider children to be their wealth. In South-East Asia the religious norms order each family to have at least one son. Catholicism also markedly affects the birth rate index. Therefore, if the function $\mu_i$ is well parametrized by statistical data, the coefficient $d_{12}$ is still an unidentified parameter. Nevertheless, the $d_{12}$ value for different global regions can be estimated from the birth rate data.

The coefficients of transition to the following age groups $d_{21}$ and $d_{32}$ are determined from the duration of the respective age group and the hypothesis of a uniform distribution of ages within the group—namely: $d_{21} = 1/15$; $d_{32} = 1/50$.

Genetically stipulated diseases and a deficit of protein in the food of children are the main causes of transitions from a younger age group to the disabled group. Thus, $d_{41} = d_{41}(G, A)$ is the age function of its arguments. Coefficients $d_{42}$ and $d_{43}$ are small compared with $d_{41}$ and determined by the sum of vital parameters of the environment, the level of medical service, and other indicators of anthropogenic medium (investment). As a first approximation, suppose $d_{4k} = \Delta_k \mu_k$ $(K = 23)$, where $\Delta_k \ll 1$.

Demographic matrix $D$ has a certain set of properties which make it possible to reveal the typical trends of the demographic process. A maximum eigenvalue is:

$$\lambda_{\max}(D) = \max\{\lambda_1; d_{33}; d_{44}\} > 0$$

where

$$\lambda_1 = 0.5(d_{11} + d_{22}) + 0.25[(d_{11} + d_{22})^2 + d_{12}d_{22} - d_{11}d_{22}]^{1/2}.$$

It follows from (1.1) that if $\lambda_{max}(D) \geq 1$, then $\lambda_{max}(D) = \lambda_1$; that is, the rate of *Homo sapiens* population growth in the neighborhood of stationary age distribution is determined by the $\lambda_1$ value. The stationary age structure is calculated as an eigenvector corresponding to $\lambda_{max}(D)$. The coefficients and modifying dependences of matrix $D$—calculated, as a rule, from the data of Stempell (1985)—give the dominating eigenvector $P_D = (0.369; 0.576; 0.05; 0.005)$. The first three components of vector $P_D$ coincide, to within second sign, with the global data on the age complement of the global population.

The matrix version of the global model demographic unit has a time step of one year. Inclusion of this unit in the chain of other global model units with an arbitrary time step $\Delta t < 1$ year is needed as an adjusting procedure. This is possible, for instance, if the demographic unit is included only at time moment multiples of 1 year. In this version spasmodic changes in population size will cause some imbalance in continued trajectories of the global system. Another procedure free of this drawback consists in the use of a prognostic equation:

$$S_{t+\Delta t} = S_t + \Delta t(D - I)S_t \tag{1.2}$$

where $I$ is an identity matrix. One can demonstrate that for the age structure vectors coinciding in direction with the eigenvector of matrix $D$, the main term of a relative error accumulated during one year with a time step $\Delta t = 1/n$ constitutes $(1 - 1/n)(\eta - 1)^2/2$. Let the right-hand side of (1.1) be denoted by $f(S, \Delta t)$. Then if $e$ is the eigenvector of matrix $D$ with an eigenvalue $\eta$, then with the time step $\Delta t = 1/n$:

$$f(e, 1/n) = e + (1/n)(D - I)e = (1 + [\eta - 1]/n)e.$$

It is apparent that $S_{t+n\Delta t} = f^{(n)}(S_t, 1/n)$ and $S_t = ef^{(n)}(e, 1/n) = (1 + [\eta - 1]/n)^n e$. But if $\Delta t = 1$, then $S_{t+1} = f(e, 1) = \eta e$, so that a relative error resulting from a subdivision of the 1-year time interval is:

$$(1 + [\eta - 1]/n)^n - \eta = n(n - 1)(\eta - 1)^2/(2n^2) + O((\eta - 1)^3).$$

In other words, the accuracy of an approximation of (1.2) rises as $\Delta t$ approaches unity.

### *1.6.1.2 Differential model of population dynamics*

The matrix model considered above enables one to use demographic statistics by age group but requires knowledge of many parameters, which leads to uncertainties in the global model. Therefore, a second version of the demographic unit is suggested which only simulates the dynamics of the total size of population. This version assumes that the impact of numerous environmental factors and social aspects on population size dynamics $G_i$ in the $i$th region is manifested through birth rate $R_{Gi}$ and mortality $M_{Gi}$:

$$dG_i/dt = (R_{Gi} - M_{Gi})G_i \quad (I = 1, \ldots, m) \tag{1.3}$$

Birth rate and mortality depend on food provision and its quality, as well as environmental pollution, living standards, energy supply, population density, religion, and other factors. Within the global model all these factors will be taken into account following the principle:

$$R_G = (1 - h_G)k_G G H_{GV} H_{GO} H_{GG} H_{GMB} H_{GC} H_{GZ}$$

$$M_G = \mu_G H_{\mu MB} H_{\mu G} H_{\mu FR} H_{\mu Z} H_{\mu C} H_{\mu O} G + \tau_{GO} \tau_{GC}^{\omega(G)}$$

where for ease of simplicity we shall omit the $i$ index attributing the relationship to the $i$th region; $h_G$ is the coefficient of food quality ($h_G = 0$ when the quality of food is perfect); coefficients $k_G$ and $\mu_G$ point to the levels of birth rate and mortality, respectively; indices $\tau_{GO}$ and $\tau_{GC}$ characterize the dependence of regional population mortality on O and C indicators of the environmental state (within the global model it is the content of $O_2$ and $CO_2$ in the atmosphere) manifesting through human physiological functions; coefficient $\omega(G)$ characterizes the extent of the influence of population density on mortality (in the present conditions $\omega(G) \approx 0.6$); functions $H_{GV}(H_{\mu FR})$, $H_{GO}(H_{\mu O})$, $H_{GC}(H_{\mu C})$, $H_{GMB}(H_{\mu MB})$, $H_{GG}(H_{\mu G})$, and $H_{GZ}(H_{\mu Z})$ describe, respectively, the impact on birth rate (mortality) of environmental factors, such as food provision, atmospheric $O_2$ and $CO_2$ concentration, living standards, population density, and environmental pollution. Functions $H_{\mu O}$ and $H_{\mu C}$ approximate the medico-biological dependences of mortality on atmospheric gas composition. Let us consider all these functions in more detail. For this purpose we shall formulate a number of hypotheses concerning the forms of dependences of mortality and birth rate on various factors.

The results of numerous studies with national specific features taken into account enable one to assume the following dependence as an approximation of the function $H_{GV}$: $H_{GV} = 1 - \exp(-V_G)$, where $V_G$ is an efficient amount of food determined as a weighted sum of the components of *Homo sapiens*' food spectrum:

$$V_{Gi} = k_{G\Phi i}\Phi + k_{GFi}\left(F_i + \sum_{j \neq i} a_{Fji} F_i\right) + k_{Gri} I_i (1 - \theta_{Fri} - \theta_{uri})$$
$$+ k_{GLi} L_i + k_{GXi}\left[(1 - \theta_{FXi}) X_i + (1 + \nu_{FXi}) \sum_{j \neq i} a_{Xji} X_j\right]$$

Here $\Phi$ is the volume of vegetable food obtained from the ocean; $F$ protein food; $L$ food from forests; $X$ vegetable food produced by agriculture; $I$ fishery products; coefficients $k_{G\Phi}$, $k_{GF}$, $k_G$, $k_{GL}$, and $k_{GX}$ are determined using the technique described by Kondratyev *et al.* (2004, Chapter 4, Section 4.5); $a_{Fji}$ and $a_{Xji}$ are, respectively, shares of the protein and vegetable food in the $i$th region available for use by the population of the $i$th region; $\theta_{FXi}$ and $\nu_{FXi}$ are shares of vegetable food produced and imported by the $i$th region, respectively, to produce the protein food; $\theta_{Fri}$ and $\theta_{uri}$ are shares of fishery spent in the $i$th region on protein food production and fertilizers, respectively.

With increasing food provision, population mortality drops to some level determined by the constant $\rho_{1,\mu G}$ at a rate $\rho_{2,\mu G}$, so that $H_{\mu FR} = \rho_{1,\mu G} + \rho_{2,\mu G}/F_{RG}$, where normalized food provision $F_{RG}$ is described by the relationship $F_{RG} = F_{RG}(t) = V_G/G(F_{RG}(t_0) = F_{RGO} = V_G(t_0)/G(t_0))$. Similarly, we assume that the birth rate depending on the living standard $M_{BG}$ of the population is described by the function with saturation so that the maximum birth rate is observed at low values of $M_{BG}$, and at $M_{BG} \to \infty$ the birth rate drops to some level determined by the value of $a_{*GMB}$. The rate of transition from maximum to minimum birth rate with changing $M_{BG}$ is set by the constants $a_{1,GMB}$ and $a_{2,GMB}$:

$$H_{GMB} = a_{*GMB} + a_{1,GMB} \exp(-a_{2,GMB} M_{BG})$$

where

$$\begin{aligned} M_{BG} &= (V/G)\{[1 - B - U_{MG} - U_{ZG}]/[1 - B(t_0) - U_{MG}(t_0) - U_{ZG}(t_0)]\} \\ &\quad \times [E_{RG}(t)/E_{RG}(t_0)], \\ E_{RG}(t) &= 1 - \exp[-k_{EG} M(t)/M(t_0)]. \end{aligned}$$

The dependence of mortality on the living standard is described by the decreasing function:

$$H_{\mu MB} = b_{1,\mu G} + b_{2,\mu G} \exp(-b_{*\mu G} M_{BG}).$$

This function shows that population mortality with an increasing per capita share of capital drops with the rate coefficient $b_{*\mu G}$ to the level $b_{1,\mu G}$.

Birth rate and mortality within certain limits are, respectively, decreasing and increasing functions of population density:

$$H_{GG} = G_{1,G} + G_{*G} \exp(-G_{2,G} Z_{GG}), \qquad H_{\mu G} = \theta_{1,\mu G} + \theta_{2,\mu G} Z_{GG}^{\omega\mu G},$$

where $Z_{GG} = G(t)/G(t_0)$.

Finally, an important aspect of *Homo sapiens* ecology is the environmental state. In this connection the problems of anthropobiocenology have been widely discussed in the scientific literature and many authors have tried to discover the requisite regularities. Without going into the details of these studies, most of which cannot be used in the global model, we shall confine ourselves to the following dependences:

$$\begin{aligned} H_{GZ} &= l_{1,G} \exp(-l_{*G} Z_{RG}), \\ H_{\mu Z} &= n_{1,\mu G} + n_{2,\mu G} Z_{RG}, \\ \tau_{GC} &= \begin{cases} \tau_{1,GC} + \tau_{2,GC}(C_a - C_{1,G}) & \text{for } C_a > C_{1,G}, \\ \tau_{1,GC} & \text{for } 0 \le C_a \le C_{1,G} \end{cases} \\ H_{\mu O} &= f_{1,\mu G} + f_{2,\mu G}/O(t), \\ H_{\mu C} &= \exp(k_{\mu G} C_a), \\ H_{GO} &= 1 - \exp(-k_{GO} O), \\ H_{GC} &= \exp(-k_{GC} C_a), \\ Z_{RG} &= Z(t)/Z(t_0), \\ \tau_{GO} &= \begin{cases} \tau_{1,GO} & \text{for } O > O_{1,G}, \\ \tau_{2,GO} - (\tau_{2,GO} - \tau_{1,GO}) O/O_{1,G} & \text{for } 0 \le O \le O_{1,G} \end{cases} \end{aligned}$$

where $C_a$ is atmospheric $CO_2$ concentration, $C_{1,G}$ and $O_{1,G}$ are safe-for-human levels of atmospheric $CO_2$ and $O_2$ content.

#### *1.6.1.3 Megapolitan zones*

A megapolitan zone is an Earth surface area with high urbanization, developed industry, and other human activity attributes concentrated in a limited territory (urban, suburban, and peri-urban areas). The number of territories like this is constantly growing in the world and their total area increases with the growing size of the global population. In this context, geographer Gottmann (1987) wrote: "... the Megapolitan concept seems to have popularized the idea that the modern cities are better reviewed not in isolation, as centers of a restricted area only, but rather as parts of 'city-systems', as participations in urban networks revolving in widening orbits." According to *European Spatial Planning* (Faludi, 2002), the primary urban unit for integration into the global economy is the trans-metropolitan area, or what Gottman (1964) refers to as a "megalopolis". Andoh and Iwugo (2002) use the term "megapolis". Characteristic examples of large megapolises are Moscow, Tokyo, New York, HoChiMinh City, etc. For example, the megapolitan total population in the U.S.A. comprises 68% of the U.S. population (Lang and Dhavale, 2005). The Moscow megapolitan zone is characterized by high concentrations of sources of anthropogenic pollution over a small area (energy enterprises, chemical industry, and automobile transport). Their share in total emissions constitutes 90.4%. The polluted air plume from Moscow can be observed at a distance of about 100 km from the city. The state of the water bodies within the megapolis is determined by the input of sewage, surface run-off as well as run-off from industrial enterprises to the main rivers Moskwa and Yausa, as well as into 70 rivers and springs around the megapolis. The concentration of chemicals in the river Moskwa both in the city and downstream varies widely depending on the season. For instance, copper ion content varies during the year from 0.004 to 0.013 $mg\,l^{-1}$ (4–13 permissible concentration levels, or PCLs) with a maximum in the spring. The content of oil products varies within 0.25 to 0.6 $mg\,l^{-1}$ (5–12 PCLs). A similar situation exists in the small river Yausa where the concentration of copper varies from 0.007 to 0.12 $mg\,l^{-1}$ and that of oil products—from 0.38 to 0.7 $mg\,l^{-1}$. The ecological service of the megapolis monitors the local concentrations of pollutants forming the respective time series of data on the state of the environment. There is a global network of megapolises, consideration of which in the global model results in fine-tuning its accuracy. Many megapolises are in the stage of formation. Khoshimin (Vietnam) is an example of a young megapolis, as are those in the adjacent provinces of South Vietnam—Dong Nai, Bin Zyong and Baria-Vung Tau. Khoshimin occupies an important place in the economy of Vietnam covering 75% of the industrial production of South Vietnam and 50% for the whole country (Bui Ta Long, 1998; Si and Hai, 1997). Development of the megapolis infrastructure includes getting services of ecological and sanitary control of the environment involved at an early stage, which makes future planning of the environment possible.

Analysis of the data on the structure of the environment of megapolises suggests that for its assessment it is possible to develop a sample system of models simulating the transport and propagation of pollutants in the atmosphere and in water bodies. The input information for this model could be both from the monitoring systems and from the global model.

The characteristic linear dimensions of a megapolis constitute tens of kilometers. For example, the Osaka–Kyoto–Kobe prefecture and New York City areas equal 11,170 $km^2$ and 54,520 $km^2$ with population densities of 1,669 and 402 persons per square kilometer, respectively. Such areas mean that—to simulate the processes of atmospheric transport of pollutants—a Gaussian-type model can be used. To assess the quality of the atmosphere in a megapolis it is sufficient to work out the composition of the Gaussian streams and at their points of intersection to summarize the concentrations of the respective types of pollutants. Population density is a key indicator for the assessment of atmospheric pollution and water quality. Water resources in many megapolises for domestic, industrial, and commercial use is becoming scarce as a result of water bodies being polluted by wastewater containing heavy metals, bacteria, etc. Flooding is a serious problem in megapolises where drainage systems are poor and there exists a relatively high water table and flat topography. All these aspects need to be taken into account in the GMNSS to make ecodynamic forecasting more precise.

### 1.6.2 Consumption society: ecological restrictions

Characterizing the consumption dynamics observed during recent decades, Starke (2004) gives China as an example. Until recently, China has been "the country of bicycles". However, a traveler who visited Beijing, Shanghai, and other places, in 1980 will now see quite another picture. By 2002, there were about 2 million automobiles in China, and during 2003 their number increased daily by 11,000 reaching 4 million. In 2003, the sale of cars increased by more than 80%, and it is expected that by 2015 the total number of privately owned automobiles will reach 150 million.

A considerable part of the global population (more than 1.7 billion people) make up the "class of consumers", about one-half residing in developing countries. During recent decades the scale of consumption in developed countries has been growing, called the "revolution of consumption". The total per capita consumption due to goods and services reached $20 trillion in 2000, exceeding the 1960 level by $4.8 trillion (1995 dollar values) with huge differences characterizing the situation in different countries. Sixty per cent of per capita consumption falls on 16% of the global population living in the U.S.A. and Western Europe, whereas the share of consumption by one-third of the global population in South-East Asia and sub-Saharan Africa constituted only 3.2% (Table 1.6).

In 1999 some 2.8 billion people lived at the minimum level to satisfy their needs ($2 per day), and about 1.2 billion at the level of extreme poverty (<$1 per day). The "class of consumers" determined as persons with an annual income >$7,000 (this level is considered in Western Europe as the limit of poverty) numbered 1.7 billion people in 2002.

**Table 1.7.** Global distribution of the class of consumers.

| Region | Quantity of the class of consumers (millions) | Share of the class of consumers (%) | Share with respect to global class of consumers (%) |
|---|---|---|---|
| U.S.A. and Canada | 271.4 | 85 | 16 |
| Western Europe | 348.9 | 89 | 20 |
| Eastern Asia and Pacific Ocean region | 494.0 | 27 | 29 |
| Latin America and Caribbean region | 167.8 | 32 | 10 |
| Eastern Europe and Central Asia | 173.2 | 36 | 10 |
| Southern Asia | 140.7 | 10 | 8 |
| Australia and New Zealand | 19.8 | 84 | 1 |
| Near East and Northern Africa | 78.0 | 25 | 4 |
| Sub-Saharan Africa | 34.2 | 5 | 2 |
| Developed countries | 912 | 80 | 53 |
| Developing countries | 816 | 17 | 47 |
| *Global* | *1728* | *28* | *100* |

As seen from Table 1.7, almost half the class of consumers is in developing countries, the share of China and India constitutes >20%, though, of course, an average Chinese or Hindu consumes far less than an average Western European.

The region where the situation is most serious is sub-Saharan Africa, where consumption in 2001 was 20% less than during the previous two decades. According to available estimates, by 2015 the number of the class of consumers will double, at least. The data in Table 1.8 illustrate the size of the class of consumers in 10 countries in 2002, and in Table 1.9 the share of family expenditure on food in different countries in 1998.

Table 1.10 contains information on various indicators about life support from the data for 2000. In 2002, about 1.12 billion families had (globally averaged) at least one TV set (of them 31% used cable TV).

A substantial feature of current domestic consumption is that it is not concerned with survival, but is connected with acquisition of luxury items (sports cars, jewelry, vacations, etc.). The respective expenditure reached gigantic levels (tens of billions of dollars). The needs of the poorest in the world could be satisfied by just a small part of this expenditure.

Naturally, the necessity to satisfy increased consumption brings forth higher requirements on the scale of use of natural resources. For instance, in the period 1960–1995, the global scale of the use of minerals increased by a factor of 2.5, metals by a factor of 2.1, wood 2.3, and synthetic materials 5.6. This growth has considerably exceeded the rate of increase in population size by its extreme non-uniformity. For instance, the U.S.A.—where population size constitutes about 5% of the global

**Table 1.8.** Quantity of the class of consumers.

| Country | Quantity of the class of consumers (millions) | Share of population size (%) |
|---|---|---|
| U.S.A. | 242.5 | 84 |
| China | 239.8 | 19 |
| India | 121.9 | 12 |
| Japan | 120.7 | 95 |
| Germany | 76.3 | 92 |
| Russia | 61.3 | 43 |
| Brazil | 57.8 | 33 |
| France | 53.1 | 89 |
| Italy | 52.8 | 91 |
| U.K. | 50.4 | 86 |

**Table 1.9.** Share of family expenditure on food.

| Country | Per capita, personal consumption expenditure ($) | Share of expenditure on food (%) |
|---|---|---|
| Tanzania | 375 | 67 |
| Madagascar | 608 | 61 |
| Tajikistan | 660 | 48 |
| Lebanon | 6,135 | 31 |
| Hong Kong | 12,468 | 10 |
| Japan | 13,568 | 12 |
| Denmark | 16,385 | 16 |
| U.S.A. | 21,515 | 13 |

**Table 1.10.** Domestic consumption.

| Country | Per capita domestic expenditure in 1995 | Electric energy | TV set | Telephone | Mobile telephone | PC |
|---|---|---|---|---|---|---|
| | ($) | (kW $hr^{-1}$/head) | (Per 1000 people) | | | |
| Korea | 194 | 81 | 68 | 6 | 4 | 7 |
| India | 294 | 355 | 83 | 40 | 6 | 6 |
| Ukraine | 558 | 2,293 | 456 | 212 | 44 | 18 |
| Brazil | 1,013 | 976 | 217 | 104 | 43 | 16 |
| South Korea | 2,779 | 1,878 | 349 | 223 | 167 | 75 |
| Germany | 6,907 | 5,607 | 363 | 489 | 621 | 556 |
| Denmark | 18,580 | 5,963 | 586 | 650 | 682 | 435 |
| U.S.A. | 21,707 | 12,331 | 835 | 659 | 451 | 625 |

one—consume about one-quarter of the global resources of fossil fuel. The U.S.A., Canada, Australia, Japan, and Western Europe—with combined populations about 15% of the global size—use every year 61% of produced aluminum, 60% of lead, 59% of copper, and 49% of steel. The annual consumption of aluminum by an average American constitutes 22 kg, in the case of an average Hindu it is only 2 kg, and of an average African less than 1 kg. Despite large reserves of re-cycled raw material, under real conditions re-cycling is still insignificant, constituting, for instance, in the case of copper only 13%. Re-cycling of municipal waste (even in ORCD countries where it is about 24%) is still inadequate.

The general conclusion drawn by the authors of the report about the state of the World (Starke, 2004) is that "in the consumption society with an abundance of food and other goods, conditions are created for the unhealthy level of consumption." The growth in consumption is facilitated by cheap energy and improved transport. The introduction of new technologies is accelerating. Although it took 38 years for the U.S.A. to increase its number of radio listeners to 50 million, the same level for TV was reached in 13 years, and in the case of the Internet it was 4 years. The need to sell goods produced has stimulated a rapid growth in expenditure on advertising reaching in 2002, on a global scale, $446 billion (in 2001 prices), which constituted almost a ninefold increase compared with 1950. At present, each credit card in the U.S.A. spends about $1,900, which is equivalent to the average per capita income, at least, in 35 poor countries.

This "unhealthy level of consumption" has an increased impact on the environment and natural resources. Table 1.11 contains the respective summary (Starke, 2004). Almost all ecosystems of the world have been subjected to various negative impacts.

As an indicator of the impact on global ecosystems, a value called an "ecological fingerprint" was introduced. It characterizes the size of the productive land area that is necessary to provide production of the needed resources and to dispose of waste. Estimates have shown (Starke, 2004) that the global mean per capita size of the biologically productive land territory should constitute 1.9 ha, and from the viewpoint of waste 2.3 ha. Of course, these average quantities mask the real heterogeneity of the size of the "ecological fingerprint" varying from 9 ha in the case of an average American to 0.47 ha for an average citizen of Mozambique. When we take these quantities into account it becomes clear that total global consumption exceeded the limits of ecological capacity of the planet as far back as the end of 1970–early 1980. Of course, life under these conditions is only possible by using natural resources—as in the case of groundwater, for example.

Excess consumption (from the viewpoint of load on ecosystems) in many countries correlates with a decrease in health indicators, resulting in an enhancement of "consumption diseases". For instance, smoking increases annual mortality in the world by 5 million people. From the data for 1999, smoking-induced expenditure on medical services and economic losses in the U.S.A. reached $150 billion, exceeding about 1.5 times the profits of the five largest manufacturers of tobacco. Also important are the consequences of obesity due to poor nutrition (in 1999 medical expenditure cost at least $117 billion). On average, against the background of material

**Table 1.11.** Trends of impacts on natural resources and the environment.

| Indicators | Trends |
|---|---|
| Fossil fuels and the atmosphere | In 2002, global use of coal, oil, and natural gas was 4.7 times higher than in 1950. The level of $CO_2$ concentration in the atmosphere in 2002 was 18% higher than in 1960, and, apparently, 31% higher than before the industrial revolution (1750). |
| Degradation of ecosystems | During recent decades, more than one-half of the Earth's wetlands (from coastal marshes to intra-continental lowlands exposed to floods) have been lost due to human intervention, about one-half of the original forests have been lost, 30% of the remaining forests have degraded. In 1999, the use of wood as a fuel and in industry had more than doubled compared with 1950. |
| Sea level | In the 20th century the World Ocean level rose by 10–20 cm (at an average rate of increase of 1–2 mm/year due to melting of continental glaciers and thermal expansion of water masses (under conditions of climate warming). |
| Soil/land surface | About 10–20% of agricultural land have degraded in many respects (loss of fertility), which has brought about a decrease (during the last 50 years) in yield by 18% on cultivated lands and by 4% on pasture. |
| Fishery | In 1999, the catch increased 4.8 times as against 1950. Creation of the modern fishing fleet has led to a 90% increase in the harvest of tuna, cod, marlin, swordfish, shark, halibut, flounder, cramp-fish. |
| Water | Over-exploitation of groundwaters has led to a decrease in their level in many agricultural regions of Asia, North Africa, the Middle East, and the U.S.A. The impact of sewage, fertilizers, pesticides, oil product remains, heavy metals, stable phosphoro-organic compounds, and radioactive substances has led to a substantial decrease in groundwater quality. |

prosperity, the "social health" of the U.S. society during the last three decades has considerably worsened, which has shown itself through an increase in the scale of poverty, suicides among young people, absence of an adequate medical service, and drastic inequality of income.

As a result of consumption continuing to increase in developed countries and, on the other hand, the serious negative aspects of the development of the consumption society begs answers in particular to the following questions (Marfenin and Stepanov, 2005; Meler and Munasinghe, 2005; Millstone and Lang, 2003; Pulido *et al.*, 2005; Davies, 2003; Nazaretian, 2004; Roels, 2005):

- Does life standard exist determined by the growth of consumption characteristic of the global class of consumers?
- Is balanced consumption possible (in accordance with the requirement for environmental protection)?

- Can the ways in which the consumption society develops be changed in order to ensure sustainable development?
- Can society satisfy the needs of all its members and consider it as the main priority?

It is clear from this that the consumption society economy based on unlimited consumption is faulty because of the unavoidable consequences for the environment. Realization of this conception would break all economic, ecological, and social limitations (thresholds). The major conclusion is that the consumption society cannot be sustained.

Now we shall consider an analysis of some of the most important systems of life support.

### 1.6.3 Freshwater problems

Freshwater problems deserve special attention, since, on the one hand, water is a key component of ecosystems and, on the other hand, about one-third of the global population is under the threat of a chronic shortage of water in a few decades time. This and other facts have determined the declaration of the U.N. Freshwater Decade (2005–2015). The following facts characterize the present state of global water resources (WCSDG, 2004):

- The total volume of water resources on Earth constitutes about 1.4 billion $km^3$.
- The volume of freshwater resources is about 35 million $km^3$ or 2.5% of the total water volume.
- Of these freshwater resources, about 24 million $km^3$ or 68.9% exist in the form of ice and constant snow cover in the mountains, the Antarctic, and Arctic.
- About 8 million $km^3$ or 30.8% are under ground in the form of groundwater (in shallow and deep basins at depths of 2,000 m), soil moisture, bog water, and permafrost.
- In freshwater lakes and rivers there are about 105,000 $km^3$ or 0.3% of the global freshwater supply.
- The general volume of freshwater supply to be consumed by ecosystems and populations constitutes about 2 million $km^3$; that is, less than 1% of the global freshwater supply.
- 3,011 freshwater biological species are included in the list of species that are either under threat or have disappeared, 1,039 of them being fish. Four of the five existing species of river dolphins and two of the three existing species of seacows as well as about 40 species of freshwater tortoises and more than 400 species of freshwater mollusk are under threat.
- Annual groundwater diversion is evaluated at 600–700 $km^3$ or about 20% of total global water diversion. About 1.5 billion people drink groundwater.
- From estimates, 70% of global freshwater was spent in 2000 on agriculture.
- The average per capita consumption of water in developed countries is about 10 times greater than in developing countries. In developed countries this

indicator varies between 500 and 800 l/day, and in developing countries from 60 to 150 l/day.

- Industry consumes about 20% of global freshwater diverted. From 57 to 69% of global water diversion are spent on energy production at hydro and nuclear power stations, 30–40% in industrial processes, and 0.5–3% on thermal energy.

Freshwater problems are very broad and diverse. Table 1.12 (WCSDG, 2004) presents a summary of the aspects of these problems on which human life depends in the context of the U.N. Millennium Declaration (UNEP, 2002, 2004).

On the whole, it should be stated that provision of sufficient pure water is of fundamental importance for achieving the goals of socio-economic development and environmental protection. Of serious concern here is the enhanced anthropogenic impact on the environment. For instance, the areas of freshwater wetlands that play an important role in natural purification of water and in water cycle formation has almost doubled during the last 20 years. Meanwhile, according to the economic assessment of the role of wetland, their losses are equivalent to $20,000 per hectare per year. About 20% of the 10,000 species of freshwater fish are either on the verge of extinction or are extinct. The number of large dams in the world has increased from 5,000 in 1950 to more than 45,000 now, which has serious ecological outcomes (Clarke and King, 2004; Davidenkov and Kesler, 2005; Evans, 2005; Giupponi *et al.*, 2006; Pulido *et al.*, 2005).

The geographical distribution of freshwater resources is non-uniform: about half the global resources are shared by six countries (Brazil, Russia, Canada, Indonesia, China, Columbia). This non-uniformity is also characteristic of individual countries. For instance, China has 7% of global freshwater resources (with a 21% share of global population), but most of the country is arid. Naturally, countries with water shortages have to use groundwater, bringing about a gradual decrease in their level. The data in Table 1.13 characterize the levels of per capita groundwater diversion in different countries as of 2000 (Kondratyev and Krapivin, 2005c).

Moreover, these data illustrate current socio-economic contrasts. About every fifth person in the developing world (they total about 1.1 billion) is subjected every day to the risk of falling ill for lack of good-quality drinking water. The main problem here is not the absence of water at all but unfavorable socio-economic conditions.

The major consumer of the freshwater of rivers, lakes, and underground sources is known to be agriculture (about 70% on a global scale and up to 90% in many developing countries). Since the enhanced use of irrigation will face a shortage of freshwater resources in the near future, the efficient use of freshwater becomes more and more important, but can be substantially increased: this refers to the use of micro-irrigation (including drip irrigation), the scale of which remains limited.

The considerable potential for water economy is connected with food production. As seen, for instance, from Table 1.14, production of 10 g of protein in the form of beef requires a water expenditure five times greater than in the case of rice. To obtain 500 food calories, this difference reaches a factor of 20. With an abundant

**Table 1.12.** Aspects of vital activity connected (directly of indirectly) with freshwater problems.

| Millennium Declaration | Problems to be resolved |
|---|---|
| *Goal 1*<br>To extirpate extreme poverty and hunger | Over the period 1990–2015 to halve the share of the population with an income of less than $1 per day and to halve the number of starving people. |
| *Goal 2*<br>To provide universal elementary education | By the year 2015 to provide the possibility for both boys and girls in the world to be elementarily educated. |
| *Goal 3*<br>To promote equality of men and women and broaden the ability of women | To liquidate inequality of boys and girls in their access to elementary and secondary education preferably by 2005 and to all levels of education by no later than 2015. |
| *Goal 4*<br>To reduce infant mortality | To reduce by two-thirds the mortality of children under 5 years over the period 1990–2015. |
| *Goal 5*<br>Improve mothers' health | To reduce by three-quarters the mortality of mothers over the period 1990–2015. |
| *Goal 6*<br>To struggle against HIV/AIDS, malaria, and other diseases | By the year 2015 to stop the propagation of HIV/AIDS, malaria, and other basic diseases and start reducing its scale. |
| *Goal 7*<br>To provide ecological stability | To take into account the principles of sustainable development in national strategies and programs and reduce the trend of ecological resource loss. By the year 2015 to halve the share of the population without access to pure drinking water, to substantially improve the life of 100 million slum dwellers, as a minimum. |
| *Goal 8*<br>To arrange a global scale partnership in the interest of development | To continue the formation of transparent, predictable, and non-discriminatory commercial and financial systems based on legal norms. To provide control, development, and reduction of poverty at national and international levels. To satisfy the special needs of the least developed countries having no access to the sea as well as small island developing countries (SIDC). In cooperation with the private sector, to take measures so that everybody can use the benefits of new technologies. |
| Plan of realization of decisions of the Summit of the Millennium | By the year 2015 to halve the share of the population without access to safe drinking water partially because of want of money (as foreseen in the Millennium Declaration) and the share of population without access to basic sanitation. By the year 2005, to work out plans for complex control of water resources and efficiency of water use. |

**Table 1.13.** Annual per capita groundwater diversion ($m^3$/person) in different countries.

| Country | Level of per capita groundwater diversion ($m^3$) |
|---|---|
| Ethiopia | 42 |
| Nigeria | 70 |
| Brazil | 348 |
| South Africa | 354 |
| Indonesia | 390 |
| China | 491 |
| Russia | 527 |
| Germany | 574 |
| Bangladesh | 578 |
| India | 640 |
| France | 675 |
| Peru | 784 |
| Mexico | 791 |
| Spain | 893 |
| Egypt | 1,011 |
| Australia | 1,250 |
| U.S.A. | 1,932 |

**Table 1.14.** Water expenditure (in litres) to produce 10 g of protein and 500 calories.

| Food | 10 g of protein | 500 calories |
|---|---|---|
| Potatoes | 67 | 89 |
| Peanut | 90 | 210 |
| Onions | 118 | 221 |
| Maize (grain) | 130 | 130 |
| Legumes | 132 | 421 |
| Wheat | 135 | 219 |
| Rice | 204 | 251 |
| Egg | 244 | 963 |
| Milk | 250 | 758 |
| Poultry | 303 | 1,515 |
| Pork | 476 | 1,225 |
| Beef | 1,000 | 4,902 |

meat diet, the average American needs 5.4 l of water per day, whereas in the case of vegetarians this amount is halved.

A serious problem is urban water supply and bringing about measures for water economy. Still more urgent is the problem of the industrial use of freshwater, which takes 22% of globally used freshwater resources (59% in developed and 10% in developing countries).

### 1.6.4 Energy production and consumption

During the period 1850–1970, global population size has more than tripled and energy production has increased 12 times. Since then (by 2002) the population size has increased by 68% and fossil fuel consumption by 73% (Starke, 2004). It is important to note, however, that there does not appear to be any rigid bond between energy production and economic growth. Measures on energy saving between 1970 and 1997 provided economic growth, despite "power inputs" of production reduced by 28%. The data in Table 1.15 illustrate the distribution of energy consumption in different regions in some countries, and Table 1.16 the global dynamics of GDP and the trend of fossil fuel consumption (Starke, 2004). Global fossil fuel consumption increased in 2002 by 1.3% (up to 8,034 million tons of equivalent oil) compared with 0.3% in 2001. The level of fossil fuel consumption increased by a factor of 4.7 compared with 1950. Currently fossil fuel constitutes 77% of global energy consumption.

As for consumption of various kinds of fossil fuel, the respective trends are rather diverse. Whereas in 2002 the global mean consumption of oil constituted only 0.5%, in China it reached 5.7%, the Middle East 2.5%, and the former U.S.S.R. countries 1.9%, which is mainly determined by a growing export. In the U.S.A. (consuming 26% of global oil) the level of consumption increased negligibly. In some countries (Japan, South Korea, Australia, New Zealand) there was a 0.6% decrease (on average), and in the countries of Latin America a decrease of 2.6%.

After the short-term but considerable decrease in coal consumption in the late 1990s, in 2002 its global consumption increased (compared with 2001) by 1.9% (2298 million tons of oil equivalent). In the U.S.A. (25% of global coal consumption) there was a 0.5% decrease, and in China (23% of global consumption) a 4.9% increase (despite a ban on the use of coal in some regions where smog and acid rain often occur).

The global mean increase in natural gas consumption constituted 2% (2,207 million tons of oil equivalent), but in the U.S.A. (27% of global consumption) there was a decrease (during the first 10 months of 2002 compared with the respective period in 2001) by 3.7%, which was determined mainly by the mild winter. The decrease in natural gas consumption was also observed in some other developed countries, reaching a maximum in Japan (10.4%), but in Norway natural gas consumption grew substantially (up to 81%). On the whole, natural gas is characterized by a very rapid increase in consumption (compared with other fossil fuels), ensuring now almost 24% of the global energy consumption (compared with 22.5% 10 years ago). This growth is explained by various factors, including the accessibility of natural gas in many countries and its less severe (compared with other fossil fuels) negative impact on the environment (McElroy, 2002; McHarry *et al.*, 2004; Munasinghe *et al.*, 2001; Nikanorov, 2005; Sedwick *et al.*, 2005).

The levels of energy consumption in various countries differ strongly (Starke, 2004). The people of the richest countries consume, on average, per capita 25 times more energy than those in the poorest countries. For up to 2.5 billion people—living mainly in Asia and Africa—the main source of energy is still wood (or other kinds of

**Table 1.15.** Regional consumption of energy from different sources in 2004 for a number of selected countries (BP, 2005).

| Region and countries | Oil ($10^6$ t) | Gas ($10^6$ t) | Coal ($10^6$ t of oil equivalent) | Nuclear energy ($10^6$ t of oil equivalent) | Hydroelectric energy ($10^6$ t of oil equivalent) |
|---|---|---|---|---|---|
| *North America* | *1,122.4* | *784.3* | *603.8* | *210.4* | *141.9* |
| U.S.A. | 937.6 | 646.7 | 564.3 | 187.9 | 59.8 |
| Canada | 99.6 | 89.5 | 30.5 | 20.5 | 76.4 |
| Mexico | 88.2 | 48.2 | 9.0 | 2.1 | 5.7 |
| *South and Central America* | *221.7* | *117.9* | *18.7* | *4.4* | *132.1* |
| Argentina | 18.7 | 37.9 | 0.7 | 1.8 | 6.8 |
| Brazil | 84.2 | 18.9 | 11.4 | 2.6 | 72.4 |
| Venezuela | 26.3 | 28.1 | 0.1 | — | 16.0 |
| Chile | 10.7 | 8.2 | 2.5 | — | 4.9 |
| *Europe and Eurasia* | *957.3* | *1,108.5* | *537.2* | *287.2* | *184.7* |
| Germany | 123.6 | 85.9 | 85.7 | 37.8 | 6.1 |
| Russia | 128.5 | 402.1 | 105.9 | 32.4 | 40.0 |
| France | 94 | 44.7 | 12.5 | 101.4 | 14.8 |
| Ireland | 89.5 | 4.1 | 1.8 | — | 0.2 |
| U.K. | 80.8 | 98.0 | 38.1 | 18.1 | 1.7 |
| Turkey | 32.0 | 22.1 | 23.0 | — | 10.4 |
| Ukraine | 17.4 | 70.7 | 39.4 | 19.7 | 2.7 |
| Uzbekistan | 6.0 | 49.3 | 1.2 | — | 1.7 |
| *Middle East* | *250.9* | *242.2* | *9.1* | — | *4.02* |
| Iran | 73.3 | 87.1 | 1.1 | — | — |
| Saudi Arabia | 79.6 | 64.0 | — | — | — |
| Kuwait | 13.7 | 9.7 | — | — | — |
| *Africa* | *124.3* | *68.6* | *102.8* | *3.4* | *19.8* |
| Algeria | 10.7 | 21.2 | 0.8 | — | 0.1 |
| Egypt | 26.7 | 25.7 | 0.7 | — | 3.3 |
| South Africa | 24.9 | — | 94.5 | 3.4 | 0.8 |
| *Asia and Oceania* | *1,090.5* | *367.7* | *1,506.6* | *118.9* | *152.0* |
| Australia | 38.8 | 24.5 | 54.4 | — | 3.8 |
| China | 308.6 | 39.0 | 956.9 | 11.3 | 74.2 |
| Japan | 241.5 | 72.2 | 120.8 | 64.8 | 22.6 |
| India | 119.3 | 32.1 | 204.8 | 3.8 | 19.0 |
| *Global* | *3,787.1* | *2,420.4* | *2,778.2* | *624.3* | *634.4* |

**Table 1.16.** GDP during 1950–2005 and global fossil fuel consumption ($10^6$ t of oil equivalent).

| Year | GGP (US$$10^{12}$, 2001) | Per capita GGP (US$, 2001) | Coal | Oil | Natural gas |
|---|---|---|---|---|---|
| 1950 | 6.7 | 2,641 | 1,074 | 470 | 171 |
| 1955 | 8.7 | 3,112 | 1,270 | 694 | 266 |
| 1960 | 10.7 | 3,516 | 1,544 | 951 | 416 |
| 1965 | 13.6 | 4,071 | 1,486 | 1,530 | 632 |
| 1970 | 17.5 | 4,708 | 1,553 | 2,254 | 924 |
| 1971 | 18.2 | 4,805 | 1,538 | 2,377 | 988 |
| 1972 | 19.1 | 4,933 | 1,540 | 2,556 | 1,032 |
| 1973 | 20.3 | 5,157 | 1,579 | 2,754 | 1,059 |
| 1974 | 20.8 | 5,174 | 1,592 | 2,710 | 1,082 |
| 1975 | 21.1 | 5,154 | 1,613 | 2,678 | 1,075 |
| 1976 | 22.1 | 5,312 | 1,681 | 2,852 | 1,138 |
| 1977 | 23.0 | 5,432 | 1,726 | 2,944 | 1,169 |
| 1978 | 24.0 | 5,573 | 1,744 | 3,055 | 1,216 |
| 1979 | 24.8 | 5,672 | 1,834 | 3,103 | 1,295 |
| 1980 | 25.3 | 5,688 | 1,814 | 2,972 | 1,304 |
| 1981 | 25.8 | 5,698 | 1,826 | 2,868 | 1,318 |
| 1982 | 26.1 | 5,664 | 1,863 | 2,776 | 1,322 |
| 1983 | 26.9 | 5,728 | 1,914 | 2,761 | 1,340 |
| 1984 | 28.1 | 5,890 | 2,011 | 2,809 | 1,451 |
| 1985 | 29.1 | 5,993 | 2,107 | 2,801 | 1,493 |
| 1986 | 30.1 | 6,101 | 2,143 | 2,893 | 1,504 |
| 1987 | 31.2 | 6,216 | 2,211 | 2,949 | 1,583 |
| 1988 | 32.5 | 6,375 | 2,261 | 3,039 | 1,663 |
| 1989 | 33.6 | 6,470 | 2,293 | 3,088 | 1,738 |
| 1990 | 34.2 | 6,492 | 2,270 | 3,136 | 1,774 |
| 1991 | 34.7 | 6,468 | 2,218 | 3,138 | 1,806 |
| 1992 | 35.4 | 6,499 | 2,204 | 3,170 | 1,810 |
| 1993 | 36.1 | 6,538 | 2,200 | 3,141 | 1,849 |
| 1994 | 37.3 | 6,663 | 2,219 | 3,200 | 1,858 |
| 1995 | 38.6 | 6,791 | 2,255 | 3,247 | 1,938 |
| 1996 | 40.1 | 6,964 | 2,336 | 3,323 | 2,032 |
| 1997 | 41.7 | 7,139 | 2,324 | 3,396 | 2,025 |
| 1998 | 42.7 | 7,202 | 2,280 | 3,410 | 2,059 |
| 1999 | 44.0 | 7,337 | 2,163 | 3,481 | 2,103 |
| 2000 | 46.0 | 7,513 | 2,217 | 3,519 | 2,192 |
| 2001 | 46.9 | 7,551 | 2,555 | 3,511 | 2,215 |
| 2002 | 47.9 | 7,601 | 2,413 | 3,580 | 2,286 |
| 2003 | 48.7 | 7,637 | 2,614 | 3,642 | 2,342 |
| 2004 | 49.3 | 7,661 | 2,778 | 3,767 | 2,425 |
| 2005 | 49.9 | 7,691 | 2,930 | 3,814 | 2,475 |

biomass). In the U.S.A. the per capita level of commercial energy consumption constituted (in weight units of oil equivalent) 8.1 $t\,yr^{-1}$, in Ethiopia this was only 0.3 $t\,yr^{-1}$. The respective extreme levels of global oil consumption are 70.2 and 0.3 barrels/day per 1,000 people, and electric power 12,331 and 22 kW/1,000 people. The fact is that the level of per capita $CO_2$ emission to the atmosphere reaches 19.7 t in the U.S.A. and only 0.1 t in Ethiopia.

However, it should be noted here that the rate of energy production increase (and, respectively, economy) is highest in Asia. For instance, since 1980 the Chinese economy has grown more than fourfold (respectively, the needs of energy increased fourfold too). One of the indicators of economic development in China is the increase in the number of personal cars from 5 million in 2000 to 24 million in 2005. The number of rich families in India with an income of $220/month increased sixfold in just 5 years, whereas the number of poor families has substantially decreased.

According to the International Energy Agency (IEA, 2005a–c) estimates, in the period 2000–2030 a global mean growth in primary energy production should take place annually, constituting 1.7% per annum, reaching the level of 15,300 million tons of oil equivalent in 2030. More than 90% of the growing need for energy (mainly in developing countries) should be satisfied by fossil fuel. However, even by 2030 about 18% of the global population will still be without such modern sources of energy as electricity. Of course, these forecasts should be considered conditional.

Of great importance is the substantial progress achieved in the use of alternative sources of energy. This principally refers to nuclear energy (Table 1.17). As seen from

**Table 1.17.** Global energy production at nuclear power stations (GW), 1960–2002.

| Year | Power | Year | Power |
|---|---|---|---|
| 1960 | 1 | 1987 | 297 |
| 1965 | 5 | 1988 | 310 |
| 1970 | 16 | 1989 | 320 |
| 1971 | 24 | 1990 | 328 |
| 1972 | 32 | 1991 | 325 |
| 1973 | 45 | 1992 | 327 |
| 1974 | 61 | 1993 | 336 |
| 1975 | 71 | 1994 | 338 |
| 1976 | 85 | 1995 | 340 |
| 1977 | 99 | 1996 | 343 |
| 1978 | 114 | 1997 | 343 |
| 1979 | 121 | 1998 | 343 |
| 1980 | 135 | 1999 | 346 |
| 1981 | 155 | 2000 | 349 |
| 1982 | 170 | 2001 | 352 |
| 1983 | 189 | 2002 | 358 |
| 1984 | 219 | 2003 | 350 |
| 1985 | 250 | 2004 | 366 |
| 1986 | 276 | 2005 | 397 |

Table 1.17, in the period 2001–2002 the production of energy at nuclear power stations (NPSs) increased by about 1.5% (>5,000 MW). The period after 1993 witnessed the most intensive growth. The number of nuclear reactors in the world (at NPSs) increased to 437 due to the launching in 2002 of new reactors in China (4), South Korea (2), and the Czech Republic (1). In 2002, the building of 6 new reactors was started in India, and 26 reactors were in the process of construction (their power totaled 20,959 MW). Seven reactors were decommissioned (the number of such reactors has now reached 106). China plans to build four new reactors generating a total of 20,000 MW, and in India the total power of NPSs should increase by 150% (up to 6,000 MW with the prospect of this growing to 20,000 MW by 2020).

In some countries nuclear energy faces difficulties, however. For instance, in Great Britain NPSs have been declared uncompetitive, and Belgium plans to stop using NPSs by 2025. At present, no new reactors are being built in Western Europe (an exception is Finland where parliament approved the building of a fifth reactor). Serious difficulties are observed for nuclear energy in the U.S.A. and Japan (Kennedy, 2003). South Korea plans to launch eight new reactors during the next 10 years.

As for other kinds of energy, the most substantial progress has been achieved in the use of wind energy, though the absolute contribution remains very small (Table 1.18). By the end of 2002, the total energy production due to wind generators constituted 32,000 MW, and global mean growth (compared with the previous year) reached 27% (31% in Western Europe). Compared with 1998, wind energy production has tripled (making it the most rapidly developing source of energy). The share of Western Europe constituted 73% of global wind energy, with the main contributions being made by Germany, Spain, and Denmark.

The increasing scale of the use of fossil fuel and related growing $CO_2$ emissions to the atmosphere is of serious concern for possible anthropogenic global climate change. The data in Table 1.19 contain information about respective observed global trends (Elliot, 2003; Fall, 2005; Schultz, 2005; Wicks and Keay, 2005; Pulido *et al.*, 2005).

Without dwelling upon the cause-and-effect feedbacks between temperature and $CO_2$ concentration—this subject has been discussed in detail elsewhere (Kondratyev, 2004a, c)—we shall only mention that despite acknowledgment of the known international documents about the need to reduce GHG emissions to the atmosphere ($CO_2$ included), the global mean $CO_2$ concentration goes on growing. There is no doubt that this trend will remain in the near future. In 2002 alone, the atmosphere received 6.44 billion t C (a 1% growth compared with 2001), and $CO_2$ concentration reached 372.9 ppm, an increase of 18% during the period 1960–2002, and from the beginning of the industrial revolution (1750) of 31%. About 24% of global emissions are the responsibility of the U.S.A., whose population constitutes 5% of the global population (per capita $CO_2$ emissions in the U.S.A. exceed those observed in India by 17 times).

One of the important features of current energy consumption is the increase in energy consumption due to various kinds of transport: mainly, privately owned automobiles, whose production reached 40.6 million in 2002, five times as many as in 1950 (Storm and Naastepad, 2001; Pulido *et al.*, 2005).. The total number of

**Table 1.18.** Global renewable energy production (with examples by region).

| Renewable energy source | 1995 | 2000 | 2003 | 2004 | 2005 |
|---|---|---|---|---|---|
| *Geothermal* (MW) | 6,833.4 | 7,972.6 | 8,402.3 | 8,911.8 | 8,937.7 |
| Argentina | 1 | 0 | 0 | 0 | 0 |
| China | 29 | 29 | 28 | 28 | 28 |
| Indonesia | 310 | 590 | 807 | 807 | 807 |
| Italy | 632 | 785 | 791 | 791 | 791 |
| Japan | 414 | 547 | 561 | 535 | 535 |
| Mexico | 753 | 755 | 953 | 953 | 953 |
| Philippines | 1,227 | 1,909 | 1,931 | 1,931 | 1,931 |
| U.S.A. | 2,816 | 2,228 | 2,020 | 2,534 | 2,544 |
| *Solar* (kW) | 198,627 | 725,832 | 1,798,664 | 2,599,247 | 3,342,632 |
| North America | 77,880 | 159,882 | 304,130 | 397,266 | 510,884 |
| Western Europe | 62,718 | 202,119 | 612,310 | 1,006,912 | 1,294,889 |
| Others | 58,029 | 363,831 | 912,224 | 1,195,069 | 1,536,859 |
| *Wind* (MW) | 4,778 | 18,450 | 40,301 | 47,912 | 59,264 |
| North America | 827 | 2,752 | 6,715 | 7,197 | 9,867 |
| South and Central America | 63 | 96 | 190 | 194 | 195 |
| Europe and Eurasia | 2,882 | 13,650 | 29,325 | 34,749 | 41,068 |
| Middle East | 16 | 18 | 72 | 101 | 101 |
| Africa | 4 | 137 | 211 | 234 | 278 |
| Asia Pacific | 986 | 1,797 | 3,788 | 5,437 | 7,755 |
| *Ethanol* (1,000 t of oil equivalent) | 8,803 | 8,256 | 12,840 | 14,030 | 16,182 |
| North and South America | 8,803 | 8,136 | 12,535 | 13,750 | 15,073 |
| Europe | 0 | 120 | 268 | 266 | 454 |
| Asia Pacific | 0 | 0 | 37 | 14 | 655 |

cars exceeds 531 million, increasing annually by 11 million. As a result, in most developed countries the use of public transport has substantially decreased, constituting only 10% (in the cities of Western Europe), 7% in Canada, and 2% in the U.S.A.

On average, about one-third of global produced energy is spent on heating, illumination, air conditioning, etc.—this energy consumption is rapidly growing with a strong contrast between countries. Domestic energy consumption in the U.S.A. and Canada is greater by a factor of 2.4 than in the countries of Western Europe, and the average difference in the levels of energy consumption between developed and developing countries reaches a factor of 9. The share of domestic energy consumption in China is about 40%, and in India it reaches 50% (and still higher in many countries of Africa), whereas in the developed world this share does not exceed 15–25%.

According to the data on HDI and energy consumption, there is a correlation between these values that is not, however, directly proportional (Starke, 2004). For the poor, even a small increase in energy consumption means a considerable increase of their living standards, but with energy consumption reaching 1 t (oil equivalent)

**Table 1.19.** Trends of global mean SAT (°C), $CO_2$ emission to the atmosphere ($10^6$ tC), and $CO_2$ volume concentration (ppm) (Karl *et al.*, 2006).

| Year | Temperature | $CO_2$ emissions | $CO_2$ concentration (ppm) |
|---|---|---|---|
| 1950 | 13.87 | 1,630 | 315.4 |
| 1955 | 13.89 | 2,043 | 315.8 |
| 1960 | 14.01 | 2,577 | 316.7 |
| 1965 | 13.90 | 3,145 | 319.9 |
| 1970 | 14.02 | 4,076 | 325.5 |
| 1975 | 13.94 | 4,615 | 331.0 |
| 1976 | 13.86 | 4,883 | 332.0 |
| 1977 | 14.11 | 5,029 | 333.7 |
| 1978 | 14.02 | 5,097 | 335.3 |
| 1979 | 14.09 | 5,389 | 336.7 |
| 1980 | 14.16 | 5,330 | 338.5 |
| 1981 | 14.22 | 5,164 | 339.8 |
| 1982 | 14.06 | 5,118 | 341.0 |
| 1983 | 14.25 | 5,101 | 341.6 |
| 1984 | 14.07 | 5,281 | 344.2 |
| 1985 | 14.03 | 5,436 | 345.7 |
| 1986 | 14.12 | 5,600 | 347.0 |
| 1987 | 14.27 | 5,731 | 348.7 |
| 1988 | 14.29 | 5,958 | 351.3 |
| 1989 | 14.19 | 6,072 | 352.7 |
| 1990 | 14.37 | 6,143 | 354.0 |
| 1991 | 14.32 | 6,252 | 355.5 |
| 1992 | 14.14 | 6,121 | 356.4 |
| 1993 | 14.14 | 6,129 | 357.0 |
| 1994 | 14.25 | 6,262 | 358.9 |
| 1995 | 14.37 | 6,402 | 360.9 |
| 1996 | 14.23 | 6,560 | 362.6 |
| 1997 | 14.40 | 6,696 | 363.8 |
| 1998 | 14.56 | 6,656 | 366.6 |
| 1999 | 14.32 | 6,522 | 368.3 |
| 2000 | 14.31 | 6,672 | 369.4 |
| 2001 | 14.46 | 6,842 | 370.9 |
| 2002 | 14.52 | 6,973 | 372.9 |
| 2003 | 14.70 | 7,303 | 375.2 |
| 2004 | 14.91 | 7,513 | 376.8 |
| 2005 | 15.15 | 7,609 | 379.3 |
| 2006 | 15.35 | 7,741 | 380.9 |

per capita per year and increasing up to 1–3 t per capita per year, the role of additional energy consumption diminishes. With still greater increases in energy consumption, its connection with HDI disappears. The countries with the level of equivalent energy consumption about 3 t per capita per year include Italy, Greece, and South Africa, whereas in the U.S.A. the per capita energy consumption is threefold greater.

To characterize life quality, the Wellbeing Index (WI) has been proposed: it classifies 180 countries by considering 87 indicators of the state of humans. The WI has two components: the Human Wellbeing Index (HWI) and the Ecosystem Wellbeing Index (EWI). The HWI includes 36 indicators of health, population, wealth, education, communication, freedom, peace, crime, and equity. The EWI includes 51 indicators of land health, protected areas, water quality, water supply, global atmosphere, air quality, species diversity, energy use, and resource pressures. The HWI is an average of indicators of health (life expectancy) and population (total fertility rate), wealth (household and national indicators), knowledge (education and communication indicators), community (freedom, governance, peace, and order indicators), and equity (household and gender equity indicators). The EWI is an average of land indicators (land diversity and quality indicators), water (use and quality indicators), air (global and national indicators), species and genera (biodiversity indicators), and resource use (energy and materials consumption indicators). In this context, Prescott-Allen (2001) writes: "The use of indicators to gauge human progress is common and well understood. Gross Domestic Product and the Index of Leading Economic Indicators are two well-known examples. Yet, most of the widely cited indicators focus exclusively on economic activity, and even the most progressive of indicators fails to account for key issues of sustainability. The Wellbeing of Nations addresses that shortcoming by combining indicators of human wellbeing with those of environmental sustainability to generate a more comprehensive picture of the state of our world."

According to the WI data, Sweden occupies first place and the United Arab Emirates comes last, though the level of energy consumption in this country is almost twice as high as Sweden (Table 1.20). Austria occupies a moderate place in energy consumption, but ranks high from the viewpoint of welfare.

This means that the present state of society cannot be considered stable from every point of view (including socio-economic and ecological conditions). The growing amount of data shows that the present structure of consumption results in

**Table 1.20.** Energy consumption and living standards in different countries (Prescott-Allen, 2001).

| Country | WI | Per capita energy consumption (rank) | Share (%) with respect to energy consumption in Sweden |
|---|---|---|---|
| Sweden | 1 | 10 | 100 |
| Finland | 2 | 6 | 112 |
| Norway | 3 | 8 | 104 |
| Austria | 5 | 26 | 61 |
| Japan | 24 | 19 | 70 |
| U.S.A. | 27 | 4 | 140 |
| Russia | 65 | 17 | 71 |
| Kuwait | 119 | 3 | 162 |
| U.A.E. | 173 | 2 | 190 |

degradation of the quality of life for many people manifested through aggravating health problems, deterioration of the environment, destruction of many ecosystems, etc. Despite the continuing improvement of technologies and measures on energy saving, under conditions of growing global population size and increasing level of per capita consumption (especially in developed countries), the prospects for global sustainable development in the 21st century are far from clear. According to the available forecasts, by 2050 the global population size will increase by more than 40% (up to 8.9 billion people). Clearly, achieving the global mean level of energy consumption of developed countries (and this level is much below that in the U.S.A.) is simply not possible. It would necessitate a more than eightfold increase in energy production in the period 2000–2050. This is an impossibility for fossil fuels—moreover, even if it were possible it would be followed by serious ecological consequences—and, on the other hand, the prospects of intensive development of alternative sources of energy in the 21st century remain unclear.

### 1.6.5 Conflict, cooperation, and the global nature–society system

Lomborg (2001) is, of course, right in rejecting apocalyptic prognoses of global ecodynamics based on exaggerated misgivings with respect to limited natural resources and the environmental state. Such views and estimates are confirmed, in particular, by the data in Table 1.21 compiled by Holdren (2003) which characterize

**Table 1.21.** Global energy resources (IEA, 2005a–c; WEO, 2006).

| | |
|---|---|
| *Non-renewable resources* (TW/year) | |
| Standard oil and natural gas | 1,000 |
| Non-standard oil and gas, except clusters of methane | 2,000 |
| Clusters of methane | 20,000 |
| Shale | 30,000 |
| Geothermal sources | |
| vapour and hot water | 4,000 |
| hot dry rocks | 1,000,000 |
| Uranium | |
| in reactors with light water | 3,000 |
| in breeder reactors | 3,000,000 |
| Thermonuclear energy | |
| heavy hydrogen limited with lithium | 140,000,000 |
| heavy hydrogen–heavy hydrogen | 250,000,000,000 |
| *Renewable resources* (TW/year) | |
| Hydroenergetics | 15 |
| Use of biomass | 100 |
| Wind energy | 2,000 |
| Solar energy | |
| on land surface | 26,000 |
| over the globe | 88,000 |

both real and potential global energy resources. Here energy units are expressed (in the case of non-renewable energy sources) in TW year, which is equivalent to 31 Exajoules (3.6 EJ = 1,000 TWh; 1 EJ = $10^{18}$ joules).

It should be added that global energy consumption in 2000 constituted about 15 TW or 15 TWh/year; this is supposed to increase up to 60 TWh/year by 2100.

Despite the optimistic data in Table 1.20, the absence of long-term prospects for further developing the present consumption society illustrated by estimates of global ecodynamics cannot be denied (Kondratyev and Krapivin, 2005a, 2006a–f). Therefore, the World Summit on Sustainable Development held in Johannesburg in 2002 emphasized the need to carry out 10-year programs to look into the possibility of reaching stable production and consumption.

It is extremely important to select in the complicated and interactive totality of the numerous processes characterizing globalization dynamics and the prospects for NSS sustainable development, a key aspect that determines the future of the global NSS. The results considered above demonstrate that this aspect is ecodynamical in nature, whereas all the others (on a global scale and with long time periods) depend on ecosystems. Despite the importance of globalization processes in the spheres of economy, finance, geopolicy, etc., all of them are subordinate. To substantiate this conclusion, one should remember that the successful functioning of the systems for global life support (i.e., life on Earth) is eventually determined by two factors:

(1) ensuring the Earth's heat balance (ecological disaster—either over-heating or over-cooling of the planet—will result of course from such extremes of temperature); and
(2) closedness of global biogeochemical cycles (clearly, the breakdown of any cycle will be catastrophic: the anthropogenic growth in $CO_2$ and other GHG concentrations in the atmosphere is of great concern).

One should emphasize the hopelessness and danger of attempting to eliminate an undesirable global imbalance (to control ecodynamics) by actively impacting nature. The only reasonable prospect relates to the Earth's inherent ability to self-regulate.

Unfortunately, accumulated information on the processes involved in NSS dynamics is far from adequate. It is necessary to plan and develop a system to globally monitor NSS dynamics that would facilitate timely recognition of dangerous situations and take measures for their prevention. At present, such a system is in an embryonic state. Also important is the development of reliable numerical simulation models of NSS dynamics which will require not only highly efficient computers but also a deeper understanding of the processes underlying the dynamics of the NSS. It is necessary to take into consideration religious and analogous aspects (Charitonova, 2004, 2006) that can influence fear of the environment by people. These problems will now be discussed in detail.

## 1.7 PROSPECTS FOR GLOBAL ENERGETICS DEVELOPMENT IN THE CONTEXT OF GLOBAL ECODYNAMICS

### 1.7.1 Resources and ecodynamics: present state and future perspectives

One of the key problems of current civilization development is the search for a compromise between the increasing needs of the growing global population and nature, whose food and energy resources are limited (Figure 1.1). The characteristic feature of the present ecological situation on Earth is the growth of instability or "system civilization crisis" the global size of which is expressed through a deterioration in the habitats of humans and animals. The difficulty of resolving all the related problems is determined by the unique character of the environment, experimenting with which is impossible. Therefore, one of the acceptable means of decision-making on the human–nature interaction is observation, analysis, and modeling. Many publications have been dedicated to analysis and resolution of the problems that have appeared, some of which can be found in our list of references (Aires *et al.*, 2005; Artiola *et al.*, 2004; Bui and Krapivin, 1998; Izrael, 2001; Kondratyev and Krapivin, 2005a, e, f; Krapivin *et al.*, 2006; Spellerberg, 2005).

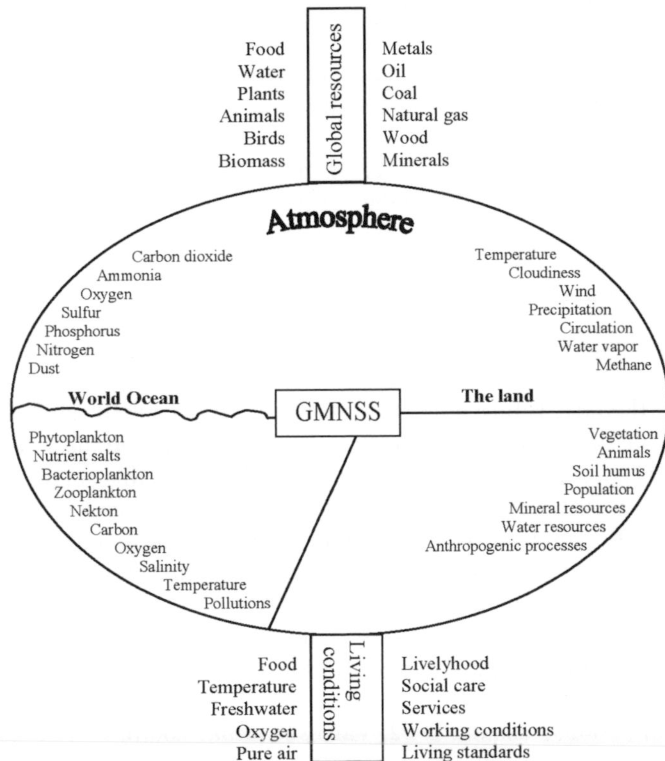


**Figure 1.1.** Key elements of the nature–society system and energy components which are taken into account when formulating a global model of ecodynamic forecast.

The global resources at our disposal are either naturally limited or vulnerable, and therefore they can change depending on human behavior. Clearly, to analyze the consequences of anthropogenic activity and attempts to predict environmental dynamics, it is necessary to take into account the current connections (both direct and indirect) between all resources. The major drawback of most climate change studies consists in ignoring elements or processes of the environment that seem now to be of secondary importance. Eventually, conditions for global population survival can only be brought about by considering the interaction of all factors whose significance can change unexpectedly. Among these factors one should take into account land resources (water, soil, plants, animals, birds), supplies of mineral resources (oil, coal, natural gas, metals, and other minerals), solar energy, and processes in the atmosphere and in the World Ocean, which can serve as energy sources. It is only the totality of all these factors and their interactions with each other and with other environmental elements that the conditions for existence of living beings are formed. Of these conditions, the most important are the content of oxygen in the atmosphere, drinking water availability, temperature, food accessibility, and a clean atmosphere.

One of the causes preventing us from getting reliable forecasts of NSS development is inadequate knowledge of environmental processes. The major sources of this inadequacy are:

- imperfection of global models and monitoring systems;
- lack of real information and fragmentary data on interactions taking place in the environment;
- some principally unknowable processes and objects;
- weak predictability of the behaviour of people and nature;
- insurmountable NSS multi-dimensionality;
- unknown physical limits of nature; and
- vagueness of the notion of sustainable development.

One of the principal problems of global ecodynamics is, of course, the future prospect of using various sources of energy (Figure 1.2). The major conclusion is that at present and in the near future non-renewable energy sources (fossil fuel and nuclear energy) will still dominate. This dictates the need to accomplish a program for ecological security, with global energy being developed in this way.

Clearly, the unreasonable use of natural resources can lead to an environment dominated by anthropogenically changed landscapes with limited resources for life. Growing instability in the environment leads to an increase in the number of natural disasters, which can considerably change habitats and change human behavior. At the same time, natural catastrophes are interactive components of global ecodynamics affecting the distribution of energy in the atmosphere and in the World Ocean as well as affecting the biogeochemical cycles of GHGs (Kondratyev *et al.*, 2006b; Ehleringer *et al.*, 2005; Fasham, 2002; Varotsos *et al.*, 2006).

The development of global processes in the environment and in socio-economics at the present stage of nature–society interaction is fraught with a multitude of limitations and consequences. The growth in population brings about a need to

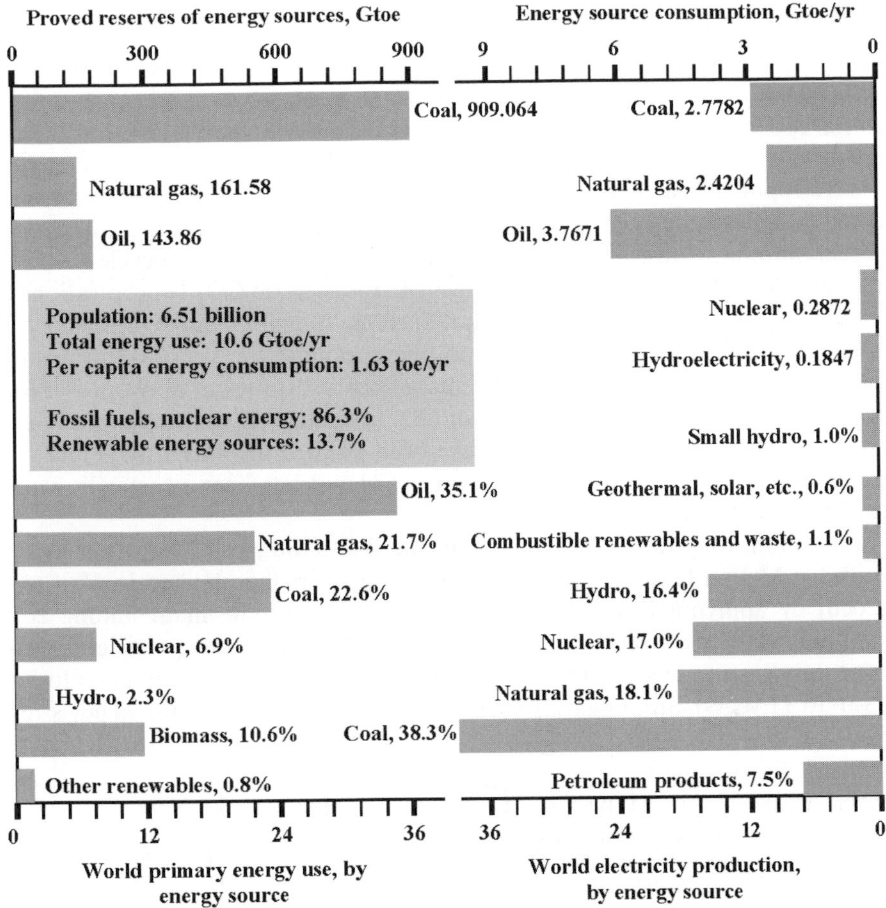


**Figure 1.2.** The present distribution of the use of energy sources (Goldemberg and Johansson, 2004; IEA, 2005b; Karekezi and Sihag, 2004). Gtoe is $10^9$ tons of oil equivalent.

increase the amount of energy consumed. Since 85% of energy production is based on hydrocarbons, the environment inevitably becomes polluted leading to subsequent ecological consequences. Moreover, the atmosphere becomes hotter, which leads to a change in habitats.

Let us now discuss the interaction between two complicated open systems: nature and society. This interaction is based on the mutual exchange of resources and information. People, using the available means, want to maximize natural resources despite the uncertain response of nature. Nature changes both due to human activity and follows natural laws of evolution towards preservation of its stability. To assess the direction of dynamics on regional and global scales, a criterion is needed. Such a criterion was proposed by Kondratyev and Krapivin (2005c). Its use in the synthesis of the NSS global model (GMNSS) will make it possible to assess the sensitivity of natural resources to the realization of anthropogenic scenarios.

### 1.7.2 Energy and the greenhouse effect

The long-term observations of $CO_2$ content in the atmosphere have shown a high stability in $CO_2$ variations during the last 11,000 years before the beginning of the industrial epoch. These variations are estimated at no more than 20 ppm. At the same time, during the last 200 years $CO_2$ partial pressure in the atmosphere increased by almost 100 ppm. Most investigators of the causes of this change believe that the observed and continuing growth in $CO_2$ content in the atmosphere is connected with fossil fuel burning. However, a thorough study of the global carbon cycle shows that the causes of the observed trends in changes of atmospheric $CO_2$ concentrations and climate are not that simple. Along with fuel burning, humans change the image of the Earth's surface, emit aerosols to the atmosphere, pollute the World Ocean, initiate forest fires, and change the structure of the global hydrological network. Unfortunately, along with the greenhouse effect of $CO_2$, a similar effect of other GHGs also takes place, whose sources and sinks have been studied insufficiently (Figure 1.3). Also the change in degassing processes on the planet has been practically ignored (Figure 1.4 and Table 1.22).

In addition, long-range correlations in the fluctuations of $CO_2$ concentrations measured at Mauna Loa, Hawaii during 1958–2004 were investigated by Varotsos *et al.* (2006) by applying Detrended Fluctuation Analysis. The main finding is that fluctuations in $CO_2$ concentrations exhibit $1/f$-type long-range persistence, which signifies that fluctuations in $CO_2$ concentrations, from small-time intervals to larger ones (up to 11 years), are positively correlated in a power-law fashion. They stressed

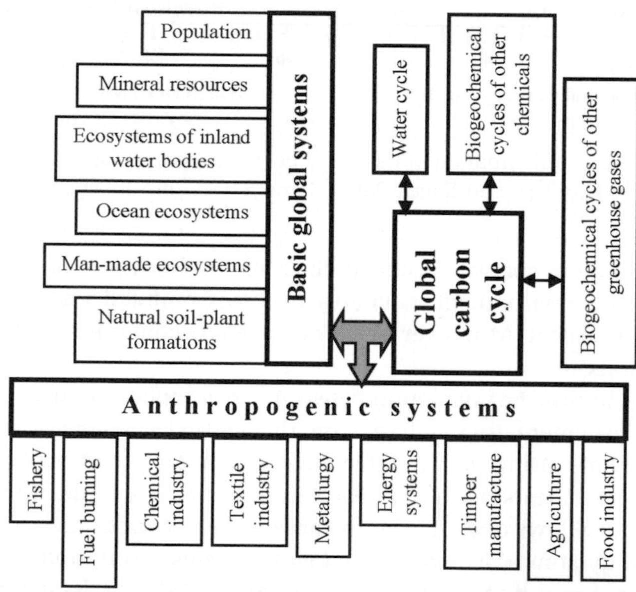


**Figure 1.3.** A conceptual scheme of the carbon cycle in the environment and the place occupied by the biogeochemical carbon cycle in the global system of energy exchange.

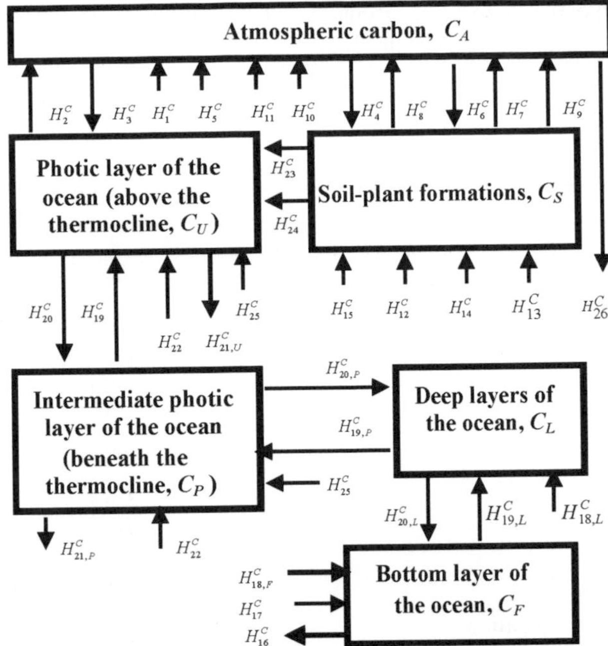


**Figure 1.4.** Block diagram of a global model of the carbon cycle with indicated unstudied and uncertain carbon fluxes. Notations are given in Table 1.22.

the point that this scaling comes from time evolution and not from the values of carbon dioxide data. They also suggested that the scaling property detected in real observations of $CO_2$ concentrations could be used to test the scaling performance of leading global climate models under different scenarios of $CO_2$ levels and to improve the performance of atmospheric chemistry transport models.

Therefore, to increase the reliability of conclusions that it is only the process of energy production due to hydrocarbon sources that plays the role in climate change, the role of other factors of anthropogenic activity should be taken into account, especially those connected with the provision of normal living conditions and in many cases of survival. About 2.7 billion people in the world live on less than \$2 per day, most of them (~2.4 billion) rely on biomass-burning, and 57% of poor people do not use electric energy and have limited access to fossil fuel energy. On a global scale, electrification has reached 72.8%, varying between 34.3% in Africa and 99.5% in countries with transitional economies. Energy production due to biomass-burning is widely practiced in African countries, covering from 56.2% in Senegal to 88.9% in Mali. This testifies to the fact that various regions have different conditions to contend with when attempting to raise the living standards of their populations and use whatever accessible energy there is with different efficiency. For instance, the countries of Latin America, because of the limited hydropotential of the continent, cannot produce more than 1 billion W per year due to HPS construction. These limits are characteristic of practically all global regions (Figure 1.5). Therefore,

**Table 1.22.** Reservoirs and fluxes of carbon as $CO_2$ in the biosphere considered in the simulation model of the global biogeochemical cycle of carbon dioxide shown in Figure 1.4.

| Reservoirs and fluxes of carbon dioxide | Identifier | Estimate of reservoir ($10^9$ t) and flux ($10^9$ t/yr) |
|---|---|---|
| Carbon | | |
| Atmosphere | $C_A$ | 650–750 |
| Photic layer of the ocean | $C_U$ | 580–1,020 |
| Deep layers of the ocean | $C_L$ | 34,500–37,890 |
| Soil humus | $C_S$ | 1,500–3,000 |
| Emission due to burning | | |
| Vegetation | $H_8^C$ | 6.9 |
| Fossil fuel | $H_1^C$ | 3.6 |
| Desorption | $H_2^C$ | 97.08 |
| Sorption | $H_3^C$ | 100 |
| Rock weathering | $H_4^C$ | 0.04 |
| Volcanic emanations | $H_5^C$ | 2.7 |
| Assimilation by land vegetation | $H_6^C$ | 224.4 |
| Respiration | | |
| Plants | $H_7^C$ | 50–59.3 |
| Humans | $H_{10}^C$ | 0.7 |
| Animals | $H_{11}^C$ | 4.1 |
| Emissions | | |
| Decomposed soil humus | $H_9^C$ | 139.5 |
| Plant roots | $H_{15}^C$ | 56.1 |
| Vital functions | | |
| Human population | $H_{12}^C$ | 0.3 |
| Animals | $H_{13}^C$ | 3.1 |
| Vegetation decay | $H_{14}^C$ | 31.5–50 |
| Sedimentation to bottom deposits | $H_{16}^C$ | 0.1–0.2 |
| Solution of marine sediments | $H_{17}^C$ | 0.1 |
| Decomposition of detritus | | |
| Photic layer | $H_{22}^C$ | 35 |
| Deep layers of the ocean | $H_{18}^C$ | 5 |
| Upwelling with deep waters | $H_{19}^C$ | 4 |
| Sinking with surface water due to gravitational sedimentation | $H_{20}^C$ | 40 |
| Photosynthesis | $H_{21}^C$ | 69 |
| Underground sink | $H_{23}^C$ | 0.5 |
| Surface sink | $H_{24}^C$ | 0.5–0.6 |
| Respiration of living organisms in the ocean | $H_{25}^C$ | 25 |
| Geological sink | $H_{26}^C$ | 2.0 |

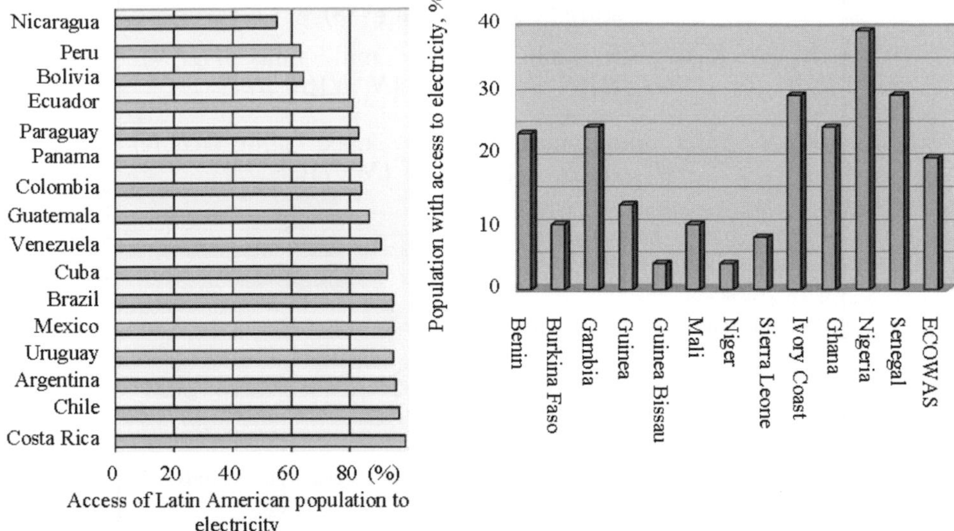


**Figure 1.5.** Characteristics of the energy indicators for some regions (Goldemberg and Johansson, 2004; IEA, 2005b; Karekezi and Sihag, 2004). ECOWAS is the Economic Community of West African States.

when developing a scenario for a region's development, it is necessary to take into account these limitations and to determine thereby what possibilities there are of negative impacts of the region on the environment.

At present there are many scenarios of $CO_2$ emissions to the atmosphere, connected with the high level of uncertainty about global ecodynamics and the absence of weighted and substantiated mathematical solutions that are able to determine a strategy for sustainable development (Figure 1.6). Nevertheless, usually among this diversity, there are three groups of scenarios that regulate the level of anthropogenic impact on the environment:

(i) rapidly developing economy, slowly growing population size, rapid introduction of new technologies, and aspiration for economic homogeneity in the world;
(ii) low increase in population size, rapid change of economic structures, introduction of clean and resource-saving technologies; and
(iii) maintaining the rate of economic development observed during the last decade.

The correspondence of the scenario of anthropogenic activity and the rate of $CO_2$ input to the atmosphere always raises doubts. Nevertheless, there is an established conception of the role of individual regions in this process with the total volume of 8–15 GtC/year emitted to the atmosphere in 2050 with subsequent stabilization at the level of 12–20 GtC/year by 2100. The rate of growth in GDP is taken into account when estimating the level of such impacts and the resulting climate change (Figure 1.7).

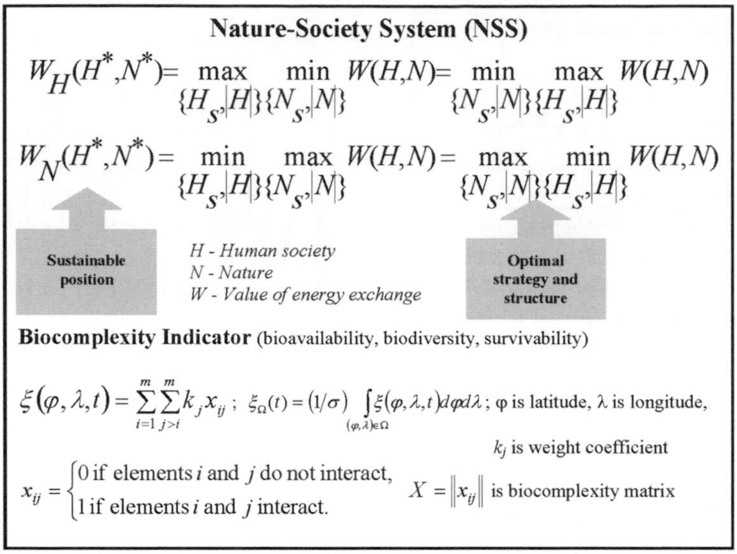

**Nature-Society System (NSS)**

$$W_H(H^*,N^*) = \max_{\{H_S,|H|\}} \min_{\{N_S,|N|\}} W(H,N) = \min_{\{N_S,|N|\}} \max_{\{H_S,|H|\}} W(H,N)$$

$$W_N(H^*,N^*) = \min_{\{H_S,|H|\}} \max_{\{N_S,|N|\}} W(H,N) = \max_{\{N_S,|N|\}} \min_{\{H_S,|H|\}} W(H,N)$$

Sustainable position

*H - Human society*
*N - Nature*
*W - Value of energy exchange*

Optimal strategy and structure

**Biocomplexity Indicator** (bioavailability, biodiversity, survivability)

$$\xi(\varphi,\lambda,t) = \sum_{i=1}^{m}\sum_{j>i}^{m} k_j x_{ij}\,;\quad \xi_\Omega(t) = (1/\sigma)\int_{(\varphi,\lambda)\in\Omega} \xi(\varphi,\lambda,t)\,d\varphi\, d\lambda\,;$$ φ is latitude, λ is longitude, $k_j$ is weight coefficient

$$x_{ij} = \begin{cases} 0 \text{ if elements } i \text{ and } j \text{ do not interact,} \\ 1 \text{ if elements } i \text{ and } j \text{ interact.} \end{cases}$$ $X = \|x_{ij}\|$ is biocomplexity matrix

**Figure 1.6.** Energy balance equations for the nature–society system and determination of its biocomplexity indicators.

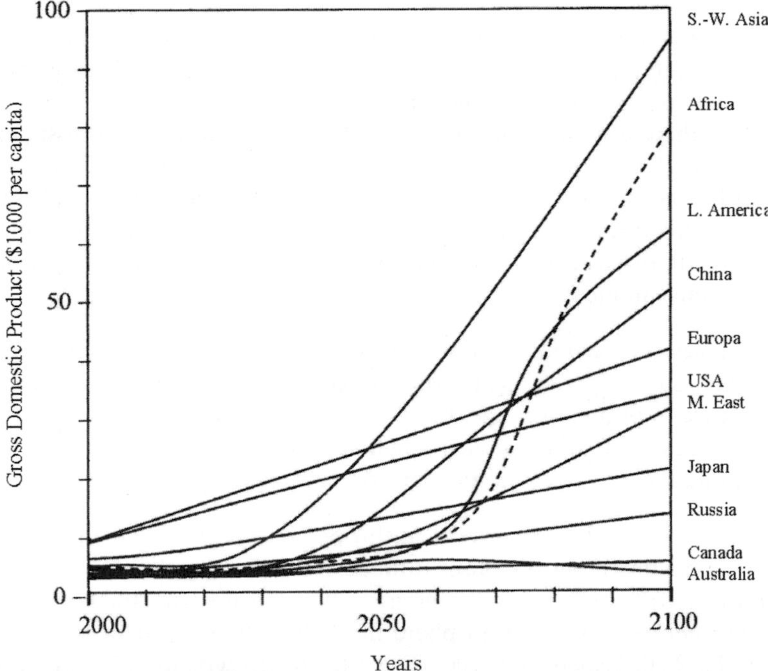


**Figure 1.7.** Predicted dynamics of GDP change in different regions by the end of the 21st century (Goldemberg and Johansson, 2004; IEA, 2005a–c; Karekezi and Sihag, 2004).

### 1.7.3 Numerical modeling results

The GMNSS makes it possible to accomplish a wide spectrum of simulation experiments over the multi-dimensional phase space of the NSS and to forecast NSS development for the rest of the century. Predictions beyond this are scarcely possible with the present level of knowledge and global databases. The uncertainties that have appeared have been analyzed earlier: they are all connected with climate modeling. Unfortunately, so far there is no universal global model which would reliably simulate the functioning of the Earth's climate system reflecting the interaction of the atmosphere, oceans, land, and society. Therefore, to use the GMNSS, scenarios have to be formulated which describe the dynamics of anthropogenic impacts on the environment.

First, we shall consider some results obtained using the GMNSS with the following *a priori* suppositions:

- Up to 2100 anthropogenic $CO_2$ emissions are described by the model of Demirchian *et al.* (2002), and emissions of other GHGs till 2050 follow the laws of the year 2000, and then they decrease by 25%.
- Extraction and consumption of non-renewable fossil fuel in the period till 2050 increase by 0.5%/year in developed countries and by 0.7%/year in developing countries and then they stabilize.
- The per capita domestic product grows in developed and developing countries at constant rates of 2.5% and 1.5%, respectively.
- The level of healthcare in developed and developing countries reduces mortality due to environmental changes by 90% and 50%, respectively, and in 2020 a means of combating AIDS will be found.
- The temperature regime of the atmosphere is formed in correspondence with the model of Mintzer (1987).
- Investment in agriculture depends on providing the world's population with food (Krapivin and Kondratyev, 2002).
- Investment in the struggle against environmental pollution for all countries until 2050 increases annually by 0.2% and then stabilizes.
- The processes of deforestation and aforestation from 2010 are balanced so that the total area of forests as of 2000 remains, and the area of tropical forests decreases by no more than 5%.
- The oil pollution of the World Ocean in the 21st century follows the laws of the last decade of the 20th century.
- Agricultural productivity grows steadily to 80% in 2100.

Within the framework of these suppositions, NSS dynamics is characterized by the following indicators:

- The ecodynamic laws of NSS functioning are characterized by derivatives of functions describing the state of NSS components never equaling zero in time periods longer than the model's time step.

- By 2100, the population size will reach 12.7 billion people, with the ratio between age groups characterized by the following average indicators: 0–15 years 32%; 16–65 years 53%; and above 65 years 15%. The disabled population will constitute 17%.
- By 2030, protein provision to the population increases by 6.8% (the contribution of the World Ocean constitutes 12.1%) and then decreases with respect to 2000 by 3.2% (the share of the World Ocean increases to 22.1%).
- $CO_2$ concentration in the atmosphere will reach 486 ppm with a respective increase in average planetary temperature by 0.87°C.
- The dynamics of the spatial distribution of atmospheric $CO_2$ sinks is characterized by the growing role of the World Ocean and land ecosystems whose share will constitute 31% and 19%, respectively, by the mid-21st century, and then by the end of the 21st century the share of the World Ocean will decrease to 26.7% with continuing growth in the role of land ecosystems up to 24.4%.
- Substitution of 5% of the area of tropical forests by urban ecosystems will lead to a reduction in the total $CO_2$ sink by 1.2%.
- Overall substitution of wet tropical forests by grasslands by 2050 will lead to global climate change in subsequent years due to breaking of the regional moisture cycle at the atmosphere–land boundary, increasing by 4.3 $W/m^3$ the longwave radiation flux outgoing from deforested territory, a 5.7% increase in the albedo of this territory, a 4.1% decrease in absorbed SWR flux, an increase in soil temperature by 1.3°C, and a decrease in precipitation and evaporation by 0.7 and 0.9 mm/day, respectively.

For educational purposes, it is of interest to consider hypothetical versions of the interaction between nature and society. We shall only consider those situations which appear when putting forward a hypothesis for the future dynamics of the most important anthropogenic factors. Note that—in contrast to many other models—the GMNSS calculates population size as a function of environmental parameters.

Numerical simulations show that the mechanisms underlying demographic process formation depend substantially on land use strategies. So, an increase in the level of food provision by 50%, which is possible due to an introduction of efficient technologies in agriculture and development of fish farming, leads to a decrease in population density by 14%. In contrast, a 10% reduction in the rate of mineral resource expenditure causes an increase in population density by 3%. The NSS is highly sensitive to possible changes in the global structures of soil–plant formation. For instance, if by 2050 the areas of forests are uniformly reduced only by 10% compared with 2000, atmospheric $CO_2$ concentration in 2100 will reach 611 ppm—that is, it will almost double. But if the areas of forests increase by 15% by 2050, then in 2100 $CO_2$ partial pressure will constitute 475 ppm.

The results of modeling NSS responses to possible anthropogenic changes of its parameters show that the proposed method makes it possible to assess the dynamics of various components of this system depending on hypotheses on the possible rates of such changes. Application of this method to reconstruct the integral NSS dynamics by its individual parameters during the period 1970–2000 has shown that the error of prediction for 10 and 20 years does not exceed 15% and 25%, respectively. Hence, the technology of global modeling based on the constant correction of the GMNSS with forecast reliability taken into account can be recommended as a constructive approach to solving the problems of global ecodynamics formulated by Kondratyev (1999a). Simulation experiments using the GMNSS with available reliable data on prehistoric NSS will make it possible to reveal the conditions that limit civilization development and to discover mechanisms for sustainable development. For this purpose, it is necessary to concentrate the efforts of both international and national research programs on GMNSS database synthesis. This requires concentrating the accumulated knowledge on GMNSS development by improving its component parts and, first of all, revealing and formalizing the factors and dynamic strategies involved in social evolution formation.

Among the many possible situations of North–South relationships, we shall try to answer the question on prospects and consequences of the use of hydrocarbons in energy production in the 21st century. Figures 1.8 and 1.9 illustrate the forecasts of air temperature changes with a constant rate of economic development in all regions and the introduction of new ecologically safe technologies. The growth in population size and climate change turns out to be below those predicted using, for instance, Hadley Centre technologies (PRECIS, 2002). Numerical modeling results show how atmospheric $CO_2$ content will change and how carbon will be distributed among its reservoirs. As can be seen, the interactive mechanisms between NSS elements in the GMNSS demonstrate more reliably the dynamics of environmental change, from

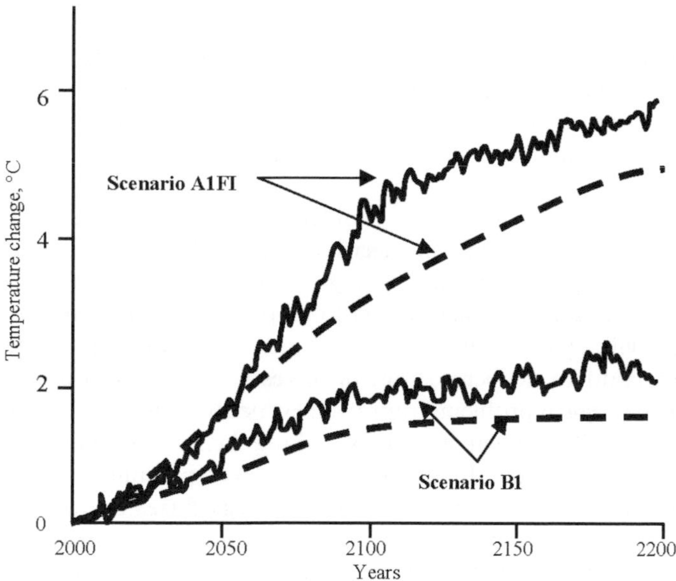


**Figure 1.8.** Comparison of temperature change predictions as deduced from realization of A1FI (intensive use of fossil fuel) and B1 (rapid introduction of new technologies to energy production) scenarios. Solid curves correspond to the forecast made with the Hadley Centre climate model (PRECIS, 2002). Dashed curves demonstrate the results of the use of GMNSS.

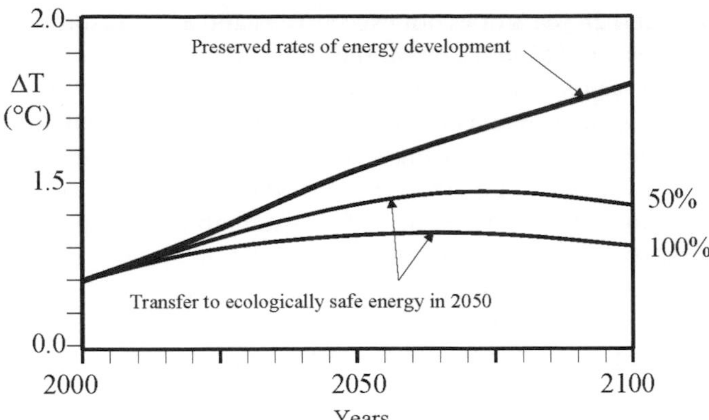


**Figure 1.9.** Air temperature change predicted using the GMNSS.

which it follows that the use of hydrocarbons in energy production in the 21st century is possible while still preserving the rate of GDP growth.

### 1.7.4 Preliminary conclusions

In connection with these aspects, there is a need for a complex analysis of the present state and prospects of further global energy development, bearing in mind the possibilities of its diversification, measures in the field of energy consumption economy, increased level of energy resource efficiency, and minimization of negative ecological consequences (transition from hydrocarbon to "pure" energy). What matters is the achievement in each country and in the world on the whole of a level of energy production which would be ecologically safe, stable, and provide the growing needs of economies.

Solution of the most important problem of the 21st century—the means of civilization development and making decisions on a global level, aimed at sustainable development by achieving acceptable living conditions for the global population—is the subject of many international and national programs on environmental studies. The observed civilization–nature conflict has moved from the spontaneous development of humankind to deliberate control of resources whose scarcity has become apparent. The ecological problems appearing everywhere on a planetary scale should motivate humankind to take up a strategy making their actions fall into step with the laws of nature. It is clear that, first of all, human development must fall into line with the laws of distribution and transformation of energy fluxes in the biosphere, which determines the boundaries of the corridor to ecologically problem-free civilization development. Realization of the fact that civilization is in a state of system crisis and has no clear behavioral strategy is a good first step but does not remove problems related to its survival.

To use this approach to search for a constructive technology to study the process of NSS global development, it is necessary to organize at an international level a special scientific center to develop the GMNSS, taking into account the available database and accumulated knowledge. One of the global goals of this center could be to work out algorithms, methods, and models which would reduce the level of uncertainty in assessments of civilization development in the future.

# 2

# Globalization and biogeochemical cycles in the environment

## 2.1 PARAMETRIZATION OF GLOBAL GEOCHEMICAL CYCLES

The interaction between abiotic factors and living organisms in the biosphere drives the continuous cycle of matter in nature. Different forms of living organisms absorb substances that are essential for their growth and the sustaining of life. Under this mechanism the products of metabolism and other complex mineral and organic compounds of chemical elements are discharged in the environment in the form of unassimilated food or dead biomass. As a result the stable sequence of global biogeochemical cycles underpinned biosphere evolution. Their disturbance within the second part of the 20th century posed humankind many principal questions related with unforeseen climate change due to the greenhouse effect, decrease in biodiversity, progressive desertification, etc. In reality, questions about what will happen to the Earth's climate and the prospects for ozone layer decrease remain unresolved in spite of enormous amounts of money spent on their study (Kondratyev and Varotsos, 2001a, b; Varotsos, 2002, 2005). At present it is clear that answers to these and other questions related to nature protection depend on finding a means of creating an effective global monitoring system based on the global model of the nature/society system (GMNSS). Simulation of the biogeochemical cycles of basic chemical elements is a fundamental part of the GMNSS (Kondratyev *et al.*, 2002; Zhu and Anderson, 2002). Exactly such an approach gives us the possibility to optimize the anthropogenic fluxes of pollutants and to determine acceptable emissions of carbon, chlorine, sulfur, fluorine, methane, and other chemical elements to the environment, when finally the ideology behind the Kyoto Protocol becomes a reality, as well as to regulate questions related with the commerce of quotas for greenhouse gas emissions (Kalb *et al.*, 2004; Dalby, 2002; Fogg and Sangster, 2003; Kasimov *et al.*, 2004; Kondratyev, 2005a).

Fundamental connections between the constituent characteristics of the biological state of the environment have not been thoroughly studied for either the land or water ecosystems. Biodiversity in the ecosystems, state, and dynamics of food chains and interactions of the biosystem with the cycles of biogenic salts are important elements of these characteristics. The following questions are key among the series of problems and questions arising when investigating global biogeochemical cycles:

(1) What physical, biological, chemical, and social processes are fundamental to the cycles of carbon, nitrogen, phosphorus, sulfur, water, and other chemical elements in space and time?
  - What mathematical correlations can be determined when the parametrization of biological processes in the computer models of biogeochemical cycles is realized?
  - What dependencies are there between biodiversity, structure of ecological chains, and biogeochemical cycles in land and water ecosystems?
  - What are the mechanisms that connect one biogeochemical cycle with another and do common principles for the parametrization of these connections exist or do they depend on the chemical element types and ecosystems studied?

(2) What form can anthropogenic intervention in global biogeochemical cycles take?
  - Which are the ways in which humankind impacts biogeochemical cycles, affects the speed of change in such cycles, impacts the spatial distributions of chemical elements, and what are the consequences of this intervention?
  - How does the change in land use strategy influence the redistribution of chemical elements in space and time?
  - Which anthropogenic pollutions are important for biogeochemical influence on ecosystems and how can these influences be forecast?

(3) What mechanisms control an ecosystem's ability to recover rapidly from impacts and what are the indicators that reflect this ability?
  - How do the introduction of new species to an ecosystem and the outbreak of new unknown diseases influence the development of biogeochemical cycles in land and water ecosystems?
  - What feedbacks between ecosystems and climate are critical and how can these correlations be parametrized in computer models?
  - Can past data about biogeochemical cycles be used for forecasting the future?
  - What features of the ecosystem influence upon their ability be restored after anthropogenic impacts?

A major problem undergoing close attention by many scientists is the global biogeochemical cycle of $CO_2$. Scientists from many countries are attempting to answer such questions as:

(1) What level of $CO_2$ concentration is likely in the future under existing and forecast rates of burning of fossil fuel?
(2) What sorts of climate change can happen due to growing $CO_2$ concentration?

(3) What sorts of consequences are likely in the future for the biosphere as a result of climate change?
(4) What can humanity do to reduce the negative consequences of climate change?

It is clear that industrial civilization has to find new sources of energy in an effort to decrease the rates of fossil fuel burning and thereby reduce external impacts on natural biogeochemical cycles. The atmosphere is an important reservoir and plays a major part in the formation of these cycles. Atmospheric chemistry and physics is changing. It demands thorough investigation of its dynamical and photochemical atmospheric processes.

In recent years the term *greenhouse effect* has been used in numerous publications on the problems of global climate change on Earth (Frances, 2004; Follett and Kimble, 2000; Houghton, 2004; Loulou *et al.*, 2005). This term indicates the set of effects arising in the climate system that are connected with a series of natural and anthropogenic processes. On the whole, the greenhouse effect concept relates to explanation of changes in the heat regime of the atmosphere. These changes arise due to the influence of some gases on the process of absorption of solar radiation by the atmosphere. Many gases are characterized by high stability and long life in the atmosphere (Table 2.1): carbon dioxide is one.

Numerous long-time observations at different latitudes show the high-level correlation between temperature and $CO_2$ content. The greatest contribution to this correlation is made by the interaction between the atmosphere and ocean. Although the atmosphere and ocean achieve a certain stability concerning $CO_2$ exchange, this balance is still regularly disturbed. The principal causes of this disturbance are the following:

- change in surface ocean temperature;
- change in ocean volume; and
- change in the regime of vertical ocean circulation.

These causes are generally characterized by the level of their influence on $CO_2$ concentration in the atmosphere. The first cause brings about a 65% change in $CO_2$ partial pressure in the atmosphere ($p_a$). The remaining 35% belongs to the

**Table 2.1.** Characteristics of the most important greenhouse gases (Mintzer, 1987). Designation: ppmv = parts per million by volume; ppbv = parts per billion by volume.

| Greenhouse gas | Life time in the atmosphere (years) | Anthropogenic emission (MtC) | Average concentration | Distribution in the atmosphere (%) | Increase in speed (%) |
|---|---|---|---|---|---|
| $CO_2$ | 3–5 | 1,585.7 (84) | 362 ppmv | 76 | 0.5 |
| $NO_x$ | 100–150 | 97.5 (5) | 308 ppbv | 6 | 0.25 |
| $CH_4$ | 11 | 175.8 (9) | 1,815 ppbv | 13 | 1.0 |
| Fluorocarbons (HCFC, HFC, PFC) | 75–111 | 31.4 (2) | 0.34–0.54 ppbv | 5 | 7 |

second and third causes. By quantity this relation is characterized by the increase in $CO_2$ partial pressure in the atmosphere by 6% per degree of the increase in temperature of the upper ocean layer. It is known that a decrease in ocean volume by 1% increases $p_a$ by 3%.

Assessment of the greenhouse effect demands consideration of the complex interconnection of all the processes of energy transformation on Earth. However, there is a certain hierarchy in the significance of various processes (from astronomical to biological) that influence the climate system within different temporal scales. But this hierarchy is not constant since the role of separate processes can vary widely in their significance for climatic variations. Consideration of these separate processes should simplify analysis of the greenhouse effect on climate. In reality the level of greenhouse effect is defined by the amount surface temperature $T_L$ exceeds effective temperature $T_e$. The temperature of the Earth's surface $T_L$ depends on radiant emittance $\kappa$. The effective temperature $T_e$ is a function of the radiativity $\alpha$ of the *atmosphere–land–ocean system*. Commonly $\kappa$ and $\alpha$ depend on many factors including atmospheric $CO_2$ concentration. There are many simple and complex mathematical models in which efforts have been made to parametrize these dependencies. Unfortunately, there is no model that can reliably satisfy adequacy conditions and describe the prehistory of the climatic trends of Earth with reliability. Nevertheless, one can assert with confidence that the greenhouse effect depends essentially nonlinearly on the difference $T_L - T_e$; that is, on atmosphere opacity especially in the long-wave range. This opacity increases with the growth of $CO_2$ concentration in the atmosphere. The strongest influence of $CO_2$ on atmosphere opacity is displayed in the long-wave area of 12–18 μm. This influence is weaker in ranges of 7–8, 9–10, 2.0, 2.7, and 4.4 μm. It is clear that the role of various ranges increases with growth of $CO_2$ concentration in the atmosphere. This means that a flow of long-wave radiation that rises up from the Earth surface will grow with intensification of $CO_2$ absorption zones. In addition, the falling flow of long-wave radiation on the Earth's surface will increase. Existing data say that a reduction in rising and an increase in falling flows of radiation are characterized by the values of 2.5 and 1.3 $W\,m^{-2}$, respectively.

Thus, to assess the level of greenhouse effect as a result of $CO_2$ influence and other greenhouse gases (Table 2.2) it is necessary to be able to forecast their con-

**Table 2.2.** Global potential of different greenhouse gases (global warming potential, GWP).

| Gas | GWP | ΔGWP (%) | Gas | GWP | ΔGWP (%) |
|---|---|---|---|---|---|
| $CO_2$ | 1 | 55 | HFC-227ea | 2,900 | 0.69 |
| $C_4$ | 21 | 17 | HFC-235fa | 6,300 | 0.75 |
| $N_2O$ | 310 | 5 | HFC-4310mee | 1,300 | 0.75 |
| HFC-23 | 11,700 | 0.96 | $CF_4$ | 6,500 | 1.15 |
| HFC-125 | 2,800 | 0.75 | $C_2F_6$ | 9,200 | 0.75 |
| HFC-134a | 1,300 | 0.34 | $C_4F_{10}$ | 7,000 | 0.87 |
| HFC-143a | 3,800 | 0.75 | $C_6F_{14}$ | 7,400 | 0.75 |
| HFC-152a | 140 | 0.28 | $SF_6$ | 23,900 | 0.30 |

**Table 2.3.** Average dry air composition.

| Gas | Molecular weight | Average concentration (%) | |
|---|---|---|---|
| | | By volume | By weight |
| Nitrogen, $N_2$ | 28.016 | 78.084 | 75.53 |
| Oxygen, $O_2$ | 32.000 | 20.946 | 23.14 |
| Carbon dioxide, $CO_2$ | 44.010 | 0.0325 | 0.046 |
| Carbon monoxide, CO | | $(0.8–5) \times 10^{-5}$ | |
| Nitrogen protoxide, $N_2O$ | 44.01 | $(2–5) \times 10^{-5}$ | $7.6 \times 10^{-5}$ |
| Nitric oxide, NO | | $10^{-6}–10^{-4}$ | |
| Nitrogen dioxide, $NO_2$ | | $10^{-6}–10^{-4}$ | |
| Sulfur dioxide, $SO_2$ | | $7 \times 10^{-7}–10^{-4}$ | |
| Ozone, $O_3$ | 48.000 | $(0–5) \times 10^{-6}–5 \times 10^{-5}$ | $(0–1) \times 10^{-5}–10^{-4}$ |
| Ammonia, $NH_3$ | | $\leq 10^{-4}$ | |
| Formaldehyde, HCHO | | $\leq 10^{-5}$ | |
| Xenon, Xe | 131.3 | $(8.7–9.0) \times 10^{-6}$ | $(3.6–3.7) \times 10^{-5}$ |
| Hydrogen, $H_2$ | 2.016 | $5 \times 10^{-5}$ | $3 \times 10^{-6}$ |
| Krypton, Kr | 83.8 | $(1.14–1.2) \times 10^{-4}$ | $(2.9–3.3) \times 10^{-4}$ |
| Methane, $CH_4$ | 16.04 | $(1.2–2.0) \times 10^{-4}$ | $(7.75–9) \times 10^{-5}$ |
| Helium, He | 4.003 | $(5.24–5.3) \times 10^{-4}$ | $(7.2–7.4) \times 10^{-5}$ |
| Neon, Ne | 20.183 | $1.818 \times 10^{-3}$ | $1.25 \times 10^{-3}$ |
| Argon, $^{40}Ar$ | 39.994 | 0.934 | 1.27 |
| Water vapor, $H_2O$ | | $\leq 4$ | |
| Radon, Rn | 222.0 | $(0.06–0.45) \times 10^{-6}$ | $6 \times 10^{-18}$ |

centration in the atmosphere taking into consideration all correlations within the global biogeochemical cycles of these gases (Watson *et al.*, 2000). This problem entails knowledge from biogeochemistry, geochemistry, soil science, ecology, agricultural chemistry, geology, oceanography, physiology, and radiochemistry. The present methods of global ecoinformatics make it possible to bring about combined use of the different knowledge accumulated in these sciences.

Certainly, study of the global cycle of chemical elements is important, not for assessment of climate change due to anthropogenic activity, but for understanding environmental dynamics and its capacity to sustain life. Really, the cycle of chemical elements in nature is closely correlated with the activity of living matter. This allows us to distinguish the geological, biogenic, and biological sub-cycles within this cycle. Tables 2.3–2.9 contain some estimates of some parts of and parameters for these cycles.

## 2.2 INTERACTIVITY OF GLOBALIZATION PROCESSES AND BIOGEOCHEMICAL CYCLES

### 2.2.1 Anomalous phenomena in the environment and climate change

Climate change shows itself at both global and regional scales. One important special feature of climate formation both at regional and global scales consists in

**Table 2.4.** Evaluation of some parameters of the global cycle of chemical elements.

| Parameter | Parameter estimation |
|---|---|
| Coefficient of molecular diffusion in the air at temperature $T_a = 0°C$ and pressure 1 atm | |
| Hydrogen | 0.634 |
| Water vapor | 0.250 |
| Oxygen | 0.178 |
| Carbon dioxide | 0.139 |
| Absolute gas constant (cal $mol^{-1}K^{-1}$) | 1.9872 |
| Atmospheric mass (t): | |
| Total atmosphere | $(5.2–5.51) \times 10^{15}$ |
| Troposphere (up to 11 km) | $4 \times 10^{15}$ |
| Number of molecules in the atmosphere | $1.8 \times 10^{20}$ |
| Organic mass in photosynthesis (billion tons per year) | 100 |
| Land vegetation (%) | 66 |
| Plankton and algae (%) | 34 |
| Balance of photosynthesis (billion tons per year) | |
| Water consumption | 130 |
| Oxygen emission | 155 |
| Number of active volcanoes | |
| Lava | 527 |
| Mud | 220 |
| Number of molecules in the atmosphere per $km^2$ | $2.1 \times 10^{35}$ |
| World metal consumption (billion tons per year) | |
| Iron | 38 |
| Aluminum, copper, zinc, lead | 2 |
| Other | 0.3 |

the presence of considerable changeability within the internal dynamics of the climate system. One of the most essential factors of its internal dynamics is ENSO (El Niño/Southern Oscillation). An episode of recurrent climate warming due to ENSO began in October–November 2002 and finished in March–April 2003. However, despite the end of this episode during the boreal spring the rise in temperature as a result of ENSO had an influence on the beginning of regional anomalies in precipitation within the enormous area of the Pacific Ocean including the formation of a zone with increased moisture along the western coast of South America and a region of moisture shortage in east Australia as well as in the southwest part of the Pacific Ocean.

The global average value of the surface air temperature (SAT) in 2003 was close to the three highest levels registered since 1880, but it was lower than that of 1998. Average global SAT in 2003 compared with the average value for the period from

**Table 2.5.** Character and origin of basic substances polluting the atmosphere.

| Pollutant character | Pollutant origin |
|---|---|
| *Gases* | |
| Carbon dioxide | Volcanic activity, living organism respiration, fossil fuel combustion |
| Carbon monoxide | Volcanic activity, internal combustion engines |
| Hydrocarbons | Plants, bacteria, internal combustion engines |
| Organic compounds | Chemical industry, waste combustion, different fuels |
| Sulfuric gas and other sulfur derivatives | Volcanic activity, sea breezes, bacteria, fossil fuel combustion |
| Nitrogen derivatives | Bacteria, anaerobic microorganisms, biomass-burning |
| Radioactive substances | Nuclear power plants, nuclear explosions |
| *Particles* | |
| Heavy metals, mineral aggregates | Volcanic activity, meteorites, wind erosion, mist spray, industry, internal combustion engines |
| Organic substances (natural and manufactured) | Forest fires, chemical industry, various fuels, waste burning, agriculture (pesticides) |
| Radioactive aerosols | Nuclear explosions |

1961 to 1990 was exceeded by 0.46°C. According to the data of thermal satellite sensing the average global temperature of the middle troposphere in 2003 was one-third compared with the average value during 1979–1998.

The season of hurricanes in 2003 was extremely active in the basin of the Atlantic Ocean, with 16 tropical storms, 7 hurricanes, and 3 powerful hurricanes. Five of these tropical cyclones caused landslides in northeastern Mexico. In 2003, New Scotland and Bermuda suffered heavily from hurricanes. A specific feature of the region of the Atlantic Ocean was the formation of five tropical storms in the Gulf of Mexico. Three

**Table 2.6.** Assessment of the annual volume of particles with radius less than 20 μm emitted to the atmosphere.

| Particle type | Particle flow ($10^6$ t $yr^{-1}$) |
|---|---|
| Natural particles, soil and rock particles | 100–500 |
| Particles from forest fires and combustion of timber industry waste | 3–150 |
| Marine droplets | 300 |
| Volcanic dust | 25–150 |
| Particles generated in gas production | |
| Sulfates from $H_2S$ | 130–200 |
| Ammonium salts from $HN_3$ | 80–270 |
| Nitrates from $NO_x$ | 60–430 |
| Hydrocarbons from vegetable aggregates | 75–200 |
| Particles as a result of manufacturing | 10–90 |

**Table 2.7.** Classification of atmospheric pollutants (Straub, 1989).

| Basic class | Subclasses | Typical elements |
|---|---|---|
| Inorganic gases | Oxides of nitrogen | Nitrogen dioxide, nitric oxide |
| | Oxides of sulfur | Sulfuric acid, sulfur dioxide |
| | Other inorganics | Carbon monoxide, chlorine, ozone, hydrogen sulfide, hydrogen fluoride, ammonia |
| Organic gases | Hydrocarbons | Benzene, butadiene, butene, ethylene, isooctane, methane |
| | Aldehydes, ketones | Acetone, formaldehyde |
| | Other organics | Acids, alcohols, chlorinated hydrocarbons, peroxyacyl nitrates, polynuclear aromatics |
| Aerosols | Solid particulate matter | Dust, smoke |
| | Liquid particulates | Fumes, oil mists, polymeric reaction products |

**Table 2.8.** Sources of atmospheric pollution.

| Pollution source | Pollutant |
|---|---|
| *Natural* | |
| Volcanoes, fumaroles, solfataras | Gases, volcanic dust, mercury vapors |
| Natural surges of natural gas and oil | Hydrocarbons |
| Mercury deposits | Mercury vapours |
| Sulfide deposits | Sulfuric gas |
| Radioactive ore deposits | Radon |
| Wind blowing from surface of seas and oceans | Chlorides, oil, sulfides |
| Underground coal fires | $CO_2$, CO, $SO_2$, hydrocarbons |
| Natural forest and steppe fires | Smoke |
| Plant transpiration | Water vapors, aromatic and other airborne material |
| *Anthropogenic* | |
| Incineration of hard and fluid organic material | $CO_2$, CO, $SO_2$, lead, hydrocarbons, mercury vapors, cadmium, nitric oxides |
| Metallurgy of black, colored, and rare metals | Dust, $SO_2$, mercury vapors, metals |
| Atomic industry | Radioactive material |
| Nuclear blasts | Radioactive isotopes |
| Cement industry | Dust |
| Building blasts | Dust |
| Forest and steppe fires | Smoke |
| Oil and gas extraction | Hydrocarbons |
| Motor transport | CO, smog, nitric oxides |

**Table 2.9.** The lifetimes of some atmospheric components.

| Component | Life time in the atmosphere |
|---|---|
| Carbon dioxide | 3–5 years |
| Carbon monoxide | 0.1–3 years |
| Water vapor | 9–10 days |
| Sulfur dioxide | 3 days |
| Ozone | 10 days |
| Hydrogen chloride | 3–5 days |
| Nitric oxide | 5 days |
| Nitrogen dioxide | 5 days |
| Nitrogen protoxide | 100–120 years |
| Ammonia | 2–5 days |
| Methane | 3 years |
| Freons | 50–70 years |

tropical storms happened outside the usual season (June–November). One formed in April and two in December. In the western sector of the Pacific Ocean in the Northern Hemisphere, storm formation was less than usual (large-scale storms were totally absent).

The summer of 2003 in some regions of Western Europe was one of the warmest with heat waves affecting mainly Central and Western Europe. Two anomalous heat waves taking place in June and July–August (especially the latter) were noticeably powerful. Droughts accompanied the heat waves an there were forest fires, which covered a considerable part of the south of France and Portugal in July and August. The summer of 2003 in Western Europe was apparently the hottest since 1540. The heat wave in France killed 11,000 people. In Germany that summer was the hottest in the 20th century and (except some regions of northern and northwestern Germany) the hottest since instrumental observations began.

The most substantial anomalous situations which began in March 2003 included:

(1) extremely intensive precipitation in the middle part of the Atlantic Ocean, in the southeast and eastern coast of the U.S.A.;
(2) extremely low SAT values and unusual snowfalls over the European territory of Russia;
(3) 546 tornados in May in the U.S.A., which was unprecedented;
(4) a long-term drought in the west of the U.S.A., where in some regions it was the fourth and fifth consecutive year of shortage of rain;
(5) heavy brush fires in the eastern part of Australia in January and powerful forest fires in the south of California in October;
(6) anomalously intensive precipitation in Western Africa and in the Sahel;
(7) return to the normal level of precipitation on the Indian sub-continent during the summer monsoon; and
(8) close-to-record extent of the "ozone hole" in the Antarctic reaching a maximum of 28.2 million km$^2$ in September 2003.

Recent years have been marked by increased interest in the study of present climate change in high latitudes of the Northern and Southern Hemispheres, mostly determined by the decision to conduct in 2007–2008 the Third International Polar Year. Major conclusions concerning Arctic climate diagnostics are concentrated on analysis of the spatial–temporal variability of the polar climate. In this context, of great interest are new results from paleoclimatic analysis of an ice core from the Russian Antarctic station "Vostok" (Vakulenko *et al.*, 2004) which demonstrated the negative correlation between changes in $CO_2$ concentration in the atmosphere and air temperature. Paleoclimatic developments are becoming a convenient way of studying the laws of present climate dynamics (Widmann *et al.*, 2004).

From the data on the Antarctic discussed in "State of the Climate" (Levinson and Waple, 2004) it follows that the last decade in this region was anomalously cold. From the late 1970s till mid-winter of 1990, the sea ice cover extent round the Antarctic continent was growing.

The monograph of Filatov *et al.* (2005) dedicated to the climate of Karelia can serve as an example of the benefit of using informational analysis on the regional features of climate. New developments dedicated to the climate of cities (Mayers, 2004) and analysis of individual long series of meteorological observations (Alessio *et al.*, 2004; Garcia-Barrón and Pita, 2004; Oganesian, 2004) have made important contributions to studies of regional climate change. A new important stage in comprehending the data of empirical diagnostics of climate was the development and application of interactive models of the climate system and a combined approach to numerical climate modeling (Kondratyev, 2004b,c; Palmer *et al.*, 2004; DeWitt and Nutter, 1989; Kolomyts, 2003; Munasinghe and Swart, 2005).

As has been repeatedly emphasized, the interactive components of the present climate system include a broad spectrum of natural and natural–anthropogenic subsystems and processes, without a complex study of which it is impossible to reliably select prevailing trends in climate change. In this connection, one should enumerate the most important ones:

- Global water cycle. Effect of "cloud" feedbacks.
- Global carbon cycle. Interaction of water and carbon cycles.
- Land use and land surface changes.
- Present trends of GHG content in the atmosphere and mechanisms of their control.
- Interaction of climate and land ecosystem productivity.
- Effect of climate regime shifts on marine ecosystems.
- Control of natural resources to neutralize the negative consequences of human activity.
- Socio-economic aspects of ecodynamics and climate and their analysis for optimization of land use strategy.
- Interactions between processes in the geosphere and biosphere and their dependence of cosmic impacts.

### 2.2.2 Climate change, forests, and agriculture

Forest and agriculture ecosystems are the environmental components that are most sensitive to climate change. The former determine many characteristics of the biogeochemical cycles of GHGs, and the latter are the basis of the interaction between humans and the environment. The problems appearing here have been thoroughly studied within many international programs on the environment and climate. They are especially emphasized in the national programs of the U.S.A. In particular, analysis of the consequences using various scenarios of possible changes of global climate has led to the conclusion that in the case of several scenarios the impact on forestry and agriculture in the U.S.A. will be economically favorable. Partially it is connected with the growth of forest productivity (due to the growth of $CO_2$ concentration) and determined by the ability of forests to adapt to climate change. As for agriculture, according to available prognostic estimates for the period up to 2060, the positive impact of the process of global warming on agriculture in the U.S.A. will be less economically favorable than follows from the earlier estimates.

Unfortunately, in view of the global and poorly studied character of correlation between climate change and behavior of vegetation cover (forest ecosystems, in particular) at present there are no reliable estimates of the consequences of climate change for their productivity. Study of the problems appearing here has only just started.

### 2.2.3 Observational data

Analysis of observational data is reduced, as a rule, to the consideration of two categories of information:

(1) SAT changes for the last 150 years (and especially during the last 20–30 years, when the increase in global mean annual mean SAT was at a maximum).
(2) Paleoclimatic changes. These attract attention because they can be compared with present climatic trends and, to some extent, as an analog to possible climate change in the future (such attempts are ongoing, though the inadequacy of paleoanalogs for future climate forecast has been repeatedly and convincingly argued).

By definition, climate is characterized by the values of meteorological parameters averaged over 30 years. For instance, climate anomalies in 1990 are determined as deviations from averages over the period 1961–1990. Analysis of spatial–temporal climate variability for individual years is also widely practiced. In particular, the World Meteorological Organization (WMO) publishes annual surveys of the global climate. In these reports attempts are being made to answer the most important questions:

- Is climate warming taking place?
- Is moisture cycle intensity changing?

- Is the general circulation of the atmosphere and ocean changing?
- Are extreme climate changes (storms, droughts, floods) intensifying?
- Is a reliable estimate of the anthropogenic contribution to climate change possible?

The 1990s, on the whole, was the warmest decade since meteorological observations began in 1860, and the year 1999 was the fifth most anomalous year regarding global mean annual SAT (+0.33°C) for the period 1860–1999—it was also fifth for the average SAT anomaly (+0.45°C) in the Northern Hemisphere, but in the Southern Hemisphere it was only tenth (+0.20°C).

The band of maximum annual mean SAT extended in 1999 from North America eastward across the Atlantic Ocean and the Eurasian continent to the equatorial band of the western sector of the Pacific Ocean. Minimum SAT anomalies were observed in a broad band of the central and northeastern regions of the Pacific Ocean (including a decrease in SAT). Analysis of observational data revealed the prevalence of positive temperature anomalies in 1999 in many regions of the globe. The most apparent anomalous situations include both warming and cooling:

(1) The cold wave observed in January brought a SAT decrease in Norway, Sweden, and in some regions of Russia to levels not observed since the late 19th century.
(2) A temperature decrease in February in Western Europe was followed, in particular, by heavy snowfall in the Alps.
(3) In western Australia the SAT decreased to values below the norm, though the extreme warming in early January led to intensive bush fires.
(4) The March temperature in Iceland reached a minimum for the last 20 years.
(5) In April, powerful heat waves formed in the northern and central regions of India, and in July and August in the northeastern and mid-western regions of the U.S.A.
(6) Unusually hot and dry weather was observed in the western part of Russia (SAT anomalies in the central and northwestern regions of the European territory of Russia exceeded 5°C).
(7) On the Australian continent, the maximum average SAT in November–December turned out to be the lowest since 1950.
(8) The second half of the year was colder (than usual) in central and southern Africa; the Sahel region was colder, wetter, and cloudier than during preceding years.
(9) Warming in the U.S.A. during the last 50 years was weaker than over the rest of the globe, with the eastern part of the U.S.A. actually getting slightly cooler.

The land and ocean surface temperature decrease in the tropics in 1999 was determined by the year-long La Niña event. The year was characterized by a great number of destructive meteorological catastrophes, especially floods. In Australia, the U.S.A., and Asia there was a multitude of tropical storms; in Europe heavy snowfall, avalanches, and storms; and again in the U.S.A. droughts and tornadoes.

The global mean annual mean SAT value in the late 20th century exceeded—by more than 0.6°C—the value recorded in the late 19th century (error in this estimate of ±0.2°C corresponds to a 95% confidence level). Analysis of SAT observational data suggested the conclusion that since 1850 there has been an irregular but substantial trend of climate warming on a global scale. This trend was very weak from the mid-19th century till 1910, and then it increased by 0.1°C over a 10-year period (from combined data on SAT and SST in the period 1910–1940 and during the last two decades). Two positive episodes of cooling were separated by an interval of slight cooling, especially in the Northern Hemisphere. In the periods from 1951–1960 to 1981–1990 the sign of the inter-hemispherical difference of temperatures changed: the Northern Hemisphere became colder than the Southern Hemisphere.

As Moron *et al.* (1998) noted, the present global warming was considered by some specialists as connected with sudden changes in the Pacific Ocean in about 1976 or with a gradual warming of the tropical band of the Pacific Ocean as well as with other regional scale phenomena. The irregular trend mentioned above and an attempt to recognize the external forcings on it (anthropogenic or natural)—complicated by the presence of the internally determined variability of the climate system—have been traditionally based on interpretation of the trend as red noise (or, later, on the idea that the trend's variability is determined by the fact that it is regularly "immersed" in noise).

The existence of such regularities has been well established and ascribed mainly (if not completely) to instability of the interactive "atmosphere–ocean" system in the tropical Pacific Ocean. Periodicities of about 4–6 years and 2–3 years connected with the ENSO event have been detected. Such regularities on scales of decadal and inter-decadal variability were more difficult to detect in view of the insufficient length of the observation series.

In this connection, Moron *et al.* (1998) undertook a detailed analysis of all available data on the spatial–temporal variability of the SAT fields of the World Ocean and for its individual regions they used multi-channel singular spectral analysis (MSSA). The main goal of analysis was to detect the laws of variability and inter-basin relationships between SST on time scales from inter-annual to inter-decadal. The length of the observational data series was sufficient for a reliable analysis of SST variability on time scales of 2–15 years, though the statistical reliability of results for longer periods is more difficult to guarantee.

In view of great interest in SST variability in the Atlantic Ocean, most attention was given to this region. The strongest climatic signal was an irregular long-term SST trend. The use of the MSSA method for data-processing for the 20th century revealed the well-known regularities mentioned above: a gradual increase in SST in both hemispheres in 1910–1940, with the subsequent increase of SST in the Northern Hemisphere till the mid-1950s; a lower SST in the Southern Hemisphere; Northern Hemisphere oceans cooling in the 1960s till the end of the 1970s; and stability and then increase of SST in both hemispheres in the 1980s with a small weakening of this trend in recent years.

The insufficient length of the series of instrumental observations makes it impossible to interpret global laws as a manifestation of more or less monotonic increase of SST or as part of long-term centennial oscillations (according to indirect data, oscillations were observed with periods from 65 to 500 years). Possible external factors of variability include: the growth in $CO_2$ concentration, change of extra-atmospheric insolation, and volcanic eruptions. A new and surprising result was detection of the fact that large-scale warming and cooling was preceded by the same SST variability near the southern edge of Greenland and (soon after that) in the central part of the Pacific Ocean in the Northern Hemisphere. This reflects the important role of high-latitude processes in the North Atlantic and possible interaction (via the atmosphere) with the Pacific Ocean.

On time scales of near-decadal variability (7–12 years) no regular oscillations have been observed that are coherent on global scales. In the Northern Atlantic, there were 13–15-year and 90-year oscillations. Near Cape Hatteras there were inter-decadal oscillations which propagated along the Gulf Stream to the zone in the North Atlantic where their phases change (similar results obtained earlier were rather contradictory). In the context of the search for inter-decadal oscillations of ENSO the data considered did not reveal any substantial maximum of SST variability with periods longer than 10 years in either the Pacific Ocean or the whole World Ocean, but in the Indian Ocean there were observed 20-year SST oscillations, especially regular during the first half of the 20th century (oscillations of this kind were observed earlier still).

Analysis of 7–8-year oscillations revealed the contradictory character of their phase in the sub-tropical and sub-polar cycles of the North Atlantic. As for inter-annual variability (2–6 years), three dominating periods were recorded: 24–30, 40, and 60–65 months. The first period is a well-known quasi-biannual ENSO component which most strongly manifests itself in the tropics of the eastern sector of the Pacific Ocean, with anomalies of constant sign propagating along the western coastline of North and South America (in other oceans such variability is negligibly small).

An important new result consists in the detection of two clearly differing low-frequency modes of oscillation, combined by the common physical nature of the quasi-4-year mode and characterized by a drastic change in periodicity in 1960—from $\sim$5 years to almost 4 years. This set of observational data suggests the conclusion that the ENSO irregularity occurs due to interaction of the internal instability of the atmosphere–ocean system in the tropical Pacific Ocean with annual change. Since the mode of quasi-biannual oscillations also exists in other oceans, but does not correlate practically with the index of southern oscillations, one should assume that a stronger quasi-4-year signal forming here can remotely propagate beyond the Pacific Ocean, whereas the opposite process is practically impossible.

A weak oscillation with periods about 28–30 months is observed in the SST field in the Southern Hemisphere Atlantic Ocean and agrees with the Hermanito event observed in the past, which is, probably, an ENSO analog. The results discussed are a stage of developments aimed at comparison of numerical modeling results with the data of observations using 2-D and 3-D models of the atmosphere–ocean system.

An important contribution to the idea of SST changes in the past has been made by analysis of the data of temperature observations in boreholes. For instance, Bodri and Čermák (1999) noted that while the amplitude of long-term SAT changes in transitions from glaciations to inter-glacial periods reached 10–15 K, during the Holocene (the last 10,000–14,000 years) changes of the order of several degrees K took place on time scales from decades to several centuries. In this connection, analysis has been made of the data on vertical profiles of temperature measured at different depths in boreholes in the Czech Republic and maps were drawn of SAT changes in the same territory taking place during 1,100–1,300 (little climatic optimum), 1,400–1,500 years, and 1,600–1,700 years (the main phases of the Little Ice Age).

Huang *et al.* (2000) discussed the results of processing the data on temperature measured at different depths in 616 boreholes in the Southern Hemisphere, which has opened up possibilities to retrieve the change of global mean temperature for five centuries. The data from 479 holes revealed global warming by about 1.0 K taking place during the last five centuries. During the 20th century, the warmest one, an increase continental surface temperature reached 0.5 K (about 80% of climate warming occurred in the 19th–20th centuries). Warming during the five centuries was stronger in the Northern Hemisphere (1.1 K) than in the Southern Hemisphere (0.8 K). On the whole, the results obtained agree with conclusions drawn on the basis of data on tree rings, though the latter demonstrate a weaker centennial SAT trend, which can be explained by special features of dendroclimatic methods.

Analysis of paleo-information on SAT obtained from the data on oxygen isotopes in Greenland ice cores for the Quaternary period has shown that long-term temperature changes are superimposed by faster changes on time scales from a millennium to 10 years (Bowen, 2000). Analysis of Antarctic ice cores revealed similar changes. In particular, in both polar regions, substantial changes of temperature have taken place in the Holocene.

Data on the foraminiferan *Neogloboquadrina pachyderma* from the northeastern region of the Atlantic (west of Ireland) has made it possible to retrieve the SST and trace the Heinrich events connected with iceberg outbreaks (Jessen, 2006).

Information on the content of carbon dioxide and methane in air bubbles contained in ice cores reflects the important role of MGCs in climate formation, but it remains unclear what came first: a change in temperature or in MGC content. For instance, it was shown that the temperature in the Antarctic changed about 4,000 years earlier than changes of $CO_2$ concentration in four interglacials.

New results of numerical modeling of the dynamics of the El Niño event, caused by variations in orbital parameters, have satisfactorily followed "Milankovich" frequencies during the last 150,000 years as well as variations on a time scale of about a millennium (Morén and Påsse, 2001; Mäkiaho, 2005). However, most surprising was detection of climate variations with a period of 1,450 years from different data for different regions of the globe, which was regularly repeated—in particular, in Greenland—during the last 110,000 years, including the last glaciation and Holocene (an increase in the amplitude of such changes was observed during glaciations). These results reflect a radical reorganization of the climate system taking place in comparatively short time periods. The Holocene looks (compared with these changes) like a period of comparatively stable climate. There is no doubt that in the absence of climatic feedbacks the growth of GHG concentration in the atmosphere should bring about climate warming. However, the real situation turns out to be much more complicated, and to understand it a reliable means of detection and quantitative estimates of the role of feedbacks are needed. Otherwise, the reliable forecast of climate change in the future is impossible. Since one of the very important sources of respective information is peat bogs, they should be thoroughly protected.

Having analyzed the data of satellite observations of SST from 1982, Strong *et al.* (2000) noted warming over most of the tropics and in the mid-latitudes of the Northern Hemisphere (with the global mean trend +0.005°C per year not exceeding the limits of observation errors). Less representative Southern Hemisphere SST data reflect the existence of an opposite—that is, cooling—trend (the problem of SST data reliability needs serious attention).

According to the data of Levitus *et al.* (2000), during the last 50 years (1948–1998) the World Ocean has substantially warmed. The upper 300-m layer has warmed most (by 0.31°C on average), whereas the temperature of the 3-km layer increased by 0.06°C. This increase in temperature of the upper layer of the ocean had preceded the SAT increase that began in 1970.

Satellite data on sea ice cover extent are important as an indicator of global climate dynamics. Gloersen *et al.* (1999) detected a statistically substantial decrease in global area of sea ice constituting $(-0.01 \pm 0.003) \times 10^6\ \text{km}^2$ for every 10 years.

Especially important are data of microwave remote-sensing, analysis of which has not revealed any substantial changes in average temperature of the lower troposphere during recent decades. This is also confirmed by results of aerological observations. From the data of Woodcock (1999a, b), the global mean SAT in October 1999 was 0.2°C below the average value for the period 1979–1999.

Santer *et al.* (2000) discussed the causes of different trends of SAT and the lower troposphere temperature. Having analyzed the SAT data for the periods 1925–1944 and 1978–1999, Delworth and Knutson (2000) came to the conclusion that the main cause of SAT changes was a combined impact of anthropogenic radiative forcing (RF) and unusually substantial multi-decadal internal variability of the climate system.

Satellite data on changes of the balance of the mass of Greenland glaciers are an important indicator of climate dynamics. The results of laser altimetry in northern Greenland, for the period 1994–1999, show that on the whole at altitudes above 2 km the ice sheet was balanced, with local changes of different signs. A decrease in glacier thickness dominated at low altitudes exceeding 1 m per year, enough to raise the World Ocean level by 0.13 mm per year (this is equivalent to about 7% of the observed rise in ocean level).

Data of observations using moisture cycle parameters still remain fragmentary. Exceptions are such publications as Russo *et al.* (2000), in which analysis was made at the Observatory of Genoa University of change in the diurnal sum of precipitation for 1833–1985. A decrease in the number of rainy days for the whole period of observations was revealed, as well as considerable growth of the rate of rain starting from 1950. During the last 30 years there was a considerable increase in the number of days with intensive precipitation.

Yu *et al.* (1999) performed an analysis of available climate data on the heat balance of the atmosphere using the results of both observations and calculations.

The atmospheric radiation budget was found from the data of satellite observations of the fluxes of outgoing short-wave and long-wave radiation and radiation fluxes at surface level retrieved from satellite data. Quantities of turbulent heat fluxes at surface level were taken from the data of observations within the COADS program, and the horizontal heat transport was calculated using respective meteorological information. To minimize random errors, spatial–temporal averaging was done: the zonally averaged components of atmospheric heat balance components for the latitudinal band 50°N–50°S as well as values for this latitudinal band were considered.

Analysis of the data discussed has shown that it is impossible to close the atmospheric heat balance because an additional $20\,\mathrm{W\,m^{-2}}$ is needed. Attempts to use different versions of the input volumes of information did not help to remove this "imbalance". Since closing of the water vapor balance using the same data was successful, one can assume that the cause of this "imbalance" is inadequacy in the estimates of the atmospheric radiation budget, manifested as underestimated solar radiation absorbed by the atmosphere.

Having analyzed the completeness and reliability of the available data of climatic observations, Folland *et al.* (2000) came to the conclusion that the existing volume and quality of data make it impossible to give adequate answers to the questions enumerated above. In this connection, anxiety is aroused because of degradation of the systems of conventional meteorological observations that have taken place in recent decades, which are also of importance for calibration of satellite remote-sensing results. Therefore, even calculations of decadal mean values of climate parameters are difficult for some regions of Africa and vast regions of the World Ocean.

Discussing the observed regularities of global climate change with time and its causes, Wallace (1998) gave top priority to consideration of the following problems:

(1) periodic climate change due to variations in extra-atmospheric solar radiation;
(2) quasi-periodic climate variability (quasi-biennial oscillations in the equatorial stratosphere being its most vivid manifestation);
(3) the ENSO event (in view of a wide range of frequencies, this event cannot be considered quasi-periodic);
(4) inter-decadal climate variations, which are to a great extent determined by the internal intra-seasonal and intra-annual variability of the climate system;
(5) climate variability on time scales from inter-decadal to centennial;
(6) analysis of statistical significance of estimates of unprecedented events and "shifts of regimes" in the light of time-dependent series of many climatic parameters; and
(7) revealing the phase relationships between climate change on time scales from inter-annual to inter-decadal.

In view of the exceptional complexity of the climate system, with its numerous degrees of freedom and multitude of feedbacks, highly regular structures and modes of climate evolution can be the exception rather than the rule. "Rough" schemes of parametrization for reconstructing the structure and evolution of climate anomalies without superfluous detail should have a higher degree of stability. Important climatic "signals" considered in solving the problems of detection and forecast of global climate change should be seen "with the naked eye". A much more complicated problem than supposed is assessment of the statistical significance of some quantitative characteristics of climate variability, especially unprecedented events and "shifts of regime" from the data on time series of limited duration (as a rule, in view of the time dependence of such series).

A future direction for studies of climatic time dependence and related catastrophic events of the type of Hurricane Katrina—which 13 years after Hurricane Andrew was the most powerful in the history of Miami (Florida), caused the U.S.A. huge economic damage, completely submerging New Orleans and destroying many building projects in late August 2005—is a search for connections between temperature variations of different scales in different water bodies of the World Ocean. For instance, Chang *et al.* (2000) and Yamagata *et al.* (2004) showed that there is a stable correlation between changes in water surface temperature in the Indian and Pacific Oceans, which especially strongly manifest themselves in the monsoon season in the Indian Ocean basin. The ENSO event favoring propagation of the sub-tropical anti-cyclone over the Western Pacific plays a marked role in stirring up feedback mechanisms. Studies of correlations that appear are successfully carried out with the use of an interactive ensemble CGCM developed at the Centre for Ocean–Land–Atmosphere Studies (COLA).

In the Pacific Ocean there are two key regions which play an important role in variability in upper water layer temperature. These are the western and central sectors of the northern Pacific Ocean. Changes taking place here affect the climatic situation in many regions of Asia and in more remote areas (Nakamura and Yamagata, 1998a, b). Therefore, study of the complicated climatic situations in the Pacific region is important for detection of latent dependences between stimulators of global climate change in the future (Field and Raupach, 2004; Watson *et al.*, 2000; Volk, 2000; Victor *et al.*, 1998; Soon *et al.*, 2003).

### 2.2.4 Climate formation factors

Discussing the prospects of developments within the CLIVAR program of study of climate variability, Bolin (1999) emphasized that "IPCC was very careful in its assessments in order to stick to conclusions known from scientific literature, which serve the basis for such assessments. The key fact is that it is necessary to distinguish between something that can be considered real and which remains uncertain. As for the future climate forecasts, many uncertainties still remain. This approach has determined the confidence in the scientific community when making concrete decisions and should be preserved in the future."

Bolin (1999) emphasized that, though the IPCC (1995) Second Assessment Report contains the statement that global climate change taking place in the 20th century is partially determined by human activity, this conclusion was formulated very carefully. Of principal importance here was evaluation of the probable contribution of random climate variations independent of human impact. Results of recent studies have clarified this question, showing that random variations in global mean SAT on time scales from decades to centuries for the last 600 years were within ±0.2°C or less—this conclusion, as mentioned above, disagrees with the data of observations, from which it follows that there were wider limits to SAT changes in the past. In this connection—as Bolin (1999) believed—skeptics about estimates of the contribution of anthropogenic warming for the last 50–75 years should be asked to explain the much stronger global warming observed in recent decades.

In the reports of the official representatives of several countries at the conferences in Kyoto and Buenos Aires and in the mass media, weather and climate anomalies—like tropical hurricanes and unusual El Niños—were ascribed to the impacts of global warming. These opinions should, however, be thoroughly tested scientifically, though the possibility of more frequent anomalous events under conditions of global warming cannot be excluded. Therefore, the development of methods of climate forecasts on regional scales, mainly bearing in mind the period 2008–2012, are especially urgent.

In Bolin's (1999) opinion, to assess the socio-economic consequences of accomplishment of measures foreseen by the Kyoto Protocol on GHG emission reduction, it is very important to develop integral models—a combination of models of climate, carbon cycle, as well as power engineering and socio-economic development—which will require much more time and much effort. In this respect, a difficult problem is to validate such models in order to analyze their reliability. The absence of a method of adequate validation means that the results of numerical modeling using integral models can be considered only as possible scenarios but not forecasts.

Characterizing climatic forcings, Hansen *et al.* (1998, 1999) pointed out that they are still not determined with sufficient accuracy for reliable climate forecast. There is reliable information about GHG content in the atmosphere bringing about a positive RF, but serious difficulties are connected with assessments of the impacts caused by such factors as atmospheric aerosol, cloud, land use change, bringing about a negative RF, which partially compensates for "greenhouse" climate warming. One of the consequences of such a compensation points to a much more significant role played by changes in extra-atmospheric insolation (solar constant—SC) as a climate-forming factor than previously supposed, based on numerical modeling based solely on GHG contribution ("greenhouse" RF due to the growth of $CO_2$ concentration since the industrial revolution was estimated at about 1.5 W m$^{-2}$).

In connection with these circumstances, Hansen *et al.* (1998, 1999) obtained new estimates of global mean RF. The data in Table 2.10 characterize the results of analytical approximation (with an error of about 10%) of various components of "greenhouse" RF (the recent detection of $SF_5CF_3$ as a substantial GHG shows that the problem of substantiation of GHG priority cannot be considered completely resolved). GHG concentrations are expressed in ppm ($CO_2$, *c*); ppb ($CH_4$, *m*); ppb (CFC-11, *x*; CFC-12, *y*), with CFC standing for chlorofluoroorganic compounds (freons). The total RF value is 2.3 ± 0.25 W m$^{-2}$. Of interest is the fact that the rate of RF increase from 0.01 to 0.04 W m$^{-2}$ per year over the period 1950-1970 during the subsequent 20 years decreased down to 0.03 W m$^{-2}$ per year in connection with the decreasing rate of the growth in $CO_2$ concentration (despite the continuing growth of

**Table 2.10.** Greenhouse radiative forcing $F$ since the industrial revolution.

| Gas | Radiative forcing |
|---|---|
| $CO_2$ | $F = f(c) - f(c_0)$ where $f(c) = 5.04\,\lg(c + 0.0005c^2)$ |
| $CH_4$ | $0.04(m^{1/2} - m_0^{1/2}) - [g(m, n_0) - g(m_0, n_0)]$; $g(m, n) = 0.5\,\lg[1 + 0.00002(mn)^{0.75}]$ |
| $N_2O$ | $0.04(n^{1/2} - n_0^{1/2}) - [g(m_0, n) - g(m_0, n_0)]$; |
| CFC-11 | $025(x - x_0)$ |
| CFC-12 | $0.30(y - y_0)$ |

$CO_2$ emissions), the reasons remaining unclear. A certain contribution was also made by reduction of the growth in $CO_2$ concentration, but once again the reason is unknown.

Radiative forcing due to the growing concentration of tropospheric ozone was estimated at $0.4 \pm 0.15\,\mathrm{W\,m^{-2}}$. A drop in stratospheric ozone could result in RF equaling $-0.2 \pm 0.1\,\mathrm{W\,m^{-2}}$. Although these changes in sign partly compensate for each other, this does not mean that they are insignificant, since variations in ozone content in the troposphere and stratosphere have substantial and different effects on formation of the vertical profile of temperature.

As for RF due to aerosol, its determination is still unreliable for lack of adequate information on real atmospheric aerosol. Numerical modeling that takes anthropogenic sulfate, organic, and soil aerosol into account for prescribing global distributions of aerosol optical thickness has made it possible to evaluate the global distributions of RF and balanced surface temperature, and then to obtain respective global mean values of changes in RF ($\Delta F$) and surface temperature ($\Delta T_s$) for purely scattering aerosol (single scattering albedo $\omega = 1$) and more realistic aerosol (Table 2.11).

Hansen *et al.* (1998) noted that the most reliable value of total RF due to aerosol was $-0.4 \pm 0.3\,\mathrm{W\,m^{-2}}$ instead of $-0.54$ given in Table 2.11, though in any case this estimate remains very uncertain due to unreliable input data on aerosol properties.

**Table 2.11.** Global mean RF for three types of anthropogenic aerosol.

| Type of aerosol | $\varpi = 1$ | | "More realistic" $\varpi$ | |
|---|---|---|---|---|
| | $\Delta F$ (W/m$^2$) | $\Delta T_s$ (°C) | $\Delta F$ (W/m$^2$) | $\Delta T_s$ (°C) |
| Sulfate | −0.28 | −0.19 | −0.20 | −0.11 |
| Organic | −0.41 | −0.25 | −0.22 | −0.08 |
| Dust | −0.53 | −0.28 | −0.12 | −0.09 |
| *Total* | *−1.22* | *−0.72* | *−0.54* | *−0.28* |

Anthropogenic RF changes that are due to cloud are undoubtedly more substantial than those due to aerosol, but they are even more uncertain. Such changes (including the impact of aircraft contrails) mainly result from the indirect impact of anthropogenic aerosol, which causes (functioning as condensation nuclei) variations in cloud droplet size distribution and optical properties. Rough estimates place "cloud" RF between $-1$ and $-1.5\,\mathrm{W\,m^{-2}}$, but this can change by an order of magnitude (depending on the input parameters used). A conditional value $-1^{+0.5}_{-1}\,\mathrm{W\,m^{-2}}$ can be assumed. Some of the increase in cloud amount observed in the 20th century may be attributed to the indirect impact of aerosol. Specification of such estimates depends on complex observational programs in different regions of the globe being accomplished.

The contribution of changes in land use to RF variations is connected with the processes of deforestation, desertification, and biomass-burning, which affect surface albedo and roughness as well as evapotranspiration. It is also important to note that changes in surface albedo are more easily determined when covered with snow rather than vegetation. Approximate estimates of ERB change due to land use evolution were $-0.2 \pm 0.2\,\mathrm{W\,m^{-2}}$.

Natural RF due to SC changes during the last century (including its indirect impact on the ozone layer) can be assumed to equal $0.4 \pm 0.2\,\mathrm{W\,m^{-2}}$. Since total RF constitutes only about $1\,\mathrm{W\,m^{-2}}$, the contribution of variability in extra-atmospheric insolation could play a substantial role. Volcanic eruptions cause RF changes from 0.2 to $-0.5\,\mathrm{W\,m^{-2}}$ (these estimates are, however, conditional). For analysis of possible anthropogenic impacts on global climate, estimates of the sensitivity of climate system to external forcings are extremely important. Hansen *et al.* (1998) assumed that the change in global mean SAT at doubled $CO_2$ concentration should constitute $3 \pm 1°C$. Since RF estimates are not reliable enough, it is best to use different scenarios of RF change. One of the developments in this sphere is the study by Tett *et al.* (1999).

Crowley (2000) estimated the contribution of various factors to climate formation (SAT change) for the last 1,000 years using the energy-balance climate model. According to the results obtained:

(1) Changes in global mean SAT during the last 1,000 years may be explained as a result of the combined impact of known RFs (in the pre-industrial epoch 41–64% of SAT changes had taken place due to extra-atmospheric insolation and volcanic activity).
(2) Global warming observed in the 20th century was mainly of anthropogenic ("greenhouse") origin, substantially exceeding the internally caused variability of the climate system.

Unfortunately, the argumentation contained in the study of Crowley (2000) is unconvincing even from the viewpoint of explaining centennial change in global mean SAT. For instance, the causes of climate cooling in the late 19th–early 20th century were not explained. The model considered by Crowley clearly could not simulate changes in regional climate. The role of North Atlantic Oscillations in climate formation was not demonstrated. Energy-balance models can not describe the dynamics of the real climate system. This conclusion also refers to results obtained using much more complicated global interactive climate models (Knutson *et al.*, 1999).

Noting that climate models with low spatial resolution (~3°–6° latitude) cannot reliably simulate (and, moreover, forecast) climate changes on regional scales, Mearns *et al.* (1999) showed that two approaches can be used to resolve such problems:

(1) statistical (with regard to observational data) scaling (changing to a higher spatial resolution) of the numerical modeling results obtained using low-resolution models; and
(2) expanding such models by including "nested" regional models at higher resolution.

In this connection, Mearns *et al.* (1999) undertook a comparison of scenarios of anthropogenic climate change (with doubled $CO_2$ concentration) calculated using the NCAR "nested" model RegCM2 and the semi-empirical method of scaling (SDS). In both cases large-scale numerical modeling was carried out using the GCM model developed by the Commonwealth Scientific and Industrial Organization (CSIRO).

The results obtained show that the RegCM2 model reveals greater spatial variability in the fields of temperature and precipitation than the SDS model, which leads, however, to a greater amplitude of the annual change of temperature than the RegCM2 or GCM models. Diurnal temperature change turned out to be weaker in the cases of SDS and GCM than RegCM2, and the amplitude of the diurnal change of precipitation varied in interval between SDS and RegCM2. Calculations using RegCM2 reproduce both an increase and decrease of probability of precipitation with doubled $CO_2$ concentration, whereas SDS gave only an increase in precipitation.

One of the causes of these differences could be the fact that the semi-empirical model SDS is only based on data from the surface level 700 gPa, whereas in the two other models the vertical structure of the atmosphere is taken into account. This comparison, however, does not allow us to say which of the results obtained "correctly" reflects the impact of forcing. To answer this question and establish the cause of these differences, numerical modeling needs to be further improved.

Just how uncertain theoretical estimates of the causes of climate change are can be illustrated by re-assessment of the role of the "Milankovich mechanism" as the main factor of the paleodynamics of climate observed in recent years.

According to the Milankowich theory (Morén and Påsse, 2001), changes in paleoclimate were determined by latitudinal re-distribution of extra-atmospheric insolation and in annual change as a result of variations of the parameters of the Earth's orbit (referring especially to glacial–interglacial cycles during the Quaternary period), which include:

(1) inclination of the rotation axis with respect to the orbital plane (fluctuating between 22° and 24.5° with the current estimate 23.4°; the average periodicity of variations is 41,000 years and mainly affects high-latitude insolation);
(2) precession of equinoctial points affecting the time of onset of equinoxes and solstices, which mainly affects low-latitude insolation (precession is characterized by dual periodicities of 19,000 and 23,000 years); and
(3) eccentricity of the Earth's orbit which changes from almost circular to strongly elliptical with a periodicity of about 95,800 years (these changes cause a modulation of precession).

Milankovich (Kunzi, 2003) supposed, in particular, that summertime low-level insolation in high latitudes is the cause of the onset of glaciation and formation of ice sheets. Summertime low-level insolation was observed at a minimum angle of orbital inclination, high eccentricity, and at an apogee in the Northern Hemisphere summer. According to Milankowich's calculations, this configuration took place 185,000, 115,000, and 70,000 years ago.

Although the concept of Milankovich has been acknowledged (but not completely) in recent years, its individual aspects have been critically analyzed. This had two reasons:

(1) a lot of new geological data from cores of sea bottom rocks and ice cores have appeared; and
(2) considerable progress has been made in numerical climate modeling.

Analysis of both these sources of information has shown that an adequate explanation of paleoclimate changes is only possible when taking other climate-forming factors into account and not just variations in orbital parameters—in particular, variations in the GHG content in the atmosphere which includes carbon dioxide.

In this connection, Palutikof *et al.* (1999) performed an analysis of new geological data on paleoclimate changes, in the context of current ideas about global climate dynamics. Data from ice cores and pollen obtained in 1990 have led to the following two important general conclusions:

(1) some observational data do not confirm glaciation cycles as determined by the Milankovich mechanism (this refers especially to data on $\delta^{18}O$ in calcite veins at Devils Hole in Nevada, which testify to opposite phases of the cycles of glaciation and theory of Milankovich); and
(2) it follows from other data that this mechanism can explain only slow quasi-periodic variations but not short-term variability (on time scales from decades to millennia), as a result of the discovery that it has happened much more often than was supposed.

This variability in global mean temperature could reach several degrees over many decades. In particular, a large-scale sudden climate cooling had happened in the

Ames interglacial (~122,000 years ago), when climatic conditions had been very close to today's. A typical example of short-term climatic variability is the Heinrich and Dansgaard/Oeschger events (Clark and Mix, 2002; Keigwin and Boyle, 2000). Such events are likely to be repeated in the future.

Many uncertainties also remain concerning the impacts of present changes in extra-atmospheric insolation on climate. Soon *et al.* (2003) and Soon and Baliunas (2003) demonstrated, for instance, the hyper-sensitivity of the climate system to changes in UV insolation, the effect of which is intensified by feedback due to the statistical stability of clouds, the effect of tropical cirrus clouds, and stratospheric ozone (the "ozone–climate" problem warrants special analysis).

Of particular interest is consideration of the interaction between biospheric dynamics as a climate system component. The significance of this problem can be exemplified by estimates of the climatic impact of deforestation in the tropical Amazon basin obtained by Bunyard (1999). In the Amazon basin (mainly in its wet tropical forests, WTFs) nature performs a number of important functions that are still inadequately considered, including the energy input from the tropics to higher latitudes, which, however, is under threat in view of the high rate of WTF destruction.

According to present assessment, every year up to 17 million ha of tropical forests are removed, with ~6 million ha in the basin of the Brazilian Amazon. By the end of 1988, 21 million ha had been deforested, and 10 years later this area reached 27.5 million ha, larger than the size of the U.K.

WTF destruction is important because of its impact on the global carbon cycle, since there is a danger of transforming the WTF zone from a sink to a source of carbon for the atmosphere. No less substantial are the ecological aspects of WTF elimination, in view of the ecological uniqueness of the tropical forests of Central and South America. According to available estimates, deforestation in the basin of the Brazilian Amazon alone (over 360 million ha in size) will result in a loss to the annual sink of carbon of up to 0.56 billion tons, and on a global scale the level of this loss may well reach 4 billion t C per year. Bearing in mind that in 1988, as a result of forest fires, the tropical forests were burning continuously over an area of about 9 million ha, then from this source alone the atmosphere could gain 1–2 billion t C.

Over virgin tropical forests, about 75% of incoming solar radiation is used in evapotranspiration. Therefore, removal of WTF will result in radical changes in energy exchange and global atmospheric circulation. Changes in local climate could be even more substantial, especially from the viewpoint of precipitation, which might well be reduced by 65%. Of key importance is the fact that the threshold level of WTF elimination determining the loss of the ecosystem for self-support remains unclear. For instance, if it is 20%, then this threshold has already been exceeded.

Bengtsson (1999) drew attention to the fact that since non-linear processes exhibit the prevailing impact on climate system variability, it is impossible to establish any simple connection between external forcings (e.g., the growth in GHG content or variability of extra-atmospheric insolation) and the response of the climate system to such forcings. By taking the unpredictability of some factors of climate into account, the difficulty of distinguishing between anthropogenic and natural variability of climate becomes apparent and even increases due to the fact that both internally and externally forced modes of climate variability are determined by the same mechanisms and feedbacks.

Although considerable progress has been recently achieved in numerical modeling of the climate system, this mainly refers to the atmosphere, which is testified to by correspondence of the results of numerical modeling of atmospheric circulation with observational data. Results of the "combined" numerical experiments indicate that 3-D atmospheric circulation in the tropics is mainly determined by the impact of boundary conditions, whereas in high latitudes the impact of atmospheric dynamics prevails. Simulation of the water cycle in the atmosphere turned out to be realistic.

Considerable progress in modeling interactions in the atmosphere–ocean system has made it possible to successfully predict seasonal and inter-annual variability and, in particular, El Niño events. Adequate consideration of the processes at the land's surface has ensured a substantial increase in the reliability of hydrological forecasts (including river run-off).

In this context, Bengtsson (1999) discussed the progress of numerical climate modeling in three directions. Successful realization of the TOGA program given the possibility to forecast seasonal and inter-annual variability of SAT. The second of these directions is connected with numerical modeling of climate change on scales of decades and longer and especially with explanation of the centennial change of global mean annual mean SAT. Apparently, stochastic forcing can be considered as a zero hypothesis in the case of long-term climatic variability.

Consideration of the impact of low-frequency climatic fluctuations on the level of the Caspian Sea shows that the long-term variability in its level is mainly connected with SST anomalies in the eastern sector of the tropical Pacific Ocean. It turns out that positive SST anomalies correlate with enhanced precipitation in the basin of the Volga watershed and vice versa. The main cause of variations in Caspian Sea level is the long-term dynamics of ENSO events, which should be considered chaotic in nature.

Important to the resolution of this problem is the study of anthropogenic climate change. Calculations have shown that a doubling of $CO_2$ concentration should result in outgoing long-wave radiation at the level of the tropopause decreasing by $3.1\,W\,m^{-2}$ and downward long-wave radiation flux in the stratosphere growing by about $1.3\,W\,m^{-2}$. Thus, the total RF at the tropospheric top level will constitute $4.4\,W\,m^{-2}$. Calculations of the resulting SAT change using 11 climate models revealed a warming between 2.1°C and 4.8°C, as well as intensification of global mean precipitation between 1 and 10%.

According to Bengtsson (1999), before 1980 centennial change in global mean SAT was characterized by the prevailing contribution of natural variability, with the anthropogenic contribution subsequently increasing. An important task of subsequent developments is to improve numerical modeling (mainly from the viewpoint of further study into the various mechanisms of feedbacks) in order to provide reliable forecasts on regional and local scales. An urgent problem of global modeling consists in the consideration of the interaction between biogeochemical cycles.

One of the most important aspects of numerical climate modeling is assessment of the contribution of anthropogenic climate-forming factors. In this connection,

Allen *et al.* (1999) discussed the possibilities of recognition, evaluation, and forecast of the contribution of anthropogenic global climate change characterized by SAT with regard to the available data of observations and numerical modeling. The latter has to be carried out by keeping in mind the internal variability of the climate system, as well as the impact of the greenhouse effect (and respective climate warming) and sulfate aerosol (the effect of climate cooling).

The four global 3-D models of the interactive atmosphere–ocean system predicted an increase in global mean SAT for the decade 2036–2046 (compared with the pre-industrial level) between 1.1 and 2.3 K. Calculation of climate system sensitivity to doubled $CO_2$ concentration gave values between 2.5 and 3.5 K. According to the HadCM2-G5 climate model developed at the Hadley Centre (U.K.), the global mean "greenhouse" warming in the period 1996–2046 should constitute 1.35 K and "sulfate" cooling 0.35 K, and thus the resulting warming will reach 1 K.

With the anthropogenic SAT increase in the 20th century equal to 0.25–0.5 K per 100 years, calculations made using a simple climate model predict a 1–2-K uncertainty in the balanced SAT forecast by the year 2040. Such estimates can, however, only be reliable by adequate consideration of the characteristic time taken for the World Ocean to adapt. A critically important aspect of prognostic estimates is the necessity to take into account possible sudden non-linear climate changes, which seriously limits the lead time of forecasts.

The most attractive prospect for assessment and forecast of anthropogenic SAT change is connected with analysis of the spatial–temporal variability of SAT fields that takes into account the impacts of the greenhouse effect and aerosol. Realization of this approach is seriously complicated, however, by the impossibility of reliably prescribing aerosol forcing on the SAT field. Another serious problem is the necessity to take into account the climatic impact of changes in content of stratospheric and tropospheric ozone.

### 2.2.5 Inconsistency in climate change studies

The problem of anthropogenic global warming is now the center of attention of not only specialists but also wide circles of the population. As our understanding of this problem increases, there appears a feeling of inconsistency about the results obtained in this field, especially during the last decade. In this connection, Mahlman (1998) carried out an overview of such results aimed at analyzing the fundamental scientific aspects of this problem, in which emphasis is placed on the role of numerical modeling and observational data in understanding current climate change. The monograph by Weber (1992) was dedicated to the same theme.

The main difficulty in understanding the causes of climate change is connected with the seeming impossibility of adequately interpreting climatic feedbacks. This principally refers to the cloud–radiation feedback, direct and indirect (through the impact on radiative properties of clouds) climatic impact of atmospheric aerosol, as well as the impact of the atmosphere–ocean interaction on climate formation.

The specific manifestation of "greenhouse" climate warming is often ignored because of the very great time constant determined by climate system inertia (temperature change of deep layers of the ocean may take centuries and even thousands of years to manifest). Schlesinger *et al.* (2000) pointed out, for instance, the existence of global SAT oscillations with a period of about 65–70 years from the observational data for the period 1858–1992. It is important that the principal differences between numerical modeling (and forecasts) of weather and climate be taken into account. In the case of numerical climate modeling, it is important to use "fine-tuning" and adjustment because of the difficulty of adequately considering the complicated totality of interactive processes and spatial–temporal scales. In this context, the use of paleoclimatic data plays a substantial role, though they cannot be considered analogs to possible climate change in the future.

Serious anxiety has been caused by the inadequacy of global observation systems and of ground observations in particular. Mahlman (1998) emphasized that the controversy about anthropogenic climate change consists in the absence of reliable quantitative estimates of the relationship between the contributions of natural and anthropogenic factors to change. This creates serious difficulties for the practical realization of recommendations contained in the Kyoto Protocol.

These conclusions have been discussed in detail before and have received universal recognition, as illustrated by the recent review of Grassl (2000). Hence, the wide use of the term "climate change" to denote anthropogenic change alone is surprising. Substitution of the notion of "climate change" (in its true meaning) with the term "global warming" is also incorrect, since both observational data and the results of numerical modeling indicate highly inhomogeneous present climate change, far from being restricted to SAT increase.

Such terminological misunderstanding is not accidental, however. It is deliberate misinformation for the sake of establishing a false conception of anthropogenic ("greenhouse") global warming, as convincingly explained by Boehmer-Christiansen (1997) and Boehmer-Christiansen and Kellow (2002), who analyzed the political motivation behind this concept.

In August 1997, the U.S. Minister of Geology V. Babbitt, addressing about 3,000 participants at the Annual Meeting of the Ecological Society of America (Albuquerque, NM, 10–14 August 1997), said that they should implement their civil obligation and help to convince the skeptical American public that global warming is both real and dangerous: "We have a scientific consensus but we have not a public consensus." In this connection, Morris (1997) carried out an overview of available scientific information in order to analyze the grounds for this opinion, since there are many specialists who do not share the apocalyptic predictions of anthropogenic global warming. Emphasis in the overview was placed on the problem of distinguishing between natural and anthropogenic climate change.

In the mid-1970s the forecasts of global cooling due to sulfate aerosol predicted, for instance, that this impact would limit an increase in global mean temperature due to the enhanced greenhouse effect of the atmosphere by less than 2°C even with an eightfold increase in $CO_2$ concentration. The warming trend observed in the 1980s attracted attention to the problem of climate warming. At the height of the summer of 1998, J. Hansen (Hansen *et al.*, 1998) declared at the U.S. Congress a 99% probability of anthropogenic global warming and its destructive consequences for ecosystems in the future as well as a consensus among specialists on this problem, though many meteorologists and climatologists did not share those views, arguing that respective developments did not permit the establishment of reliable cause-and-effect relationships between anthropogenic GHG emissions and observed climate change.

The political motivation for support of the global "greenhouse" warming conception determined the massive growth in governmental financing of such developments in the U.S.A. in the period 1990–1995 from \$600 million to \$1.8 billion.

As for the contributions of different factors to the formation of global climate, Morris (1997) emphasized the importance of the combined consideration of contributions of "greenhouse" warming and solar activity, which during the period of instrumental observations constituted 0.31°C and 0.41°C, respectively. According to these estimates, a doubling of $CO_2$ concentration would lead to global warming between 1.26 and 1.33°C. With such climate changes, there are no grounds to expect the catastrophic consequences often predicted, such as the rise of the World Ocean level and an increase of epidemic diseases. However, there is no doubt that one should expect an increase in agricultural yield. On the whole, the consequences of climate warming should be positive. Although reliable forecasts of the ecological consequences of GHG emissions are still impossible, it is apparent that their reduction would strongly affect the world economy. Therefore, the main conclusion is that one should not try to prevent climate change that is still unreliably predicted but adapt to it. However, this conclusion is disputed in numerous publications (Hansen *et al.*, 1998; Houghton, 2000; Hulme *et al.*, 1999; Wigley, 1999; Steffen *et al.*, 2002; Smith, 2001; Smil, 2002; Saul, 2005; Pielke *et al.*, 2003).

In view of the complexity of the climate problem relevant scientific publications usually emphasize the serious uncertainty of available estimates and favor instead the continuation of studies, but in contrast this is fraught with serious consequences for ecological policy. As Boehmer-Christiansen (1997) noted, "ecological" bureaucrats and experts as well as well-organized representatives of the fuel–energy industry are interested in proving the existence of anthropogenic warming and want to obtain the needed scientific support. These three groups—with common interests—support non-governmental ecological organizations and have taken steps to thoroughly consolidate them after the Rio conference.

The main targets of "green" pressure are government departments which, in view of their responsibilities, as a rule either do not want or cannot bring about the needed measures—such as ecological taxes, subsidizing development of renewable energy sources, nuclear energy, public transport, energy efficiency, and others. The World Bank (WB) and Global Ecological Fund (GEF) are also lobbied to support the many expensive developments recommended by numerous experts and specialists in the sphere of finance. These experts are the most vigorous advocates of measures to prevent supposed undesirable climate change, and the WB and respective U.N. organizations increasingly use their recommendations—in particular, those connected with so-called "Joint Implementation".

This system is based on such fundamental documents as the International Framework Convention on Climate Change, the scientific bases of which are the IPCC Reports. To achieve these goals there are many components that need to be considered: scientific proof to substantiate the threats; the "green movement" and its emotive views; rhetoric and "principle of precaution"; development of new technologies aimed at reducing GHG emissions; and bureaucratic bodies that make the necessary plans and develop strategies.

Only after that can politicians, responsible for finding the financial support to transform plans into concrete actions, start acting. In this context Boehmer-Christiansen (1997) discussed various circumstances concerning "scientific provision". Clearly, the IPCC Reports are fundamental documents if the anthropogenic impact on climate is really dangerous. However, the difficult problem is how to determine "danger" and who should be responsible for it (taking socio-economic and political factors into account as well).

There are still many uncertainties. In particular, when discussing the necessity to reduce GHG emissions the emphasis is placed on decreasing the scale of use of coal, especially in electric power production (while the conditional "factor of $CO_2$ emissions" for natural gas constitutes 15, in the case of coal it is very variable reaching 25 and more). The "attack" on coal was not, however, connected with ecological motives. The question of ecology did not rise at all until the mid-1980s, when the price of oil and gas dropped (especially in 1986). To resolve the economic problems that cropped up, ecological allies were needed to provide the competitive ability of "pure" energy for the sake of "sustainable development".

It is no mere chance that the IPCC was set up as a forethought in 1995, it was really planned in 1987 and started functioning in 1988. The governments of various countries supported plans to reduce GHG emissions not for ecological but for other reasons: enhancement of national nuclear energy (Germany); increase in the export potential for electric power produced by nuclear power stations and gas (France, Norway); increase in the size of financial support, etc. Naturally, those countries where electric power depends on coal are the most skeptical about "greenhouse" warming.

Having analyzed the role of various international organizations and programs in relation to global climate change, Boehmer-Christiansen (1997) emphasized: "Climate policy cannot be understood without a deeper analysis of the role of science and scientific understanding of the coalition of non-ecological interests (both commercial and bureaucratic) which serve as a driving force of development of events on international scales. Where this coalition will lead us remains unclear, so far."

Constructive prospects of resolving this problem are connected with development and application of complex models to assess possible changes in climate and socio-economic development. In the detailed overview of methods and results of numerical modeling of global climate change with regard to the dynamics of socio-economic processes, prepared by Parson and Fisher-Vanden (1997), the main aspects of so-called integral assessments (IA) were discussed. The main goal of such developments is to substantiate recommendations for people making decisions about ecological policy.

Four concrete goals include the following problems:

(1) assessment of possible responses to climate change;
(2) analysis of the structure of scientific bases of modeling and characteristic uncertainties of the results obtained;
(3) comparative estimates of possible risks; and
(4) analysis of scientific progress achieved.

In recent decades, two approaches prevailed in the development of IA models: (1) use of assessments obtained by interdisciplinary groups of experts; (2) formal numerical modeling. The former approach is characteristic of IPCC efforts and developments within the Montreal Protocol, whereas the latter approach was realized by individual specialists. Emphasis was placed on the use of IA models to analyze possible impacts of climate change on the development of energy and economy in the context of $CO_2$ emissions to the atmosphere.

General assessment of the results obtained was that IA models cannot be used to substantiate highly specialized measures in view of the insufficiently detailed character of such models. However, the use of such models is important for assessment of possible uncertainties and, hence, the expediency of making any decisions. In this case it is important that considerable uncertainties concerning the system as a whole or recommendations on the needed ecological policy might turn out to be not those uncertainties that are most important for understanding the processes involved in changes taking place in the environment or characterized by the greatest variability.

In considering the data for concrete models, it is possible to range uncertainties by their significance, but this ranging depends on the specific character of the models. Although the use of IA models has made it possible to substantiate some correlations of the dynamics of socio-economic development and variability of the environment, the relevant results should be considered only preliminary. So far, the use of IA models has made only a small contribution to estimation of comparative risks and to obtaining the answer to the fundamental question: To what degree and in which respects is possible climate change most substantial?

In view of this, the results of the use of IA models to substantiate an adequate ecological policy have been, so far, rather restricted. This refers especially to the purely didactic estimates based on the use of simple models. The main areas for IA model improvement include: overcoming the insufficient understanding of probable impacts and the possibilities of adaptation; poorly substantiated or missing descriptions of social and behavioral processes in developing countries; and very limited ideas of unlikely but radical climate change. Despite these and other unresolved problems, the urgency of further efforts in development of IA models is obvious.

In conclusion, we emphasize once again that improvement of the global observation system is paramount. The urgency of this problem has been illustrated in Demirchian *et al.* (2002) where the complicated spatial–temporal SAT variability at high latitudes of the Northern Hemisphere has been demonstrated, as has the groundlessness of the "greenhouse" warming conception, according to which under conditions of the Arctic the effect of the growth of GHG concentration should manifest itself especially strongly.

## 2.3 LONG-RANGE TRANSPORT OF POLLUTANTS

Numerous observations of chemical compound content in the atmosphere above different territories have shown that there exists indisputable evidence of their movement over long distances. The following results can be mentioned as key stages in these studies (Kravtsov, 2002; Matthies and Scheringer, 2001):

- Ozone from North America prevails in ozone distribution above the North Atlantic during summer.
- Emissions of chemical elements in Asia impact on chemical processes above the northwest waterways of the Pacific Ocean.
- Sulphur oxidized above the west Pacific comes from sources located in Asia.
- Pollutants are transferred from North America to the waterways of the North Atlantic.
- Transport of pollutants between continents is controlled by meteorological fronts and is bound to influence the movement of air masses.
- Ozone balance in the troposphere above North Atlantic depends on $NO_x$ and VOC content in the atmosphere above North America.
- Ozone layer formation in the free troposphere above the North Atlantic depends highly on aircraft emissions of $NO_x$.
- Intercontinental transport of CO due to wild fires and biomass-burning exists in the Northern Hemisphere.
- The positive summer and negative winter correlation between CO and $O_3$ concentrations are detected above the North Atlantic.
- Long-range transport of dust particles from Asia and Africa to other territories is documented in a series of international experiments.
- There is a lot of evidence to support chemical element accumulation in organisms, soils, and reservoirs when pollutants arrive from other regions.
- Extraction and exploitation of oil deposits in Siberia proves the movement of a wide spectrum of heavy metals and organic compounds to the tundra and taiga.

Observations of the long-range transport of atmospheric pollutants, realized within the framework of IGAC, have given us the possibility to detect the impact caused by them, such as:

- The total flow of ozone to the North Atlantic from anthropogenic sources located in North America during summer is 1.0–1.6 gmol $day^{-1}$ and is detected at distances from 1,500 to 300 km.
- The amount of $NO_x$ delivered to the low troposphere of the North Atlantic during summer contributes to an increase in ozone content of 1–4 ppbv $day^{-1}$.

Thus, the long-range transport of contaminants and gases is important for spatial and temporal variability in the concentration of aerosols and gases in the atmosphere. Although most aerosols remain in the planetary boundary layer, the streams of desert dust and smoke from biomass-burning can in part be raised to the free troposphere and transported over large distances, even between continents. For example, the long-range transport of pollutants on the west coast of North America from Asia has been the subject of numerous analyses of chemical compounds found in sedimentation, snow, fish, and birds. Pesticides, nitrates, sulfates, heavy metals, polychlorobiphenyls of Asian origin were detected in the Pacific Ocean environment. That sort of long-range transport was confirmed in April 1998 during a severe dust storm in China, clouds to the west of which registered on satellite images. Detailed study of the transport and transformation of aerosols and gases between continents was realized in summer 2004 during the INTEX-NA within the international research program ICARTT where measurements from the Aqua, Terra, and Envisat satellites were used jointly with NASA aircraft laboratories DC-8 and J-31 (Bull *et al.*, 2005; Sinreich *et al.*, 2006). Basic attention was given to assessment of regional atmosphere quality, the effects of intercontinental transport of chemical contaminants and calculations of the atmosphere–radiation balance. ICARTT-2004 included coordinated multi-agency international field experiments to investigate the direct and indirect radiative effect of aerosols over northeastern North America and the downwind north Atlantic Ocean.

The high concentration of aerosol particles registered in the atmosphere above the Amazon and central Brazil during the dry season connected with the intensive biomass-burning and wildfires coincided recently with the long-range transport of these atmospheric pollutants. According to assessments made by Freitas *et al.* (2005) in South America more than 30 Tg of aerosol particles enters the atmosphere each year in this way. Most aerosols can remain in the atmosphere for approximately one week. In addition, CO, VOC, NO, $NO_2$ and other gases enter the atmosphere.

According to investigations by Prins *et al.* (1998) the smoke clouds of area 4–5 million $km^2$ register on satellite images even in the optical range. Such an area is capable of changing the global radiation balance and strongly influences the biospheric water cycle. Assessment of the consequences of the smoke from large-scale territories is only possible by means of global models such as GOCART, MOZART, or RAMS (Chin *et al.*, 2002). As was shown by Freitas *et al.* (2005), the classification of smoke source types is important. There exist two main types: (1) controlled biomass-burning and (2) uncontrolled forest fire. Biomass-burning introduces to the atmosphere on average 1,700 $gCO_2\,kg^{-1}$ and 60 $gCO\,kg^{-1}$, and from forest fire these fluxes total 1,524–1,670 $gCO_2\,kg^{-1}$ and 70–140 $gCO\,kg^{-1}$, respectively. The burning of dry vegetation on grasslands or savannas gives $65 \pm 20\,gCO\,kg^{-1}$.

The ITCT Project—part of the IGAC Program (Bates *et al.*, 2006)—studied the tropospheric chemistry and atmospheric transport of ozone, small particles of dust, greenhouse gases and their components. The main goal of this study was to understand how one continent can influence the air quality above another continent and what chemical transformations take place during intercontinental transport of atmospheric pollutants (Singh and Jacob, 2000). The choice of method for assessment of atmospheric air quality is significant for analogous investigations. Riccio *et al.* (2006) noted in connection with this that to minimize the effect of unavoidable uncertainty in air quality modeling it makes sense to use a Bayesian approach that permits spatial–temporal interpolation of experimental data at high reliability.

The micro-physical, optical, and radiative characteristics of aerosols are modified during the long transfer from the source region to the distant territory. Information on aerosol transformation effects and on randomization processes during long-range transport is very important for:

- quantitative evaluation of radiative forcing from aerosols;
- estimation of heterogeneous processes; and
- identification of characteristics (such as optical thickness) that are important for synoptic and climatic studies.

Numerous measurements of atmospheric optical thickness have been made in a series of international and national environmental projects. The use of passive satellite sensors guarantees the data averaged by the altitude above the pixel of spatial resolution. Therefore, aircraft measurements can be used to ascertain the vertical distributions of tropospheric aerosol and other atmospheric characteristics as well as to extend knowledge on the optical thickness of the atmosphere. Three experiments—ITOP, SHADE, and SAMUM—were successful in this, giving a possibility to study aerosol transformation during long-range transport of smoke and desert dust. The ITOP experiment took place in summer 2004 and had a general aim to investigate the transport and transformation of particulate and gas phase pollutants as they travel from North America to Europe. During this campaign, rich CO plumes reaching Europe were captured by different instruments (Bousserez *et al.*, 2005). During the SHADE in September 2000 filter samples on board a Met Office C-130 aircraft were collected (Formenti *et al.*, 2003). The main study area was over the Atlantic Ocean between Sal Island (Cape Verde) and Senegal. Dust was encountered in the altitude range from 0.5 to 4.5 km. The SAMUM focused on measurement and analysis of the effect of mineral dust from the Sahara Desert on the atmospheric radiation budget and was dedicated to understanding the micro-physical and optical properties of desert dust and the impact of desert dust on the global climate system. Measurements were made by using the flying laboratory aboard a Falcon-20, permitting the aerosol and trace gas concentration above northern Paris in July–August 2004 to be obtained. During these measurements aerosols coming from Canada and Alaska across the North Atlantic were detected. These aerosols were the result of biomass-burning. The Falcon-20 equipment permits:

- registering size of particles, from 4 nm to 100 μm;
- registering the variability of particles by size;
- assessing the aerosol absorption coefficient;
- determining the optical properties of non-spherical particles; and
- accumulating particles of diameter less than 4 μm for subsequent chemical analysis.

Atmospheric aerosols influence the Earth's radiative balance directly via their interaction with solar and Earth radiation and indirectly via cloud condensation and ice particles in the atmosphere. Since the polar regions have a sufficiently clear atmosphere even a negligible increase in aerosol concentration can substantially disturb radiation fluxes in these regions. Investigations realized within the ACE-1 have shown that even distant regions of the Arctic are not protected from global dispersion of Asian dust (Stone, 1998). Detection of chemical extraneous contaminants in the Arctic atmosphere is yet another complex global question.

One of the problems arising from the long-range transport of aerosol is assessment of the consequences of possible introduction of extrinsic microorganisms to the environment. They can generate undesirable and uncontrolled changes in the ecosystem. But, portable microscopic water droplets or particles can—along with solid matter content microorganisms—get into them or on their surface from agricultural soils, industrial sources, wild fires, dust storms in the desert, or ocean surface disturbances.

The winds transport throughout the globe trillions of tons of dust, which is synchronously increased with the development of desertization processes that accompany frequent droughts and scaled changes in hydrological structures.

Water droplets in the atmosphere can contain dissolved ions and organic compounds, and solid particles of atmospheric dust can condense these droplets. There exists a lot of data about the influence of aerosols on the climate (Kondratyev *et al.*, 2006a), but their role in influencing biological processes due to long-range transport of microorganisms has not been studied in depth. As a consequence a series of unsolved problems has arisen:

- Can microorganisms transported across the atmosphere influence ecosystem evolution (as an intrusive element)?
- How can microorganisms survive in extreme conditions in other territories under appropriate radiation, temperature, dryness, and nutrient salt loss in the atmosphere, snow, and ice, for example, in high latitudes?
- Do conditions exist in which pathogen infections can remain for a long time in dust clouds and carry with them risks for people or the biological community?
- Is there sufficient information and data to justify the anxiety about the long-range transport influence on possible environmental change in regions remote from civilization?
- What uncertainties inhibit reliable assessment of the importance of long-range atmospheric transport and how can its true contribution to environmental changes be measured in distant regions?

From the biological perspective, it is clearly important to study the type and amplitude of chemical compounds that can spread across the atmosphere over long distances. This leads to a further set of questions:

- How many chemical contaminants of biogenic type can get into the atmosphere and move with it?

- Can these aggregates be detected in the atmosphere, precipitation, or snowfall?
- Which of these chemical contaminants are capable of generating trophic effects after long-range transportation?
- Can chemical aggregates of biogenic type transported over long distances cause changes in water ecosystems and in the territories with a low level of biogenic supply (e.g., in mountain areas)?
- What type of microorganisms can be transported across the atmosphere to distant areas and how resistant are target ecosystems to their invasion?

Transfer of fine-grained aerosol dust from the Gobi Desert (Mongolia) to North America is a common process in spring. Silicon, iron, aluminum, and calcium are included in the dust content, and particle sizes range from 2 μm to 4 μm. This dust can only be differentiated from volcanic ash and aircraft discharge with great difficulty. Creation of a means of identifying these types of dust particles is important in connection with the dangerous long-range transport of microorganisms. But Asian dust transported thousands of kilometers can be hazardous for the health of people living in other territories. Therefore, knowledge of the character and origin of dust present in the atmosphere at any given moment is vital for human health. This is the reason Beyer and Matthies (2001) studied the potential of long-range transport (LRTP) as an indicator of the exchange processes of chemical contaminants between different environments.

To control the global transport of CO and $CH_4$ in the low troposphere NASA launched the Terra satellite in 1998 (EOS AM-1)/MOPITT which had an orbital altitude of 705 km and the ability to scan the 600-km band under a spatial resolution of $22 \times 22$ km. Choi and Chang (2006) analyzing the data received discovered that the high $CO_2$ concentration on the east coast of Japan was caused by long-range transport of CO from Siberian regions where there was biomass-burning and from industrial regions of China. CO levels above east China exceeded 35 ppb due to biomass-burning in Myanmar (Burma) and Indo-China. The CO concentration above the east shore of Japan at the time of such forest fires was $217 \pm 18$ ppb and at other times $186 \pm 15$ ppb giving an average difference of 31 ppb. Received estimations also confirmed that the average speed of increase in methane concentration in the atmosphere is close to 1% per year, but the motives for this increase were not established. Overcoming uncertainties in the biogeochemical cycle of chemical contaminants was undertaken by Inomata *et al.* (2006). Measurements of $SO_2$, $O_3$, sulfur, and volatile sulfur compounds above the shores of the Japan Sea were realized in 1995 and 1996 at an altitude range between 0.5 km and 5.5 km. These measurements show that concentrations of particles and short-living sulfurous gases ($CS_2$, $H_2S$, $SO_2$) sharply differ by season owing to changes in airflow trends and meteorological conditions. For example, mineral dust particles detected on 23 April 1996 had a desert origin and were brought by westerlies from Asia. In contrast, measurements taken on 28 December 1995 show a low concentration of particles and short-living sulfurous gases.

Lin *et al.* (2005) assessed the role of long-range transport of pollutants in the formation of air quality above Taiwan based on data from the monitoring network

**Figure 2.1.** The structure of global environments, sources, and sinks of chemical contaminants that take part in biogeochemical cycles (*www.adirondackphotos.info/large/forest-picture.jpg*, *www.hebdo.bf*, and *www.tan.org.au/Images/Photos/Dubai%20Arabian%20Sands.JPG*).

TEPA in 2000 and 2001 during winter monsoons. It was shown that long-range transport added 30 μg of pollutants to each cubic meter of atmospheric air. Under such conditions, the partial pressures of CO and $SO_2$ are increased by 230 ppb and 0.5 ppb, respectively. The content of dust in the air from Asia equaled $71 \pm 34\,\mu g\,m^{-3}$. On the whole, the influence of long-range transport more sharply manifested itself in coastal zones than in cities in the interior of the country.

These and analogous observations highlight the need to eliminate uncertainty in the estimations of biogeochemical processes by constructing a global model of the biogeochemical cycle of chemical compounds that is capable of integrating existing knowledge and suggesting regimes for observation (Kondratyev *et al.*, 2006a). In addition, such a model would facilitate parametrization of the interaction of biogeochemical cycles and their influence on the climate system via the dynamics of greenhouse gas concentrations in the atmosphere (Schulze, 2000). The biogeochemical cycles of chemical elements must not be considered as separate from other global ecodynamic processes, particularly those processes that take place in socio-economic areas. Energy fluxes in these areas controlled by nature and people directly influence the power of local and regional fluxes of gases and aerosols that—due to the atmosphere and hydrosphere—acquire global scales (Dearden and Mitchel, 2005). Therefore, the structure of global environments presented in Figure 2.1 indicates the

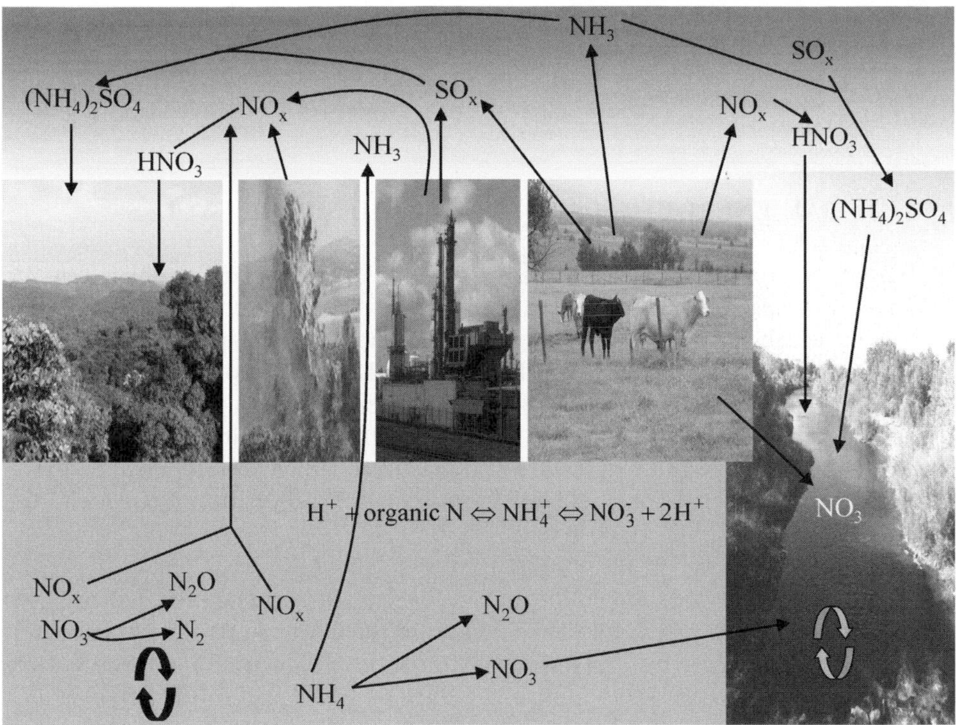


**Figure 2.2.** Schematic illustration of the role of anthropogenic nitrogen in the life of natural ecosystems (Matson *et al.*, 2002).

necessity of considering both the spatial and temporal heterogeneities that are appropriate to physical, chemical, biological, socio-economic, and even political processes in the GMNSS.

The impact of human activity on natural biogeochemical cycles reached a global scale in the 20th century, making the task of assessing the consequences to and reliable prognosis of global ecodynamics more complex. Figure 2.2 shows the spreading processes of anthropogenic intervention in the nitrogen cycle. Anthropogenic processes are responsible for doubling excess nitrogen flux to the land and aquatic ecosystems practically everywhere on the globe. As a result, a change in photosynthesis caused natural nitrogen and carbon cycles to deviate significantly from their stable states. The observed reduction in biodiversity and increase in speed of nitrogen transportation from the land to the hydrosphere are the results of this phenomenon. As follows from Figure 2.2 anthropogenic influence to the natural nitrogen cycle leads to ecosystem supersaturation by nitrates and hence to their salting and eutrophication, with acid rain destroying land vegetation. Such a chronic sequence of influences on natural ecosystems generates a chain of changes in biogeochemical cycles in practically all chemical elements, and stimulates reduction in the C : N ratio in land ecosystems. Undoubtedly, the scale of the consequences depends on the structure and acidity of soil, type of vegetation cover, topography, and regional climate (Cordova *et al.*, 2004; Matson *et al.*, 2002; Charman, 2002; Grigoryev and Kondratyev, 2001; Hulst, 1979; Koniayev *et al.*, 2003).

## 2.4 GLOBAL ENERGETICS AND BIOGEOCHEMICAL CYCLES

Globalization processes and their present tendencies pose many problems in the search for a strategy to control the waste products of different sectors of the economy. Regions in which NSS dynamics appear to be unstable cause special anxiety. Particularly, the uncertainty in future development of oil and gas imports from reservoirs in the Middle East and North Africa because of massive growth in the economies of China and India leads to uncertainty in the prediction of future fluxes of anthropogenic $CO_2$. It is expected that if present trends in economic development continue in these regions $CO_2$ emission could well increase by 52% in 2030 compared with the current level (WEO, 2002, 2004, 2005).

The need for energy in the Middle East and North Africa is regulated by the requirements of a growing population and by the growth of economics. The need for primary energy in this region by 2030 could more than double. At the same time oil extraction could increase by 75% and gas by three times. The quota of this region to oil extraction will increase from 35% to 44% by 2030. All this will be possible as a result of investments of $56 billion per year to energy infrastructure development in the countries of the Middle East and North Africa. These sums will be provided by the G8 countries to satisfy their own energy needs. Thus, the development of these regions are closely connected with levels of investments and prices, consideration of which introduces uncertainty in global $CO_2$ emissions (Cozzi, 2003).

A special IPCC Report (IPCC, 2005) analyzing long-term environment changes gives a set of assessments and makes some conclusions about the tendencies and causes of these changes:

- Global energy consumption and related $CO_2$ entering the atmosphere in the 21st century continue to increase. Fossil fuels deliver 86% of world energy and are the source of 75% of anthropogenic emission of $CO_2$ to the atmosphere. The world economy consumed in 2002 as a whole 341 Exajoules (EJ) including 149 EJ of oil, 91 EJ of natural gas, and 101 EJ of coal.
- Global consumption of primary energy increased on average at a rate equal to 1.4% during 1990–1995 and at 1.6% from 1995 to 2001. The rates of energy consumption increase are distributed between different sectors of the economy with the following shares: 0.3–0.9% in industry, 2.1–2.2% in transport, 2.1–2.7% in construction, and −0.8 to −2.4% in agriculture.
- About 38% and 17.3% of electric power is generated from coal and natural gas, respectively. Oil combustion gives 9% of electric power, and about 16.8% of electric power is produced by nuclear power plants, 17.5% and 1.6% of electric power is delivered by hydroelectric plants and other renewable sources of energy.

Some 2,250 MtC were emitted to the atmosphere in 2001 by electric power production. It is expected that coal will supply 36% of global energy needs up to 2020.

- Average $CO_2$ emission increased by 1% during 1990–1995 and by 1.4% from 1995 to 2001. The greatest input to $CO_2$ emission came from electricity generation. Here the contribution of different sectors of the economy was: 1.7–2.0% industry, 2.0–2.3% building, and −1.0 to −2.8% agriculture.
- Many experts expect a stable level of 550 ppm $CO_2$ can be reached by 2100 by reducing emissions between 7 and 70%. This uncertainty is determined by the wide variety of estimates of $CO_2$ sources and sinks, both natural and anthropogenic.
- Total emission of $CO_2$ under fossil fuel combustion for 2001 was assessed as 24 $GtCO_2$ (6.6 GtC) per year. Here 47% of this emission was from developed countries, 13% from countries with transient economics, and 25% from developing countries in the Pacific sector of Asia. Input of separate sources was: 263 MtC $yr^{-1}$ automobile industry, 1,173 MtC $yr^{-1}$ construction and manufacturing, 1,150 MtC $yr^{-1}$ travel and transport, and 520 MtC $yr^{-1}$ residential areas.
- Input of greenhouse gases to variation of the Earth's radiation balance is 20% for $CO_2$ and 6% for nitrogen oxides.

Tables 2.12–2.14 show the distribution of $CO_2$ emission by economic sector and region. It is expected that the growth of $CO_2$ emissions by total volume will change

**Table 2.12.** Distribution of $CO_2$ emissions due to the production of energy by the economic sector (MtC) in 2002 (IEA, 2002).

| Region | Indicator of the economic sector | | | | | | | | |
|---|---|---|---|---|---|---|---|---|---|
| | PEHP | VP | OES | MAC | T | CPS | RO | S | Σ |
| Countries with a transitional economy | 307.3 | 107.5 | 29.3 | 143.3 | 87.1 | 15.9 | 85.9 | 35.1 | 811.4 |
| Western sector of OECD | 298.7 | 36.1 | 60.9 | 197.3 | 284.4 | 47.8 | 135.1 | 26.3 | 1086.6 |
| U.S.A. | 622.3 | 36.9 | 74.4 | 179.8 | 470.0 | 61.6 | 101.5 | 11.7 | 1558.2 |
| Pacific sector of OECD | 139.9 | 23.8 | 17.0 | 82.3 | 94.1 | 26.0 | 20.7 | 9.8 | 413.6 |
| Central regions of South-East Asia | 254.3 | 28.4 | 37.7 | 145.7 | 123.4 | 13.9 | 50.7 | 10.8 | 664.9 |
| Asian countries with planned economies | 366.0 | 10.3 | 37.8 | 267.3 | 67.0 | 19.8 | 60.5 | 32.4 | 861.1 |
| Middle East | 77.1 | 1.8 | 32.4 | 52.7 | 46.9 | 4.5 | 24.8 | 30.7 | 270.9 |
| Africa | 75.9 | 4.3 | 11.0 | 37.6 | 39.2 | 1.4 | 12.2 | 9.5 | 191.1 |
| Latin America | 61.1 | 10.1 | 36.7 | 76.3 | 108.2 | 4.9 | 22.1 | 11.3 | 330.7 |
| *Total* | *2202.6* | *259.2* | *337.2* | *1182.3* | *1320.3* | *195.8* | *513.5* | *177.6* | *6188.5* |

PEHP = production of electric power and heat for the population; VP = vehicle production; OES = other energy sources; MAC = manufacturing and construction; T = transport; CPS = commerce and population service; R = residential; OS = other sectors; Σ = total energy.

**Table 2.13.** List of basic stationary $CO_2$ sources annually emitting more than 0.1 Mt $CO_2$ (WEO, 2002).

| Source | Volumetric concentration of $CO_2$ in emitted gas (%) | Volume of emitted $CO_2$ (MtC) | Contribution to $CO_2$ (%) | Average value of $CO_2$ emission per source (MtC per source) |
|---|---|---|---|---|
| Coal | 12–15 | 2,181.4 | 59.69 | 1.08 |
| Natural gas | 3–10 | 205.5–207.4 | 5.65 | 0.21–0.28 |
| Petroleum | 3–8 | 89.1–178.7 | 2.43–4.89 | 0.15–0.35 |
| Other fossils | — | 16.7 | 0.45 | 0.21 |
| Hydrogen | — | 0.8 | 0.02 | 0.35 |
| Cement industry | 3–20 | 218.0–254.6 | 5.97–6.97 | 0.22–0.34 |

from 24,409 Mt$CO_2$ in 2002 to 33,284 Mt$CO_2$ in 2015, and will reach 38,790 Mt$CO_2$ in 2025, exceeding the level of 1990 by 81%.

Despite the Kyoto Protocol after its ratification by Russia on 16 February 2005 coming into force for more than 55 countries, its recommendations on $CO_2$ emission reduction have not made accurate prognosis of the energy development in the coming decades easier. Nevertheless, if all signatories to the Kyoto Protocol follow its recommendations the average rate of $CO_2$ emission increase due to fossil fuel combustion during 2002–2025 will be equal to 2%. The prognosis can be improved to a certain extent by considering the data in Table 2.15, where a comparison between all economies and the most representative regions is given by an index of their economic development. To calculate the effect of such a reduction on GDP efficiency it is necessary to take into account population growth, GDP increase, and the human development index (HDI). In fact, an ideal model of globalization is to bring about HDI alignment of all countries from surplus resource redistribution. However, such a model is far from reality and therefore existing predictions of increase in $CO_2$ emissions vary widely in estimations of rates and volumes. Table 2.16 indicates the current heterogeneity in the world distribution of resources, and consequently the role played by separate regions in environment improvement.

Rates of changes in anthropogenic carbon flux intensity in the future could vary depending on technological achievements, economic policy, and change of priorities in energetic policy. The expected increase in science–technical progress is anticipated in the future to decrease continuously anthropogenic $CO_2$ fluxes to the atmosphere. This means that achievement of a high HDI level in countries with transitional and developing economics can be realized without violent changes in climate due to anthropogenic causes. As seen in Table 2.16 the U.S.A. not signing the Kyoto Protocol is a serious setback. The U.S.A. put forward its own initiative (RGGI) according to which elaboration of a reduction strategy for greenhouse gases is planned. This strategy will only satisfy U.S. national interests, though. This initiative assumes the development of methods for the management of greenhouse gas emission by trading carbon quotas between states.

**Table 2.14.** Distribution of $CO_2$ emissions by economic sector and region with prognosis to 2025 (IEA, 2005b).

| Region/Country | $CO_2$ emissions in 2002 as a result of energy production (MtC $yr^{-1}$) | | | | Prognosis of total volume of $CO_2$ emission (MtC $yr^{-1}$) | | $\sigma$ |
|---|---|---|---|---|---|---|---|
| | Due to oil combustion | Due to gas combustion | Due to coal combustion | Total emission | 2010 | 2025 | |
| U.S.A. | 617.3 | 328.7 | 565.6 | 1,565.6 | 1,853.3 | 2,339.1 | 1.7 |
| Canada | 74.9 | 43.7 | 42.1 | 160.7 | 191.3 | 236.6 | 1.7 |
| Mexico | 69.9 | 22.7 | 6.3 | 98.9 | 120.8 | 173.5 | 2.5 |
| Western Europe | 522.1 | 221.9 | 225.4 | 969.4 | 1,055.2 | 1,134.4 | 0.7 |
| Japan | 179.5 | 40.2 | 102.5 | 322.2 | 337.7 | 360.4 | 0.5 |
| Australia/New Zealand | 38.0 | 16.9 | 67.5 | 122.4 | 145.4 | 181.1 | 1.7 |
| Russia | 97.3 | 212.0 | 106.6 | 415.9 | 500.0 | 629.5 | 1.8 |
| Other CIS countries | 61.5 | 101.4 | 76.8 | 239.7 | 298.9 | 397.3 | 2.2 |
| Eastern Europe | 52.5 | 37.2 | 108.5 | 198.2 | 237.2 | 305.7 | 1.9 |
| China | 182.8 | 19.4 | 705.5 | 907.7 | 1,583.6 | 2,524.0 | 4.5 |
| India | 76.2 | 13.1 | 190.7 | 280.0 | 189.1 | 603.6 | 3.4 |
| South Korea | 65.8 | 13.1 | 44.0 | 122.9 | 157.4 | 231.1 | 2.8 |
| Other Asian countries | 205.2 | 73.0 | 106.3 | 384.5 | 535.0 | 845.6 | 3.5 |
| Middle East | 219.7 | 124.9 | 27.3 | 371.9 | 504.4 | 730.6 | 3.0 |
| Africa | 103.0 | 38.0 | 92.3 | 233.3 | 317.8 | 464.8 | 3.0 |
| Brazil | 75.1 | 7.1 | 11.2 | 93.4 | 125.4 | 207.1 | 3.5 |
| Other countries of Central and South America | 119.4 | 48.4 | 8.7 | 176.5 | 248.6 | 350.8 | 3.0 |
| Developed economies | 1,556.0 | 674.0 | 1,009.3 | 3,239.3 | 3,703.6 | 4,425.4 | 1.4 |
| Transitional economies | 211.1 | 350.5 | 291.8 | 853.5 | 1,036.3 | 1,332.2 | 2.0 |
| Developing economies | 1,047.5 | 336.6 | 1,186.3 | 2,570.4 | 3,861.2 | 5,957.9 | 3.7 |

$\sigma$ = average annual percentage $CO_2$ emission change for the period 2002–2025.

An important globalization aspect is the impossibility of assessing reliably the development of the processes involved in the working of the NSS, without analyzing all the correlations that exist in the real world. Hence, the numerous international programs can only have significance as a means of data and knowledge accumulation. A specific dependence can be seen between energetic and biogeochemical cycles.

Energy production now and in the near future is dependent on the use of natural gas, oil, and coal. Processing of these fossil fuels in industry and other sectors of economies leads inevitably to the emission of many other gases to the atmosphere that have a higher potential for atmosphere heating (Figure 2.3). Prognosis of energy development in the future depends on reliable parametrization of socio-economic

**Table 2.15.** Regional and national relationships between $CO_2$ emission and gross domestic product (GDP) (IEA, 2005a).

| Region/Country | Ratio between $CO_2$ emission (t) and GDP ($USm, year 2000 value) | | | | | Average annual rate of change (%) | |
|---|---|---|---|---|---|---|---|
| | History | | | Forecast | | | |
| | 1970 r. | 1990 r. | 2002 r. | 2010 r. | 2025 r. | 1990–2002 | 2002–2025 |
| U.S.A. | 1,117 | 701 | 571 | 501 | 393 | −2.1 | −1.6 |
| Canada | 1,046 | 691 | 612 | 562 | 481 | −1.7 | −1.0 |
| Mexico | 351 | 452 | 377 | 340 | 255 | 0.2 | −1.7 |
| Western Europe | 695 | 471 | 377 | 333 | 264 | −1.9 | −1.5 |
| Japan | 627 | 348 | 359 | 310 | 259 | −1.7 | −1.4 |
| Australia/New Zealand | 1,094 | 702 | 721 | 667 | 544 | −1.3 | −1.2 |
| Russia | 837 | 820 | 850 | 635 | 445 | 0.0 | −2.8 |
| Other CIS countries | 1,211 | 1,843 | 1,346 | 926 | 602 | 0.3 | −3.4 |
| Eastern Europe | 1,454 | 1,198 | 679 | 549 | 372 | −2.3 | −2.6 |
| Asia | 890 | 637 | 470 | 434 | 300 | −2.0 | −1.9 |
| China | 2,560 | 1,252 | 605 | 570 | 375 | −4.4 | −2.1 |
| India | 286 | 346 | 324 | 272 | 185 | 0.4 | −2.4 |
| South Korea | 791 | 698 | 680 | 555 | 454 | −0.5 | −1.7 |
| Middle East | 506 | 894 | 951 | 833 | 621 | 2.0 | −1.8 |
| Africa | 522 | 609 | 595 | 549 | 431 | 0.4 | −1.4 |
| Central and South America | 481 | 408 | 414 | 407 | 314 | −0.5 | −1.2 |

processes in the GMNSS. According to EIA (2005) the world is characterized by several clusters that differ in their economic development:

- Countries with developed market economies covering 15% of the world population (North America, Western Europe, Australia, and New Zealand).
- Countries with transitional economies covering 6% of the world population (Eastern Europe and countries of the former Soviet Union).
- Countries with developing economies (78% of world population):
  - developing Asia (53%);
  - Africa (14%);

**Table 2.16.** Global consumption per nation or region.

| Region/Country | Percentage of world population | Percentage of world resource production | Consumption of world resources (%) |
|---|---|---|---|
| U.S.A. | 5 | 6 | 40 |
| OPEC | 15 | 10 | 40 |
| Russia | 3 | 25 | 5 |
| Third World | 77 | 59 | 15 |

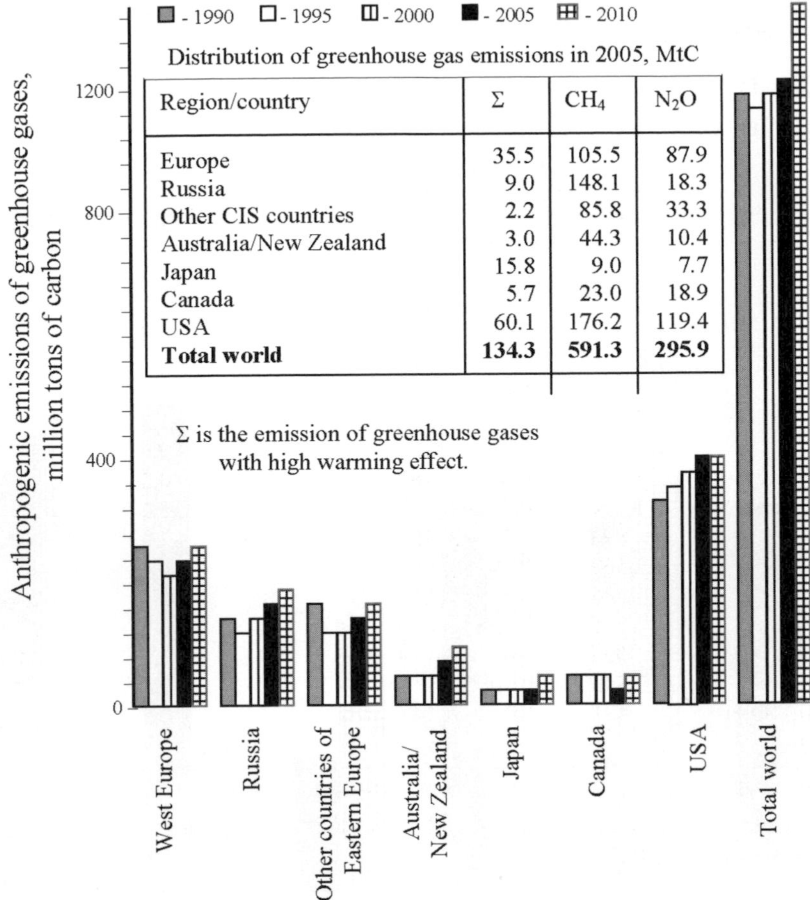


| Region/country | Σ | $CH_4$ | $N_2O$ |
|---|---|---|---|
| Europe | 35.5 | 105.5 | 87.9 |
| Russia | 9.0 | 148.1 | 18.3 |
| Other CIS countries | 2.2 | 85.8 | 33.3 |
| Australia/New Zealand | 3.0 | 44.3 | 10.4 |
| Japan | 15.8 | 9.0 | 7.7 |
| Canada | 5.7 | 23.0 | 18.9 |
| USA | 60.1 | 176.2 | 119.4 |
| **Total world** | **134.3** | **591.3** | **295.9** |

**Figure 2.3.** Regional distribution of non-$CO_2$ greenhouse gas emissions from developed countries projected to 2010 (EPA, 2001, 2005, 2006).

- Middle East (4%);
- South America (7%).

In other aspects, countries can be selected by groupings (Table 2.17):

- Countries that ratified, accepted, supported, or approved the Kyoto Protocol.
- Countries of the EU.
- G8 countries.
- Three countries of NAFTA.
- Thirty countries of OECD.
- Eleven countries of OPEC.
- Eight countries of developing Pacific basin.
- Seven countries of the Persian Gulf.

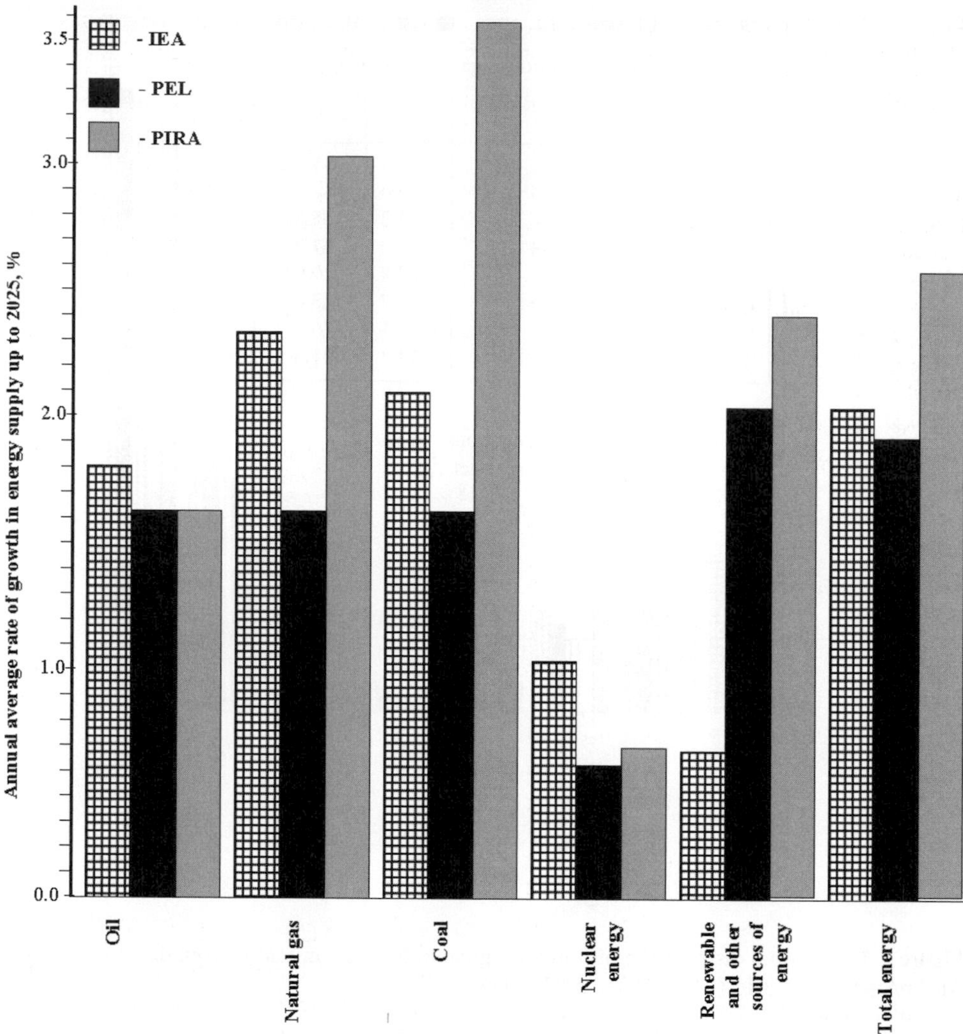


**Figure 2.4.** Forecasts of rates of average annual increase in energy supply made by IEA, PEL, and PIRA (IEA, 2005a–c).

It is clear that coordinated strategies for energy development exist within each of these clusters, but there are contradictions. Optimality of such a structure can be found by means of the $(V, W)$-exchange model described by Kondratyev *et al.* (2006c). A method developed in EIA (2005) for prognosis of energy development up to 2025 gives a possibility to make considered estimations of the need for energy resources, the realizing of which leads to achievement of $CO_2$ atmospheric concentration equal to 550 ppm by the end of the 21st century (Figure 2.4). To some degree, this is connected with priority redistribution between sectors of economies. There is expected to be a significant increase in the role of the transport sector on the

**Table 2.17.** List of regions and countries in which primary energy consumption exceeds 0.5% of total energy generated in the world at the beginning of the 21st century (BP, 2005).

| Region/Country | Energy consumption (Mt of oil equivalent) | | | | | $\Delta_1$ | $\Delta_2$ |
|---|---|---|---|---|---|---|---|
| | 2000 | 2001 | 2002 | 2003 | 2004 | (%) | (%) |
| *North America* | *2,736.3* | *2,681.5* | *2,721.1* | *2,741.3* | *2,784.4* | *1.6* | *27.2* |
| U.S.A. | 2,310.7 | 2,256.3 | 2,289.1 | 2,298.7 | 2,331.6 | 1.4 | 22.8 |
| Canada | 289.8 | 289.9 | 296.7 | 302.3 | 307.5 | 1.7 | 3.0 |
| Mexico | 135.8 | 135.3 | 135.3 | 140.4 | 145.3 | 3.5 | 1.4 |
| *South/Central America* | *450.7* | *452.0* | *454.4* | *460.2* | *483.1* | *5.0* | *4.7* |
| Venezuela | 61.9 | 65.2 | 66.1 | 60.3 | 67.6 | 12.1 | 0.7 |
| Argentina | 58.9 | 57.8 | 54.3 | 58.6 | 62.0 | 5.9 | 0.6 |
| Brazil | 176.9 | 174.1 | 177.9 | 180.0 | 187.7 | 4.3 | 1.8 |
| *Europe and Eurasia* | *2,830.4* | *2,855.5* | *2,851.5* | *2,908.0* | *2,964.0* | *1.9* | *29.0* |
| Belgium/Luxembourg | 66.4 | 64.0 | 64.9 | 68.6 | 70.3 | 2.4 | 0.7 |
| France | 254.9 | 257.8 | 256.7 | 259.6 | 262.9 | 1.3 | 2.6 |
| Germany | 330.5 | 335.7 | 330.0 | 223.1 | 330.4 | −0.5 | 3.2 |
| Italy | 176.4 | 177.2 | 175.9 | 181.0 | 183.6 | 1.5 | 1.8 |
| Kazakhstan | 41.0 | 42.3 | 44.1 | 47.6 | 52.8 | 10.9 | 0.5 |
| The Netherlands | 86.4 | 88.3 | 89.0 | 90.4 | 95.3 | 5.4 | 0.9 |
| Russia | 636.0 | 637.5 | 646.6 | 656.9 | 668.6 | 1.8 | 6.5 |
| Spain | 129.2 | 133.0 | 134.7 | 141.2 | 145.5 | 3.1 | 1.4 |
| Sweden | 48.6 | 52.1 | 48.5 | 46.2 | 48.4 | 4.8 | 0.5 |
| Turkey | 76.3 | 71.5 | 75.1 | 80.7 | 85.3 | 5.7 | 0.8 |
| Ukrane | 136.7 | 135.9 | 129.1 | 139.0 | 142.8 | 2.7 | 1.4 |
| England | 224.0 | 227.4 | 221.6 | 225.4 | 226.9 | 0.6 | 2.2 |
| Uzbekistan | 51.6 | 55.3 | 56.7 | 51.3 | 53.2 | 3.9 | 0.5 |
| *Middle East* | *395.8* | *413.2* | *438.7* | *454.2* | *481.9* | *6.1* | *4.7* |
| Iran | 119.6 | 125.0 | 140.0 | 147.6 | 155.5 | 5.4 | 1.5 |
| Saudi Arabia | 113.6 | 117.4 | 121.0 | 128.8 | 137.2 | 6.6 | 1.3 |
| UAE | 40.5 | 42.9 | 47.2 | 49.1 | 51.3 | 4.4 | 0.5 |
| *Africa* | *276.8* | *280.0* | *287.2* | *300.1* | *312.1* | *4.0* | *3.1* |
| Egypt | 47.6 | 49.4 | 49.5 | 52.0 | 53.8 | 3.6 | 0.5 |
| South Africa | 108.4 | 107.0 | 110.9 | 117.3 | 123.7 | 5.4 | 1.2 |
| *Asia and Oceania* | *2,389.7* | *2,497.0* | *2,734.9* | *2,937.0* | *3,198.8* | *8.9* | *31.3* |
| Australia | 111.0 | 112.9 | 116.6 | 114.7 | 119.0 | 3.8 | 1.2 |
| China | 766.0 | 837.9 | 1,034.9 | 1,204.2 | 1,386.2 | 15.1 | 13.6 |
| India | 320.4 | 324.2 | 338.7 | 350.4 | 375.8 | 7.2 | 3.7 |
| Indonesia | 95.2 | 101.4 | 104.4 | 104.2 | 109.6 | 5.2 | 1.1 |
| Japan | 515.9 | 514.8 | 506.6 | 504.9 | 514.6 | 1.9 | 5.0 |
| Malaysia | 45.7 | 48.0 | 51.3 | 56.2 | 60.3 | 7.2 | 0.6 |
| Pakistan | 42.0 | 42.9 | 43.8 | 46.0 | 47.2 | 2.6 | 0.5 |
| South Korea | 191.1 | 195.9 | 205.0 | 211.8 | 217.2 | 2.6 | 2.1 |
| Taiwan | 85.4 | 86.8 | 91.0 | 94.7 | 97.8 | 3.4 | 1.0 |
| Thailand | 62.4 | 63.6 | 69.2 | 75.7 | 81.5 | 7.7 | 0.8 |
| 25 EU countries | 1,655.1 | 1,680.9 | 1,665.7 | 1,697.5 | 1,718.8 | 1.3 | 16.8 |
| OECD countries | 5,359.4 | 5,327.4 | 5,360.0 | 5,414.5 | 5,503.3 | 1.6 | 53.8 |
| World | 9,079.8 | 9,179.3 | 9,487.9 | 9,800.8 | 10,224.4 | 4.3 | 100 |

$\Delta_1$ = increase in energy consumption in 2004 compared with 2003; $\Delta_2$ = % total world energy consumed.

environment. In 1990 transport equaled 20% of consumed energy, but by 2095 this will approach 40%. However, at the same time progress in industry will bring about a reduction in consumed energy from 38% in 1990 to 17% in 2095. It is expected that the role of such sectors as electric power production and construction will not significantly change. The forecasts given in Figure 2.4—as realized by different experts—are characterized by high uncertainty about the initial data from biogeochemical cycles in the GMNSS. Hence it can be seen that synthesizing a global model for economic development has special attention for models of biogeochemical cycles and for the assessment of their role in formation of the environmental state.

# 3

# Numerical modeling of the nature/society system

## 3.1 GLOBAL ECOINFORMATICS AS A SCIENCE FOR NATURE/SOCIETY SYSTEM MODELING

On a planetary scale, all living beings in the biosphere are closely interconnected by the ways in which the mechanisms that regulate energy fluxes and cycles of substances are organized: a single biocybernetic system of the highest rank. Within the continents and oceans—the structural units making up the biosphere—processes of energy and substance transformation take place automatically. Land biogeocenoses are characterized by distributed productivity, a function in many territories that is under the control of humans and therefore depends on development of scientific–technical progress. The World Ocean now provides about 1% of resources consumed by humankind and practically remains an element of the biosphere not under human control. This low level is connected with the insufficiently studied production processes in the oceans.

Anyway, the interconnection of human living standards and natural processes has recently become of fundamental importance. Study of the biosphere as a complicated hierarchically organized unique system has become an urgent problem for humankind because human life depends completely on the state of the biosphere. In this study, a central role is played by system ecology—the science of using numerical modeling methods and computers to study how biospheric ecological systems function. Development in this direction has led to new ideas in the sphere of global change studies as a result of realization of numerous anthropogenic projects, and as a result various scientific disciplines have appeared, such as Geographic Information Systems (GIS), global modeling, Geoinformation Monitoring Systems (GIMS), survival theory, and systemology (Fleishman, 1982; Nitu *et al.*, 2004). However, each of these directions has limited possibilities to study the dynamics of natural sub-systems with different spatial–temporal limitations. Therefore, there is a need to develop an interdisciplinary direction of science that combines GIS, global modeling, GIMS, expert systems, and takes into account the socio-economic aspects of nature protection activity at the same time. Ecoinformatics has become such a direction (Arsky *et al.*, 1992, 2000).

### 3.1.1 Development of ecoinformatics

Many international conferences were organized towards the end of the 20th century, whose subject matter was closely connected with the problems of ecoinformatics and with certain features of Earth sciences that need to be considered in order to solve problems about environmental protection and to give recommendations about how to organize global ecological monitoring. Analysis of this subject matter shows that it is only in Russia that a brand new approach has been proposed aimed at bringing about a complex information system that both describes biospheric and climatic processes and takes into account anthropogenic activity trends at the same time. The main idea of this approach consists in creating information technologies which would permit several models to be developed despite the fragmentary and distorted information about processes taking place both in nature and in society (Borodin *et al.*, 1997; Bui and Krapivin, 1998; Golfeld and Krapivin, 1997; Kolokolchikova *et al.*, 1997; Krapivin *et al.*, 1997).

The First International Symposium on Ecoinformatics was held in Zvenigorod, Moscow Region, in December 1992 (Mkrtchian, 1992). It was here that the problems facing human survival were first discussed by considering the socio-ecological aspects of current scientific–technical progress. It was shown that progress toward resolving these problems is only possible by using a systematic interdisciplinary approach. The scientific approach outlined attracted the attention of many specialists in the studies of global natural change from different countries. In Russia, this new scientific direction was supported by the Ministry of the Environment and Natural Resource Protection and took the form of the national program "The ecological security of Russia": this outlined the ecoinformatic problems to be resolved by the Institute of Ecoinformatics of the Russian Academy of Natural Sciences. This research was continued in other countries in many scientific centers: Universities of Alaska and Dillard in the U.S.A. (Kelley *et al.*, 1992a, b; Krapivin *et al.*, 1996), U.S. National Laboratory of Marine Studies (Krapivin and Phillips, 2001a, b), Hokkaido University in Japan (Aota *et al.*, 1993) as well as in Canada, Bulgaria, Romania, and Vietnam (Krapivin *et al.*, 1998a, b; Nitu *et al.*, 2000a, b).

In accordance with the principles of the ecological doctrine of the Russian Federation and the conception of a single governmental system for the ecological monitoring of Russia to ensure the ecologo-economic security of Russia, it is necessary to develop a multi-purpose information–observing system that has the task of data collection on environmental state, trends, and level of change due to economic activity, and is able to provide the needed ecological information for all levels of economic management and to assess the ecological expediency of realization of large-scale anthropogenic projects. The governments of other countries are faced with a similar problem. At present, many scientists and numerous research centers of the world are attempting to resolve this problem. For instance, within the framework of the International Biospheric Program "Global Change", in 1992 the U.S.A. started work on synthesizing a simulation system for prognostic estimation of the dynamics of Arctic Basin pollution (Krapivin and Phillips, 2001b). The underlying idea of this program is to create a technology of Arctic aquageosystem modeling with no fewer than five spatial levels of digitization taken into account, starting at the 1-m scale and ending at the global scale. This system of simulating how the Arctic Basin functions is aimed at monitoring the strength of the mutual effects of climatic, biogeochemical, ecological, and hydrological factors over the entire hierarchy of spatial scales of the Arctic natural system. As seen from the interim results of this program, the use of ecoinformatic methods should make it possible to successfully resolve the problem.

The basic idea of ecoinformatics consists in creation of a model—universal in its thematic content—which describes the interaction of natural and anthropogenic processes. Modeling social relationships and how they change depending on the environmental state is most complicated and insufficiently studied. The most complicated task for global modeling is the problem of prediction of these processes on a global scale in order to work out optimal behavioral policies at a governmental level. Nevertheless, ecoinformatics can suggest an approach that may well solve this problem too.

### 3.1.2 The basic model of ecoinformatics

Nature, $N$, and human society, $H$, constitute a single planetary system. Therefore, separating them when developing global or regional models should be considered a conditional step. Systems $N$ and $H$ have hierarchical structures $|N|$ and $|H|$, goals $\underline{N}$ and $\underline{H}$, behaviors $\overline{N}$ and $\overline{H}$, respectively. From the mathematical point of view, interactions between systems $N$ and $H$ can be considered as a random process $\eta(t)$ with the unknown distribution law, representing the level of tension in the interaction of these systems or assessing the state of one of them. The goals and behaviors of the systems are functions of the indicator $\eta$. There are ranges of $\eta$ in which the system's behavior can be antagonistic, indifferent, and cooperative. The main goal of system $H$ is to reach a high living standard with guaranteed long-term survival. Similarly, the goal of system $N$ can be defined in terms of survival. System $N$ behavior is determined by objective laws of co-evolution. In that sense, a selection of $H$ and $N$ is conditional and can be interpreted as a separation of the multitude of natural processes into controllable and non-controllable. Without dwelling upon the philosophic aspects of this separation, we shall consider systems $H$ and $N$ as symmetrical—in the sense of their description given above—and open. System $H$ disposes of technologies, science, economic potential, industrial and agricultural productions, sociological structure, population size, etc. The interaction between systems $H$ and $N$ leads to a change in $\eta$, the level of which affects the structure of vectors $\underline{H}$ and $\overline{H}$. There exists a threshold for $\eta_{max}$ beyond which humankind ceases to exist but nature survives. Asymmetry of systems $H$ and $N$ causes a change in the goal and strategy of system $H$. Apparently, under present conditions of interaction of these systems, $\eta \to \eta_{max}$ at a rapid rate and therefore some components of vector $\underline{H}$ can be attributed to the class of "cooperative". Since the present socio-economic structure of the world is represented by a totality of countries, we shall consider a country as a functional element of system $H$. The function $\eta(t)$ reflects the result of the interaction of countries between themselves and nature. The totality of the results of these interactions we shall describe by matrix $B = \|b_{ij}\|$, each element of which has a concept of its own:

$$b_{ij} = \begin{cases} + & \text{for cooperative behavior,} \\ - & \text{for antagonistic relationships,} \\ 0 & \text{for indifferent behavior.} \end{cases}$$

A country $H_i$ has $m_i$ possible ways to achieve goal $\underline{H}_i$. In other words, it uses a row of strategies $\{\overline{H}_i^1, \ldots, \overline{H}_i^{m_i}\}$. The weight of each strategy $\overline{H}_i^j$ is determined by the magnitude $p_{ij}$ $(\sum_{j=1}^{m_i} p_{ij} = 1)$.

The resulting quantity of the parameter $\eta$ is a function of these characteristics and, on the whole, the situation at any moment is described by the game theory model.

### 3.1.3 Global models

Objective assessment of environmental dynamics $N = \{N_1, N_2\}$ is possible by making certain assumptions using models of the biosphere $N_1$ and climate $N_2$. Such models have been developed by many authors and accumulated experience covers examples of point, regional, box, combined, and spatial models. This experience makes it possible to move toward synthesizing a global model of a new type covering the key relationships between the levels of hierarchy of natural processes on land and in the World Ocean. Relationships between the global model components are provided by parametric interface of its units. The units of climate and hydrosphere are basic, since the main circulation of substance and energy is realized through their components. A multiplicity of spatial digitization in these units provides an adaptive flexibility of the global model with coordination of data fluxes between them. As a result, the model's structure is independent of the structure of a global database, and hence does not change when the latter changes. Versions of the forms of space digitization include:

- random setting of regions and water basins;
- division of the planetary surface into sites with constant steps in latitude and longitude;
- superposition of this grid only on land and the shelf zone of the World Ocean with point presentation of its pelagic zones; and
- taking the administrative structure as the basis of NSS digitization.

This universality of space digitization is possible due to presentation of the global model in the form of a set of units connected only through inputs and outputs by the principle of open systems. In other words, let $A_s(t)$ be the content of chemical matter or energy of the element $A$ in a medium $S$ at a time moment $t$. Then, following the law of conservation of substance and energy, we write the balance equation:

$$dA_s/dt = \sum_j Q_{js} - \sum_i Q_{si} \tag{3.1}$$

where fluxes $Q_{js}$ and $Q_{si}$ are, respectively, input and output. Summing up is made by external media $i$ and $j$ interacting with the medium $S$.

Biogeocenotic processes in the global model consist of photosynthesis, dying off of plants, their respiration, and their growth. A change in biomass $P$ is approximated by the dynamic equation:

$$dP/dt = \min\{\rho P, R\} - M - U - V \tag{3.2}$$

where $\rho$ is the maximum production/biomass ratio for the plant type considered, $R$ is the real productivity of plants, $M$ and $U$ are the quantities of dying off and expenditure on energy exchange with the medium, $V$ is the loss of biomass for anthropogenic reasons.

Distribution of the types of soil–plant formations over land has been well studied. Data on assessment of productivity, biomass supplies, and dead organic matter on land and in the ocean make it possible to parametrize the components of Equations (3.1) and (3.2), with the whole structure of energy exchange in the environment taken into account. The equation of the balance between energy and matter in water ecosystems is written as:

$$\partial P_i/\partial t + Y = R_i - H_i - T_i - M_i - \sum \gamma_{ij} R_j, \tag{3.3}$$

where $\|\gamma_{ij}\|$ is the matrix of the efficiency of trophic bonds in the water ecosystem, and $Y$ is the hydrodynamic component.

The monographs by Kondratyev *et al.* (2002, 2003a, b, 2004, 2006a–c) describe in detail the global model of the nature–society system (GMNSS) which takes into account the interaction of NSS components shown in Figure 3.1.

### 3.1.4 Survivability criterion

In a general case, the state of systems $H$ and $N$ can be described by vectors $x_H(t) = \{x_H^1, \ldots, x_H^n\}$ and $x_N(t) = \{x_N^1, \ldots, x_N^m\}$, respectively. The combined trajectory of these systems in $n+m$-dimensional space is described by the function $\eta(t) = F(x_H, x_N)$, which is determined by solving Equations (3.1)–(3.3) and other ratios from the global model. The form of $F$ is determined from the laws of co-evolution, and therefore there is a wide field here for studies in different spheres of knowledge. Available estimates of $F$ indicate a relationship between the notions "survivability" and "stability". According to Ashby (1956), the dynamic system is in a "living state" in the time interval $(t_a, t_b)$ if its determining phase coordinates are within "permissible limits": $x^i_{H,\min} \leq x^i_H \leq x^i_{H,\max}$; $x^j_{N,\min} \leq x^j_N \leq x^j_{N,\max}$. Since systems $H$ and $N$ have a biological basis and limited energy resources, one of these boundary conditions is unnecessary—that is, for the components of the vector $x = \{x_H, x_N\} = \{x_1, \ldots, x_k\}$, $(k = n+m)$ conditions $x_{\min} \leq \eta = \sum_{i=1}^k x_i$ should be

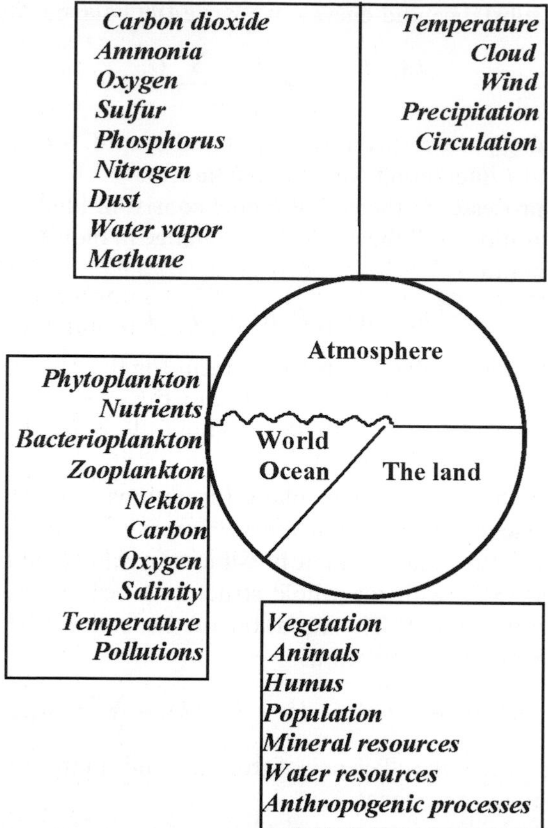


**Figure 3.1.** List of NSS elements taken into account in the GMNSS.

satisfied. This simple scheme includes requirements of both consideration of total energy in the system and the diversity of its components.

Of course, the notion of system survivability is more capacious and informative. In system ecology many authors use this term bearing in mind the stability and integrity of the system, meaning the system's ability to resist external forcings. In other words, survivability is measured by the trend of the system to suppress large oscillations in its structure and components, returning the system to its equilibrium state. Thus, system survivability is supposed to mean its ability to actively withstand external forcings, to preserve its characteristics (i.e., probabilities of the states in which it remains able to work), and to function with certain applied methods and under certain conditions of its exploitation.

The conceptual scheme of a new version of the global model covering the vital components of the environment is now being developed by many scientists; in particular, at the Potsdam Institute for Climate Impact Research (Boysen, 2000). A new version of the global model based on the use of open systems theory is also being worked out in Russia (Kondratyev *et al.*, 2004; Krapivin and Kondratyev, 2002). The structure of connections between the model's units foresees an information exchange between them via their inputs and outputs which ensures not only the use of the evolutionary selection among these units, but also accomplishment of the structural synthesis of the whole global model. Its elements are either represented by a set of earlier models or described by observational data series on their characteristics. This structure of the global model ensures its independence of the procedure of changing some units and their interconnection, such that the relationship between global model units is ensured by the parametric compatibility of all its units and nothing else.

The spatial structure of the environment is described by a geographical grid with steps $\Delta\varphi$ in latitude and $\Delta\lambda$ in longitude. With a homogeneous surface the atmosphere is divided into two layers—a mixed layer $h_1$ in height and the upper layer $h_2$ thick; the oceans are represented by a multi-layer structure with selection in each cell $\Delta\varphi \times \Delta\lambda$ of upper layers $\Delta z_1$ thick down to the thermocline, then $\Delta z_2$ thick to a photically effective depth $Z(\varphi, \lambda)$ and the deep ocean. Inside a cell pixel $\Delta\varphi \times \Delta\lambda$ all components are assumed to be homogeneous. Biotic, biogeochemical, and demographic processes are described by balance finite-differential equations, while anthropogenic, climatic, and socio-economic processes are represented either by sets of scenarios or by models of evolutionary type (Bukatova and Makrusev, 2004; Bukatova and Rogozhnikov, 2002). The computer version of the global model gives the possibility to realize numerical experiments on the consideration of numerous problems in the study of biospheric processes. The user can choose the domain of studies, adjusting the whole model to the chosen scenario. There is a possibility here to choose between a standard set of scenarios or to synthesize a new scenario both in the sphere of anthropogenic activity and in analysis of other processes in the biosphere and climate system. The choice of territory for analysis of the environmental processes taking place over this territory is realized by indicating in the menu a concrete object or, with the help of a special procedure, by formation of the inner structure of subject-thematic identifiers $\{Y_k\}$, with the use of which a model is synthesized either for all the environmental elements of the territory or for the whole biosphere. Table 3.1 contains a list of some standard identifiers of the global model.

**Table 3.1.** List of basic semantic structures (identifiers) of the global model.

| $Y_s$ | Description of identifier |
|---|---|
| $Y_1$ | Contouring the territory $\Omega$: $y_{ij}^1 = 0$ at $(\phi_i, \lambda_j) \notin \Omega$ and $y_{ij}^1 = 1$ with $(\phi_i, \lambda_j) \in \Omega$ |
| $Y_2$ | Topological structure of the territory |
| $Y_3$ | Types of hydrological elements |
| $Y_4$ | Types of soil–plant formations |
| $Y_5$ | Types of sources of environmental pollutants |
| $Y_6$ | Relief of the territory |
| $Y_7$ | Types of pollutants and scales of intensity of chemical element emissions |
| $Y_8$ | Types of anthropogenic landscapes |
| $Y_9$ | Socio-economic structures |
| $Y_{10}$ | Climatic situation |

For instance, the identifier $Y_4$ specifies the global distribution of soil–plant formations by the model geographic grid $\Delta\varphi \times \Delta\lambda$ ($1^\circ \times 1^\circ$ and $4^\circ \times 5^\circ$ are the most widespread), and the identifier $Y_5 = \|y_{ij}\|$ makes it possible to concretize the structure of allocation of the sources of environmental pollution. This procedure is realized through decoding of the semantic structure of pixel $\Delta\varphi \times \Delta\lambda$ with the coordinates of the center $(\varphi_i, \lambda_j)$ by the scheme suggested to the user by the system itself.

The database and the base of knowledge of the global model contain information and laws in *default mode*. This information is accessible to the user for analysis and correction.

Development of the methods of ecoinformatics as applied to global processes in nature and society is dictated by the present trend toward deterioration of living conditions, reduction of biodiversity on the planet, and depletion of non-renewable resources; the processes of energy reproduction are confronted with the inability of the environment to perform all its functions. Parametrization of biospheric, climatic, demographic, and socio-economic processes is the subject of many exact sciences, and the urgency of obtaining efficient technologies to study all these processes as soon as possible is undoubted. However, the anthropogenic component of global processes contains many inadequately parametrized elements. Primary among which are problems of the interdependence of these processes and the political structure of human society.

## 3.2 PROBLEM OF GLOBAL CHANGE

For many centuries of human history, humans in the Earth's biosphere have remained only one of several million species. Whereas the common activity of these millions has determined and now determines the "face" of planet Earth, the contribution of *Homo sapiens* to this activity before the industrial epoch had been negligible. Though human existence and their ecological niche depended on biospheric evolution, they had not however affected this evolution.

The problem of the combined evolution of humans and nature surfaced at the beginning of the industrial period, when the impact of the global population on natural processes became noticeable. The rate of this impact rapidly increased in time, and now the existence of life on Earth is the subject of question. Anthropogenic processes in the biosphere have become comparable with those of natural processes. The landscape is rapidly changing. Forested areas are reduced, and the areas of cultivated lands grow. The share of forest fires due to anthropogenic activity in different regions varies from 10 to 75%. The tropics covering 30% of the Earth's surface (4.6 billion ha) contain 42% of global forests, and the rate of their destruction is estimated at 12 million ha per year. This has been caused by the growth of population size in tropical countries by 2.4% per year, on average, which leads to the annual increase in ploughed land of 100 million ha. Similar trends are also observed in other regions of the globe. On a global scale, the loss of forested area by 2000 constituted about 0.5 billion ha. Scientists dealing with studies of the environment are seriously concerned about the growing rate of the human impact on global biogeochemical processes by introducing to the biosphere chemical elements and breaking the natural cycles of their transformation. The negative balance of organic matter in the geological shell of the Earth caused by human economic activity leads to breakdown of the gas composition of the atmosphere, especially to the growth of GHG concentration. The quantitative characteristics of these processes are well known (Demirchian and Kondratyev, 1998a, b, 1999; Golitsyn, 1995). To assess the trends in current biosphere dynamics and to understand the level of danger for the environment from human activity, many scientists, based on accumulated data on environmental parameters and understanding the unique character of natural systems, are trying to develop global models and with their use forecasting the possible consequences of anthropogenic activity. For the development of such complex models, one has to formalize climatic, biotic, geochemical, economic, and social processes. As Monin and Shishkov (1990) noted, the level of adequacy of these studies is determined by the level of simplification of real processes in model presentations. Using a model of the biosphere in the experiment instead of the biosphere itself makes it possible to get answers to questions about the consequences of realization of anthropogenic projects for natural system reconstruction. At the present level in the development of numerical modeling of global processes and databases, the similarity between the observed behavior of the biosphere and model forecasts can be achieved by monitoring natural systems.

The problem is that—based on the available technical means for environmental monitoring and using modern means of data-processing—one could synthesize a technology of prognostic diagnostics of global processes which would give reliable estimates of dynamic processes both in nature and in society. The complexity of this problem is the driving force behind numerous scientific approaches to development of new non-traditional methods of assessing biospheric survivability; that is, the biosphere's ability to safeguard survival of its major element—human beings.

## 3.3 A NEW INFORMATION TECHNOLOGY IN GLOBAL MODELING

An experience of global modeling based on traditional methods of numerical modeling shows that wishing to raise the accuracy of the developed models, scientists are faced with numerous contradictions caused by the limited technical means of collection and processing of monitoring data. It is obvious that the creation of a global model adequate to the biosphere and climate system of the Earth will run into numerous problems. On the one hand, complete consideration of all the parameters of the biosphere and climate system will lead to an effect called the "multidimensionality curse". On the other hand, simple models that take into account few parameters are inadequate and practically worthless. Moreover, in many spheres of science, such as physics of the ocean, geophysics, sociology, and climate, development of an adequate model is principally impossible due to the impracticability of obtaining complete information. Difficulties in the synthesis of climate models are well known, occurring as they do mainly due to the non-stationary processes under study.

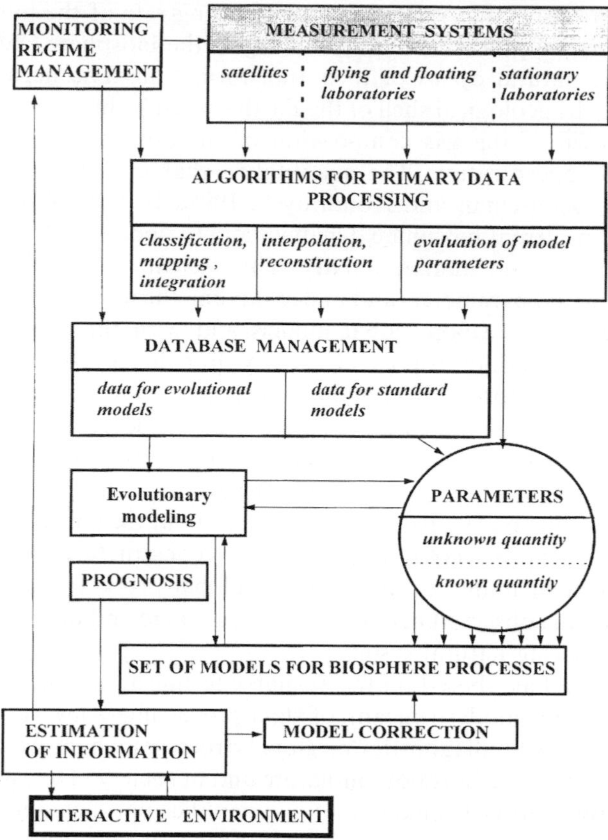


**Figure 3.2.** GMNSS interactive adjustment and control of the geoinformation monitoring regime.

An approach to global modeling developed in the late 20th century overcomes these difficulties by combining two approaches to the synthesis of global model elements. The first, traditional approach is based on balance equations and motion equations. The second approach uses evolutionary modeling, game theory, and scenarios. However, the models developed turn out to be based on expert estimates and, as numerical experiments show, are not very good at studying the stress states of the environment. The way out of these difficulties is to apply a new approach and develop models of poorly parametrized processes. This was proposed by Fogel *et al.* (1966) and developed further by other scientists (Bukatova and Makrusev, 2004; Bukatova and Rogozhnikov, 2002; Nitu *et al.*, 2000a). This approach is based on evolutionary modeling technology. Application of evolutionary technology provides for categorization of the whole system by a class of subsystems with variable structures, making it adaptable to changes in the natural process or entity under observation. Figure 3.2 demonstrates the position of this technology in a principal scheme of geoinformation monitoring. A constructive mechanism for realization of this scheme can be provided by GIMS technology (Krapivin *et al.*, 1997), the main principles of which are:

- combination, integration, and coordination of databases;
- optimization of environmental control systems;
- agreement and compatibility of information fluxes;
- centralization of access to information;
- coordination of spatial scales; and
- use of a single system of classification.

So, let a real object $A$ have an unknown functioning algorithm; the pre-history of its life observed at time moments $t_1, \ldots, t_n$ is all that is known. Evolutionary technology proposes synthesizing a global model by formation of a succession of partial models $\{A_j^i\}$, changed in the process of adaptation to the pre-history of all environmental sub-systems. The models $A_j^i$ are modernized here both parametrically and structurally. The procedure of the evolutionary selection of models provides a regime of data collection on the environmental parameters practically independent of the requirements of traditional statistical analysis and of information uncertainty. The method of evolutionary modeling removes the difficulty of parametrization of many poorly modeled processes by synthesizing a combined model whose structure is adapted to the pre-history of the complex of biospheric, climatic, and socio-economic components. The form of combination of the types of models is diverse and depends on the spatial–temporal completeness of global data archives. Moreover, an adaptive scheme makes it possible to solve the problem of determination of the structure of these archives. In fact, data for units of traditional structure are formed as schematic maps and tables of the coefficients of model equations. In this case all the cells of schematic maps should be filled in, which requires a large number of measurements. Data for evolutionary models can be fragmentary both in time and in space. There is no doubt that the relationship between two structures of data depends on the criterion of global model adequacy.

## 3.4 GLOBAL CLIMATE CHANGE MODELING

There are many climate models. Among them are models developed at the Hadley Centre, Lawrence Laboratory, U.S. National Center for Atmospheric Research, Potsdam Institute for Climate Impact Research (PIC), and others (Svirezheva-Hopkins and Schellnhuber, 2005). Below we shall concentrate on studies of climate model development. These studies are connected with a complex analysis of the processes on the globe responsible for the functioning of the climate system. At the end of the 20th century this theme became important in connection with several decisions of the U.N. on GHG emission reduction. In particular, one of these decisions is the Kyoto Protocol, the result of the International Convention on Climate signed in 1997 by many countries. After this, there was a series of scientific works in which the unfounded character of these decisions has been proved (Danilov-Danilyan and Losev, 2000; Gorshkov and Makarieva, 1999; Kondratyev, 2000). Earlier publications generally mentioned that statements about anthropogenic climate change need to be thoroughly checked (Gorshkov, 1995).

The basic goal of climate studies undertaken at many centers is to understand the participation of the many observed global changes in the formation of climate trends. Of course, the increase in $CO_2$ concentration in the atmosphere is clearly noticeable and attracts the attention of specialists from various scientific spheres. Scientists dispute the scale of climatic consequences of the doubling of this concentration. There are many prognostic estimates of responses of the Earth's climate system to atmospheric $CO_2$ doubling. The conclusions of scientists dealing with climate studies attract worldwide attention and need analysis.

Russel *et al.* (2000) noted that accumulated experience necessitated a substantial change in conceptual approaches to studies of the processes involved in the interaction between the climate system and human society. To make successful prognoses, one should move on from primitive linear models of feedbacks in the NSS to large-scale spatial parametrizations of the laws of interaction of natural and socio-economic components. The main thing here is to identify "strategic variables" which substantially enhance and favor the propagation of disturbances and interactions through the ecosphere/anthroposphere system. For instance, one such variable is the intensity of respiration processes in ecosystems. In principle, all factors critical to the self-regulation mechanisms of the climate system should be attributed to this series of variables. However, relevant data are fragmentary and their synthesis into a purposeful system, in the authors' opinion (Russel *et al.*, 2000), would bring about a reduction of uncertainties in climate forecasts. For such a reduction, some authors (Schröter *et al.*, 2005) put forward the idea of making a complete inventory of the planet Earth, including human civilization, by developing formalized academic constructions reflecting reality at different levels of accuracy. Such a reconstruction of the NSS was discussed in the work by Moiseyev *et al.* (1985) and Kondratyev (2005b). Of course, this stage in global process formalization should transform into a constructive form of description by creation of a computer version of the global model (Kondratyev *et al.*, 2004; Kondratyev and Krapivin, 2005a).

The Goddard Institute for Space Studies (GISS), based on these principles and using approved methodologies from physics, informatics, and other sciences, is trying to construct simulation models of environmental sub-systems on different spatial–temporal scales and on this basis to find the causes of global changes taking place in the environment (Russel *et al.*, 2000). The models developed of global atmospheric circulation make it possible to consider a set of spatial scales—the main ones being $8° \times 10°$, $4° \times 5°$, and $2° \times 2.5°$ in latitude/longitude—with a possibility to choose between 9, 18, 23, and 31 levels in the vertical. One of the latest versions of the climate model GISS GCM (Friend and Kiang, 2005) includes description of a 3-D structure of World Ocean circulation, atmospheric chemistry, and global aerosol transport.

Important in this respect are the many results of numerical modeling using coupled models which describe the interaction of various media of atmosphere–ocean, atmosphere–land, and other types. Studies accomplished at GISS cover the period 1950–2099 with consideration of different variations in anthropogenic emissions of $CO_2$ and aerosols. Comparison of the obtained SAT trends with those recorded has shown a high reliability of climate forecast with the exception of summer months in the Northern Hemisphere, when the greenhouse effect was weak.

### 3.4.1 Subjects of climate studies

The climatic component of the GMNSS is most difficult in global model synthesis, since it is characterized by numerous feedbacks, most of which are unstable. These feedbacks include *ice–albedo, water vapor–radiation, cloud–radiation,* and *aerosol–radiation.* Functioning of the Earth's climate system is determined by the state of the atmosphere, oceans, cryosphere, continental surface with land biota, lakes, rivers, groundwater, and various anthropogenic structures. Therefore, development of a climate model requires consideration of numerous factors whose role in climate in many cases has been little studied. Attempts at complex descriptions of the Earth's climate system using numerical approaches have not yet given results which could be used in the GMNSS. We cannot even be sure that realization of the goals of the Global Carbon Project (GCP) will clarify this problem. One thing is sure, though, and that is without parametrization of climate all problems of the greenhouse effect will remain unsolved.

There are two approaches to climate model synthesis. One approach is based on inclusion of biospheric components in climate models that have already been created or developed. The other approach consists in development of a unit within the numerical model of the biosphere which would simulate the dependences of biospheric components on climate parameters. In the first case, unstable solutions of respective systems of differentiated equations appear problematical, which hinders prognostic estimates of global environmental change. In the second case, there is a possibility to obtain stable forecasts of environmental change, but their reliability depends on the accuracy of parametrization of correlations between elements of the biosphere and climate. The second approach has the advantage of incorporating models which can be described at a scenario level. Detailed analysis of the problems facing climate modeling and assessment of the present state can be found in such publications as Kondratyev (1999a) and Marchuk and Kondratyev (1992). Here some models of individual components of the NSS were discussed, which correspond to the second approach. Among them are models of atmospheric general circulation, atmosphere–ocean interaction, sensitivity of climate parameters to boundary conditions on the Earth's surface, and interactions of biogeochemical and climatic processes (Chen *et al.*, 2000; Lin *et al.*, 2000).

Climatic impact is a physical–chemical–biological system that is unlimited in its degree of freedom. Therefore, any attempt to model such a complex system faces insurmountable difficulties. This circumstance explains the plethora of parametric descriptions of individual processes in this system. For a global model with a time step of digitization up to one year, an acceptable approach is to use two versions. The first version consists in combined use of correlations between partial processes involved in climate formation over a given territory together with climate scenarios.

The second version is based on the use of global monitoring data, which are the basis for the formation of data series about climatic parameters, with territorial and temporal attachment, and are used to retrieve a complete pattern of their spatial distribution. One of the wide-spread correlation functions is the dependence of mean temperature $\Delta T_g$ of the atmosphere on atmospheric $CO_2$ content:

$$\Delta T_g = \begin{cases} 2.5[1 - \exp\{-0.82(\xi - 1)\}], & \xi \geq 1; \\ -5.25\xi^2 + 12.55\xi - 7.3, & \xi < 1 \end{cases}$$

where $\xi$ is the ratio of current $CO_2$ content in the atmosphere $C_a(t)$ to its pre-industrial level $C_a(1850)$.

According to Mintzer (1987), there is a possibility to extend consideration of the temperature effect of other GHGs:

$$\Delta T_\Sigma = \Delta T_{CO_2} + \Delta T_{N_2O} + \Delta T_{CH_4} + \Delta T_{O_3} + \Delta T_{CFC\text{-}11} + \Delta T_{CFC\text{-}12},$$

where

$$\begin{aligned} \Delta T_{CO_2} &= -0.677 + 3.019 \ln[C_a(t)/C_a(t_0)], \\ \Delta T_{N_2O} &= 0.057[N_2O(t)^{1/2} - N_2O(t_0)^{1/2}], \\ \Delta T_{CH_4} &= 0.019[CH_4(t)^{1/2} - CH_4(t_0)^{1/2}], \\ \Delta T_{O_3} &= 0.7[O_3(t) - O_3(t_0)]/15, \\ \Delta T_{CFC\text{-}11} &= 0.14[CFC\text{-}11(t) - CFC\text{-}11(t_0)], \\ \Delta T_{CFC\text{-}12} &= 0.16[CFC\text{-}12(t) - CFC\text{-}12(t_0)]. \end{aligned}$$

The value $t_0$ was identified in 1980, when GHG concentrations first became known.

Simple but adequate climate models can be specified by taking into account the characteristic times for feedbacks to occur. The time range of response delay within the NSS is diverse and should be taken into account when assessing the consequences of changes within one or several climate sub-systems. In particular, cold supply in the Antarctic ice sheet is so large that to raise its temperature up to 0°C the average temperature of the World Ocean would need to be lowered by 2°C; that is, transferred from its current state $T_0 = 5.7$°C to the state $T_0 = 3.7$°C. The inertia of this operation would involve hundreds of years. The observed rate of climate warming for anthropogenic reasons does not have so much energy to spend.

Experience of modeling the Earth's climate demonstrates that the desire of many authors to accurately and completely take into account all possible feedbacks and elements of the climate system leads to complicated mathematical problems, solution of which would necessitate a huge amount of data—in most cases solution of the equations involved is unstable. Therefore, the use of these complicated models as a GMNSS unit leads inevitably to a negative result—that is, the impossibility of synthesizing an efficient model. The most promising approach is to combine climate models with global monitoring data. The way in which this combination can be brought about is very simple. Current ground and satellite systems involved in climate-forming process control cover a certain number of cells $\{\Omega_{ij}\}$ of the Earth's surface. Over these cells, measurements are made of temperature, cloud, the content of water vapor, aerosols and gases, albedo, and other parameters of energy fluxes. By using simple climate models and methods of spatial–temporal interpolation it is possible to retrieve from these measurements a complete pattern of climate parameter distribution over the territory $\Omega$.

As mentioned above, extensive studies on creation of efficient climate models are being undertaken at many research centers in the U.S.A., Japan, Russia, Germany, and elsewhere (Dai *et al.*, 2004; Kondratyev, 2004b; Svirezheva-Hopkins and Schellnhuber, 2005). In most cases these studies concentrate on analyzing integrated systems covering not only the interaction of natural subsystems but also the NSS interaction (Ganti, 2003). The main conception of such approaches is based on analysis of the properties of global NSS sub-systems in order to create a base of knowledge about their interaction at different spatial levels. To do this genuinely observed processes and expert estimates are used. Close attention is paid to the current analysis of natural extreme events and assessment of the risk to civilization as viewed by politicians, businessmen, and the Earth's population. The question about the danger for humankind of emerging trends in global climate change can only be answered by simulating feedbacks in the environment. The world population needs to have answers to such questions as, for instance, whether a doubling of $CO_2$ concentration in the atmosphere can cause global warming by 1.5–4.5°C or what volumes of GHG emissions will be harmless during the coming 200 years. There is a multitude of such questions, and finding the correct orientation for scientific studies to take is a major problem, solution of which will make it possible to increase the efficiency of scientific effort and to save time in getting answers to the most critical questions. But the speed at which this can be done is never going to be fast enough for the world population.

Important here are international projects aimed at creating reliable information technologies for prediction of the consequences of anthropogenic intrusions into natural processes, especially into the biogeochemical cycles of GHGs. These projects include the Coupled Model Intercomparison Project (CMIP) which plans a detailed study of interactions in the atmosphere–ocean–land system by using satellite monitoring data. Considerable progress has been achieved in this direction at NCAR (Boulder, Colorado) where the CCSM (Community Climate System Model) has been developed—this describes many environmental processes both in the past and present. Such a past-to-present transition in parametrization of climate trends makes it possible to resolve the problem of evolutionary adjustment of the model and to remove many sources of errors connected with uncertainty in assessment of environmental parameters.

The scale of global changes in natural systems depends in many respects on the role of surface ecosystems in regulation of the biogeochemical cycles of GHGs as well as in provision of humankind with food, freshwater, materials for light industry, and other conditions of life support. To assess this role, a global model of land vegetation cover was proposed by Kondratyev *et al.* (2004), on the basis of which the processes involved in the interaction between hydrological and ecological land cycles were studied as well as simulation procedures constructed to parametrize the growth of forests and vegetation control in agriculture. All these studies correlate with the global climate change problem, and their results should be taken into account in climate models.

Understanding global change mechanisms is impossible without considering the laws of development of social systems and the principles underlying the human–nature interaction against the background of the socio-political structure on the globe. Methods and technologies to study the problems that crop up used by many authors include modeling the long-term processes of macroeconomic development and, on the basis of historical studies, revealing the mechanisms involved in human adaptation to environmental changes (Kondratyev *et al.*, 2003c).

### 3.4.2 Climate models

Works on synthesis of computer models describing climate system dynamics are being undertaken at many scientific centers in Russia, the U.S.A., Japan, Germany, and elsewhere. Among the best known models one should mention a series of Hadley Centre developments where models are created, as a rule, as 3-D presentations realized using modern supercomputers. The main configuration of these models includes (Jenkins *et al.*, 2005):

- atmospheric general circulation models (AGCMs);
- ocean general circulation models (OGCMs);
- atmosphere–ocean global circulation models (AOGCMs); and
- regional climate models (RCMs).

There is also a complex climate system model (CCSM) (Kiehl *et al.*, 2003) and a parallel climate model (PCM) (Dai *et al.*, 2004). Extensive climate studies are made within the framework of the PCMDI—a program started in 1989 at the California Institute—and at the Goddard Institute for Space Studies where a series of GCMs, including 2-D energy balance models (EBMs) has been developed (Aires *et al.*, 2005).

The CCSM climate model developed at NCAR is one of the most improved models with a huge set of functions to predict climate processes on regional and global scales. In particular, this model enables one to assess the variability of precipitation and temperature depending on the volumes of $CO_2$ and aerosols emitted to the atmosphere as well as to analyze the consequences of changes such as volcanic eruptions. However, despite substantial progress in modeling climate trends, this model and similar ones have a serious drawback consisting in the absence of units in them to parametrize the interactions between natural and anthropogenic processes. Studies to overcome this drawback are being undertaken at the Institute of Radioengineering and Electronics and at the St. Petersburg Centre for Ecological Safety, Russian Academy of Sciences (Krapivin and Kondratyev, 2002). By comparing these directions—i.e., the American and Russian ones—the principal differences between them can be discovered. Whereas the authors of climate models choose parametric descriptions of climate processes as initial base structures which have some biospheric elements in them, the Russian direction proceeds from the postulate that the initial cause of global change is a complicated hierarchy of spatial–temporal interactions of NSS elements, and climatic trends are formed as a consequence of these interactions. Here the work of a group of American NASA specialists headed by Sellers *et al.* (1996) should be mentioned as a third direction. This group has developed a simplified global model SiB2 (Simple Biosphere Model-2) completely oriented toward space monitoring data. Unfortunately, these three groups are undertaking their studies individually. Clearly, these approaches supplement each other and form an effective technology for prognostic geoinformation monitoring. Nevertheless, the process of interaction has recently begun between Russian specialists and foreign groups of global "modelers" on the initiative of academician Kirill Ya. Kondratyev. This will enable one to take a new qualitative step in the sphere of global modeling. This cooperation is based on a principally new conception of assessing global change (Kondratyev *et al.*, 2003a; Kondratyev and Krapivin, 2005a; Kondratyev, 2005b), by developing a complex system of models of global biogeochemical, biogeocenotic, climatic, and socio-economic processes with their orientation toward maximum use of data from geoinformation monitoring systems.

The problem of sustainable development in the 21st century with continuing growth of population size has the potential to be solved by analysis of correlations between living standard, living conditions, social level of development, and economic parameters. The totality of these correlations is called "the syndrome of global change". By making a theoretical analysis of syndromes, many authors are studying available information about the interaction of human civilization with the environment with a view to classifying the mechanisms of trends of development in two types: leading to description of stable trends and leading to stability (Neumayer, 2003; Nicholls, 2004). By introducing qualitative indicators of syndromes, with the help of analysis of data on the development of agriculture, mineral resource supplies, and other civilization parameters, one can classify countries by their contribution to global change, and by sensitivity of the global environment to anthropogenic processes in a given territory. Of course, this analysis is useful when deciding on the structure a global model will take.

In his recent work, Kondratyev (1999a) outlined the priorities of the global dynamics of socio-economic development and the environment. It is apparent that to make cardinal decisions of the Kyoto Protocol type it is necessary to develop a global model with a broad cooperation of specialists in natural and societal sciences as well as in politics and other spheres of human activity. Available global models can serve the basis for this cooperation. In this connection, Moiseyev (1995) wrote: "I want to call the 20th century not the century of catastrophe as it is sometimes called, but the century of warning. Events of the present century have made it possible to look beyond the horizon—we saw a reality that lies ahead of everybody, the whole humankind. Our experience and events are really warning us. But at the same time, they give us a chance because we understood—there is time still enough to do many things. But to do this, Collective Decisions and Collective Will are needed! I am sure that the world goes to a rational society in which, with a multi-coloured palette of cultures needed to provide the future for *Homo sapiens*, a unity will be established without national boundaries, national governments, and confrontation."

The U.N. Convention on Climate and the Kyoto Protocol have pointed scientists toward development of models and methods mainly aimed at studies of the global $CO_2$ cycle—this, however, considerably distorts the basic ideology of global studies and leads to creation of ineffective technologies of global change assessment. The strategic lines of basic ideology contained, for instance, in the GCP is to create carbon cycle models based on a simplified idea of biospheric structure.

An important aspect of the problem of consequences of anthropogenic impact on the environment is consideration of processes at a regional level. At each level of spatial hierarchy there appear and disappear correlations between population activity and respective responses of nature. The river basins of Europe were chosen as a concrete region for study (Schröter *et al.*, 2005). Hydrological and ecologo-hydrological models of different levels are used as instruments in these studies. These models are based on assessment of European water resources and use large-scale parametrizations of the European river network. Special attention was paid to the basin of the river Elba ($\sim$100,000 km$^2$). Scenarios of urbanization considered for this basin made it possible to pose the problem of synthesizing a more complicated model of the use of land areas.

Such work in the sphere of water resource modeling is used in other regions—for instance, in Brazil. Unfortunately, the approach proposed by Schröter *et al.* (2005) depends on concrete configurations of the basin. It would be better to consider more general and universal technologies of regional water balance modeling.

Climate change and agricultural productivity are interdependent components of the global process of environmental transformation. According to FAO estimates, in 1998 the global yield of wheat was 590.8 Mt (China 109.7; U.S.A. 69.3; India 65.9; Europe 103.9). The PIC Working Group which studies the interactions of agro-economic processes with the climate system, based on analysis of the mechanical model, established the dependence of agricultural productivity of European countries on $CO_2$ concentration in the atmosphere. The scenarios of changes of temperature and precipitation considered made it possible, using the model, to evaluate likely trends in the agro-economic complex of the European Community countries. Three hypotheses were proposed about the impact of $CO_2$ concentration growth on photosynthesis. It was shown that, in the absence of this impact, temperature and precipitation would change by $+3^\circ$K and $-14\%$, respectively, with the consequent losses in agriculture constituting 40%. If increased $CO_2$ concentration stimulates photosynthesis, but does not reduce the transpiration of plants, then agricultural yield would be reduced by 8%. However, if as a result of $CO_2$ increase photosynthesis grows and transpiration decreases, then yield would grow by 13%. Of course, these estimates depend on the reliability of subjective assumptions introduced in the model and probably play an important role in the final results. Nevertheless, the study of such scenarios is an important stage toward synthesis of a general global model, since they highlight weak points in the global base of knowledge.

Apart from the importance of forests as a source of building material, they play an important role in stabilization of the environment, providing water balance stability over forested territory and creating combinations of atmospheric parameters favourable to human health. In Europe, for instance, the problem of predicting the fate of forests is urgent from economic, ecological, and social points of view. The fate of forests in Brazil and Russia is primarily important for the state of the biogeochemical cycle of carbon and level of the greenhouse effect. Therefore, many international projects foresee the development of forest models to reflect correlations between forested territories and global climate change. Moreover, these models will take into account the biodiversity level, carbon supplies, functions of forests in groundwater dynamics, and other processes connected with forests. The division of forested territories in Europe as controlled or wild is also taken into account. In Europe most studies are made within the framework of such European projects as ther LTREF (Long-Term Regional Effects of Climate Change on European Forests). Computerized GIS technology is mainly used.

It should be noted that, under present conditions, underestimation of the role of forests in global climate change has led to inadequate conclusions contained in the Kyoto Protocol. According to Gorshkov and Makarieva (1999), a forest as a biotic regulator of the environment manifests its mechanisms of regulation differently, depending on whether it is natural or created technogenically. This important conclusion means that the structure of the forest ecosystem model needs to be revisited and should serve as a stimulus to correct development of the respective unit of the global model.

Development of an NSS global model will deal with a complicated set of hierarchical causes and effects as a result of human activities concerning the use of both renewable and non-renewable resources. Forrester's (1971) model and the like have already tried to formulate these relationships. It is clear that a spatio-temporal distinction should exist between regional and global changes. Outside this distinction, transition is inevitable to a qualitative change in human–nature relationships. The task is to find this distinction and to give recommendations on the structure of the anthropogenic unit to be used in the global model. One should not make a mistake here which will inevitably lead to a conflict between the social status of the problem and preservation of environmental parameters within the boundaries that guarantee life.

The most complete and constructive analysis of the environmental problems of various scales in the context of special features of the socio-economic and political factors was made by Kondratyev (1999a). It is clear that modeling the real and hypothetical ecological risk undertaken by many scientists should in time result in parametric distinctions that are able to give a quantitative estimate of various manifestations of ecodynamics.

## 3.5 BIOCOMPLEXITY PROBLEM IN GLOBAL ECOINFORMATICS

The interaction of various elements and processes in the climate–biosphere–society global system has recently attracted the attention of many scientists. Attempts to evaluate and predict the dynamics of this interaction have been made by scientists in various scientific directions. One such attempt was the "Biocomplexity" program announced by the U.S. National Science Foundation within which plans for

2001–2005 were made to study and understand relationships between biological, physical, and social systems and trends in current environmental change. Within the framework of this program, biocompexity refers to phenomena that result from the dynamic interactions among various components of the Earth's diverse environmental systems.

Biocomplexity derives from biological, physical, chemical, social, and behavioral interactions between environmental sub-systems, including living organisms and global population. As a matter of fact, the notion of biocomplexity in the environment is closely connected with the laws of biospheric functioning as a unity of its forming ecosystems and natural–economic systems of various scales, from local to global. Therefore, to determine and assess biocomplexity, a joint formalized description is needed of the biological, geochemical, geophysical, and anthropogenic factors and processes taking place at a given level of the spatio-temporal hierarchy of scales.

Biocomplexity is the indicator of all systems of the environment connected with life. The manifestation of this is studied within the theory of the stability and survivability of ecosystems. It should be noted that the functioning of biocomplexity includes indicators of the degree of mutual modification of interacting systems, and this means that biocomplexity should be studied by taking both spatial and biological levels of organization into account. The complexity of this problem is determined by complicated behavior of the object under study—especially the human factor due to which the number of stress situations in the environment is constantly growing.

Humankind has accumulated much knowledge about environmental systems. This knowledge can be used to study biocomplexity by synthesizing a global model which reflects the laws of interaction between environmental components and enables one, without detriment to it, to assess the "efficiency" of realization of different scenarios of human society development. It is this problem that forms the basis of all questions posited by the Biocomplexity program mentioned above.

We will now attempt to formalize the notion of biocomplexity, and a technology for its study is proposed.

### 3.5.1 Biocomplexity indicator

Studies of the processes of human–nature interaction are aimed, as a rule, at understanding and assessing the consequences of this interaction. The reliability and accuracy of such assessments depends on criteria taken as the basis for conclusions, examinations, and recommendations. At present, there is no single method to choose such criteria since there is no single scientifically substantiated approach to ecological normalization of economic impacts on the environment. The accuracy of ecological examination of the functioning and planned production as well as representativeness of global geoinformation monitoring data depends on the choice of such criteria.

The processes taking place in the environment can be presented as all the interactions between its subsystems. Humankind is just one of its elements; hence, it is impossible to simplistically divide the environment, for instance, into biosphere and society. Everything on Earth is correlated and interconnected. The problem is to find such mechanisms for description of these correlations and interconnections which would reliably reflect dynamic trends in the environment and answer the questions formulated in the Biocomplexity program:

(1) How does complexity between biological, physical, and social systems appear and change within the environment?
(2) What are the mechanisms behind the spontaneous development of numerous events in the environment?
(3) How do environmental systems with living components, including those created by humankind, respond and adjust to stress situations?
(4) How does information, energy, and matter move within environmental systems and across their levels of organization?
(5) Is it possible to predict the system's adaptability and forecast future changes in it?
(6) How does humankind affect and respond to biocomplexity in natural systems?

Many important and significant questions can be added to those above. For instance, what should the level of complexity and multiplicity be of satellite systems for environmental observations for their information to be sufficient for reliable assessment of biocomplexity, if only at the moment of obtaining this information. Also of importance is optimal allocation of the means of geoinformation monitoring at different levels of its organization.

Environmental biocomplexity is partially a characteristic (level) of the interaction of its systems. In other words, one can introduce the scale $\Xi$ of biocomplexity, which changes from the position when all interactions in the environment stop (are broken) to a level when they correspond to the natural process of evolution. In this case we obtain an integral indicator of the environmental state on the whole, including bioavailability, biodiversity, and survivability. This indicator reflects the level of all kinds of interaction between environmental components. With the biological interaction connected with relationships of the type "predator–victim" and "competition for energy resource" there is a minimum level of food concentration when it becomes practically inaccessible and the consumer–producer interaction ceases. The chemical and physical characters of environmental component interaction also depend on sets of critical parameters.

This emphasizes that biocomplexity refers to categories which are difficult to measure empirically and express quantitatively. However, we shall try to move from a purely verbal tautological reasoning to formalized quantitative estimations. For transition to numerical gradations of the scale $\Xi$ we postulate that between two values of the scale indicator there are relationships of the type $\Xi_1 < \Xi_2$, $\Xi_1 > \Xi_2$, or $\Xi_1 \equiv \Xi_2$. In other words, there is always a value of this scale $\rho$ which determines the biocomplexity level $\Xi \to \rho = f(\Xi)$, where $f$ is some transformation of the notion of biocomplexity into quantity.

Let us try to find a satisfactory model which will transform a verbal portrait of biocomplexity into notions and indicators subjected to a formalized description and transformation. With this aim in view, we select $m$ element-subsystems of the lowest level in the NSS, the interaction between which is determined by the binary matrix function: $A = \|a_{ij}\|$, where $a_{ij} = 0$, if elements $i$ and $j$ do not interact; $a_{ij} = 1$, if elements $i$ and $j$ interact. Then any point $\xi \in \Xi$ is determined as the sum:

$$\xi = \sum_{i=1}^{m} \sum_{j>i}^{m} a_{ij}.$$

Here, of course, an ambiguity occurs which it is necessary to overcome and complicates the scale $\Xi$, for instance, by introducing weight coefficients for all elements of the NSS. The character of these coefficients depends on the nature of elements. Therefore, we select in the NSS three basic types of elements: living, vegetation, and non-living elements. Living elements are characterized by density calculated in the number of species per unit area (volume) or by biomass concentration. Vegetation is characterized by the type and share of the occupied area. Non-living elements are divided by the level of their concentration related to an area or volume of space. In a general case, some characteristic $k$ is ascribed to each element $i$, which corresponds to its significance. As a result, we obtain a specification for calculation of a formula in transition from the notion of biocomplexity to the scale $\Xi$ of its indicator:

$$\xi = \sum_{i=1}^{m} \sum_{j>i}^{m} k_j a_{ij}.$$

It is clear that $\xi = \xi(\varphi, \lambda, t)$, where $\varphi$ and $\lambda$ are geographic latitude and longitude, respectively, and $t$ is the current time. For some territory $\Omega$ the biocomplexity indicator is determined as an average:

$$\xi_\Omega(t) = (1/\sigma) \int_{(\varphi,\lambda) \in \Omega} \xi(\varphi, \lambda, t)\, d\varphi\, d\lambda,$$

where $\sigma$ is the area of a territory $\Omega$.

Thus the indicator $\xi_\Omega(t)$ plays the role of an integral indicator of NSS complexity, reflecting the individual character of its structure and behavior at each time moment $t$ in space $\Omega$. In accordance with the laws of natural evolution, a decrease (increase) in the $\xi_\Omega$ value will trace an increase (reduction) in biodiversity and the ability of natural–anthropogenic systems to survive. Since a decrease in biodiversity opens up biogeochemical cycles and leads to an increase in the load on non-renewable resources, the binary structure of matrix $A$ shifts toward the growing application of resource-exhausting technologies, and the vector of energy exchange between sub-systems of the biosphere–society system moves to a state when the level of its survivability lowers.

### 3.5.2 Model of nature/society system biocomplexity

The NSS consists of element sub-systems $B_i$ $(i = 1, \ldots, m)$ whose interactions are formed in time depending on many factors. NSS biocomplexity is the sum of the structural and dynamic complexity of its components. In other words, NSS biocomplexity is formed in the process of interaction of its parts $\{B_i\}$. In time, sub-systems $B_i$ can change their state, and hence the topology of their connections will change. The evolutionary mechanism of adjusting sub-systems $B_i$ to the environment makes it possible to propose a hypothesis that each sub-system $B_i$, independent of its type, has a structure $B_{i,S}$, behavior $B_{i,B}$, and goal $B_{i,G}$ such that $B_i = \{B_{i,S}, B_{i,B}, B_{i,G}\}$. The goal $B_{i,G}$ of sub-system $B_i$ is to achieve certain states preferable to it. The expediency of the structure $B_{i,S}$ and the purposeful behavior $B_{i,B}$ of sub-system $B_i$ is estimated by efficient achievement of the goal $B_{i,G}$.

Since the interaction of sub-systems $\{B_i\}$ is connected with chemical and energy cycles, it is natural to assume that each sub-system $B_i$ organizes geochemical and geophysical transformations of substance and energy such that a stable state can be preserved. The formalized approach to this process consists in supposition that exchanges of some quantities $V$ of spent resources take place between sub-systems $B_i$ in the NSS structure for some quantities $W$ of consumed resources. Let this process be a $(V, W)$-exchange. The goal of sub-system $B_i$ is the most profitable $(V, W)$-exchange—that is, a minimum amount of $V$ gets as much $W$ as possible, which is a function of the structures and behaviors of interacting sub-systems: $W = W(V, B_i, \{B_k, k \in K\})$, where $K$ is the number of sub-systems contacting sub-system $B_i$. Let $B_K = \{B_k, k \in K\}$. Then the interaction of sub-system $B_i$ with its environment $B_K$ results in the following $(V, W)$-exchanges:

$$\begin{aligned} W_{i,0} &= \max_{B_i} \min_{B_K} W_i(V_i, B_i, B_K) \\ &= W_i(V_i, B_{i,\text{opt}}, B_{K,\text{opt}}); \\ W_{K,0} &= \max_{B_K} \min_{B_i} W_K(V_K, B_i, B_K) \\ &= W_K(V_K, B_{i,\text{opt}}, B_{K,\text{opt}}). \end{aligned}$$

As can be seen, there is some smearing of the goal of sub-system $B_i$ in determining levels $V_i$ and $V_K$. Since the limiting factors have a functioning nature, in this case it is natural to assume the presence of some threshold $V_{i,\min}$, which after reaching, the energy resource of the sub-system is no longer spent on extraction of an external resource—that is, with $V_i \leq V_{i,\min}$ the sub-system $B_i$ moves on to the regime of regeneration of the internal resource. In other words, with $V_i \leq V_{i,\min}$ the biocomplexity indicator $\xi_\Omega(t)$ decreases due to broken bonds between sub-system $B_i$ and other sub-systems. In a general case, $V_{\min}$ is a structural function of graduated type—that is, the transition of $a_{ij}$ from the state $a_{ij} = 1$ to the state $a_{ij} = 0$ does not for all $j$ take place simultaneously. In reality, in any trophic pyramid "predator–victim" relationships cease with decreasing concentration of victims below a critical level. In other cases, the interaction of sub-systems $\{B_i\}$ can cease depending on different combinations of their parameters. A formalized description of possible situations of sub-system $\{B_i\}$ interaction is possible within a simulation model of NSS functioning.

### 3.5.3 Nature/society system modeling

Two approaches to a formalized description of NSS functioning are possible. The first approach is based on evolutionary modeling technology and consists in presentation of each sub-system $B_i$ in the form of a cybernetic graph, the points of which correspond to the multitude of its states. The use of this approach requires a detailed inventory of NSS components and their parametrization. In fact, a graph is drawn with two levels of organization: global and sub-system. At the global level the points of the graph correspond to sub-systems $\{B_i\}$, and at sub-system level the points of a local graph reflect the states of sub-systems $\{B_i\}$. The bonds in the graph of the global level are described by the $(V, W)$-exchange process between $B_i$ and $B_K$, and at the sub-system level they are responsible for transitions of $B_i$ between its possible states. In principle, the chain of selection of sub-levels in the global graph can go on for ever by choosing more and more partial sub-systems.

The process of synthesizing an evolutionary-type NSS is based on its prehistory, which can be presented as a series of states of several or all of its sub-systems. In fact, the process consists of two stages: mutation and quality assessment. This process is realized at a global level beginning with some *a priori* project of a model of the biosphere–society system in the form of $B^k$. The transition to a new version of the model $B^{k+1}$ is realized by changes in structure $B_S$ and behavior $B_B$. The efficiency of this transition is estimated by the function $C(B^{k+1})$, which characterizes the quality of $B^{k+1}$—for instance, by its deviation from prehistory. The structure of the model $B^k$ is prescribed by the graph $\Im = (X, \Phi)$, where $X = \{B_1, \ldots, B_m\}$, $\Phi = \{\Phi_{B_1}, \ldots, \Phi_{B_m}\}$, and $\Phi_{B_i}$ is the multitude of points on the global graph in which the sub-system $B_i$ is reflected. At the same time a graph of the model $B_{i,k}$ is drawn. The transition to new realizations of models $B^{k+1}$ and $\{B_{i,k}\}$ is brought about by adding new or removing old points, by adding or removing the bonds between the points. In this procedure, limitations are possible in trends and laws on a real-time scale. For instance, there is no sense in considering the substitution of some terrestrial covers for others over time periods of several years, or substitution of water surface for land, etc. In other words, absurd and unreal decisions are excluded from the spectrum of possible states and bonds.

Another wider spread and developed approach to NSS modelling consists in synthesizing the complex of numerical descriptions of partial processes involved in the interaction between sub-systems $\{B_i\}$. For this purpose, one should select representative NSS levels when detailing its structure, and then describe the bonds between them numerically. Following Krapivin and Kondratyev (2002), we select these NSS levels: $B_1$ population below 16 years, $B_2$ population from 16 to 60 years, $B_3$ population above 60 years, $B_4$ disabled people, $B_5$ wild animals, $B_6$ domestic animals, $B_7$–$B_{26}$ various types of soil–plant formations, $B_{27}$ mineral resources, $B_{28}$ soil humus, $B_{29}$ nitrogen compounds, $B_{30}$ carbon dioxide, $B_{31}$ oxygen, $B_{32}$–$B_{34}$ water in liquid, frozen, and vapor states, $B_{35}$ phytoplankton, $B_{36}$ bacterioplankton, $B_{37}$ zooplankton, $B_{38}$ nekton, $B_{39}$ dead organic matter, $B_{40}$ phosphorus compounds, $B_{41}$ sulfur compounds, $B_{42}$ methane, $B_{43}$ capital, $B_{44}$ heavy metals, $B_{45}$ oil hydrocarbons, $B_{46}$ radionuclides, and $B_{47}$ other pollutants.

Considering the studied processes of matter and energy transformation in the structures of the NSS, the following balance equation between $B_i$ ($i = 1, \ldots, 47$) can be written:

$$\frac{\partial B_i}{\partial t} + \theta_\varphi \frac{\partial B_i}{\partial \varphi} + \theta_\lambda \frac{\partial B_i}{\partial \lambda} + \theta_z \frac{\partial B_i}{\partial z} = \sum_{i+} - \sum_{i-} + k_\varphi \frac{\partial^2 B_i}{\partial \varphi^2} + k_\lambda \frac{\partial^2 B_i}{\partial \lambda^2} + k_z \frac{\partial^2 B_i}{\partial z^2},$$

where $(\varphi, \lambda, z)$ are spatial coordinates, $\sum_{i+}$ and $\sum_{i-}$ are the receiving and spending parts of the balance of the element $B_i$, $(\theta_\varphi, \theta_\lambda, \theta_z)$ and $(k_\varphi, k_\lambda, k_z)$ are indicators of the rates of motion and mixing of elements within the sub-system $B_i$. These components are specified for each sub-system separately with regard to the physical, biological, and chemical processes that follow the realization of structural, purposeful, and behavioral changes. Of course, the bases of knowledge behind the concrete sciences responsible for study of $B_i$ characteristics should be used. In this case description of the $(V, W)$-exchange is replaced by a set of numerical descriptions of relationships characterizing the intensity and structure of the bonds of the element $B_i$ with other elements. For instance, the "predator–victim" bond is characteristic of relationships between elements $B_{37}$ and $B_{36}$, $B_{38}$ and $B_{37}$, $B_i$ ($i = 1, \ldots, 4$) and $B_5$, $B_6$, etc. This bond is described using the model:

$$\sum_{i+} = kB_i \left[ 1 - \exp\left( - \sum_{j \in S_i} k_{ij} \overline{B}_{ji} \right) \right]; \qquad \sum_{i-} = \sum_{j \in \Gamma_i} C_{ij} \sum_{j+}$$

where $\overline{B}_{ji} = \max\{0, B_j - B_{j,\min}(i)\}$, $B_{j,\min}(i)$ is the minimum biomass of the $j$th element beyond which its consumption by the $i$th element ceases, $S_i$ is the nutrient spectrum of the $i$th element acting as a predator, and $\Gamma_i$ is the spectrum of nutrient subordination of $B_i$:

$$C_{ij} = k_{ji} B_i \Big/ \sum_{m \in S_j} k_{jm} B_m$$

According to this model, at $B_j \leq B_{j,\min}(i)$ in matrix $A$ an element $a_{ij}$ turns from state $\langle\langle 1 \rangle\rangle$ into state $\langle\langle 0 \rangle\rangle$ (the energy bond in the trophic graph gets broken), and NSS complexity changes.

Similar relationships are written for all the elements $\{B_i\}$, including those belonging to the anthropogenic component. It is clear that, as a result, we obtain a rather cumbersome system of non-linear equations whose analytical analysis is impossible. Though interest in the analytical results of biocomplexity studies is undoubted, in this case the study of biocomplexity dynamics can only be done in computer experiments.

### 3.5.4 Models in biocenology

Some of the important GMNSS units parametrize the functions of land ecosystems. Creation of such parametrizations is the sphere of numerical biocenology, which has become an independent scientific direction due to the work of the many authors who combine the potentials of ecological science and numerical modeling in their studies. This combination has resulted in re-orientation of many of the postulates of traditional ecology, which has led to the creation of an information interface between ecology and other sciences and to an increase in the role of complex interdisciplinary studies. Numerical biocenology further entails the problem of complexity and generality in the development of ecological laws (Ingegnoli, 2002; Legendre and Legendre, 1998; Maguire *et al.*, 1991; Tiezzi *et al.*, 2003; Troyan and Dementyev, 2005).

Current models cover such problems as parametrization of succession of the vegetation community, description of production processes as a function of environmental factors, modeling competition within the vegetation community for common resources, modeling natural–anthropogenic systems, modeling the spatio-temporal dynamics of ecological systems, and many others.

The problem of modeling succession in the vegetation community is currently being solved by many authors who have identified the processes involved. There are two fundamental treatments of these processes based on the conceptions of endoecogenesis and exoecogenesis. The most settled opinion of the character of dynamic phenomena in phytocenoses is that it is a single process of vegetation development under the influence of external and internal factors. In this case the process of succession includes such factors as soil development, layered division of the vegetation community, change in height of plants and their accumulated biomass, microclimate formation, change in species diversity, and the relative stability of the community. Classification of the forms of vegetation community dynamics and factors affecting it are important for further synthesizing a numerical model. For example, Vasilevich (1983) introduced the following succession categories: endogenic (autogenic), exogenic (allogenic), continuous, and post-disruptive (post-catastrophic). In endogenic succession the source of changes comes from the vegetation itself, and through environmental changes succession is affected. There are no stages in continuous successions. In the course of post-disruptive succession, changes in community take place from unstable to stable states. Exoecogenesis is characterized by external factors predominating.

A development in succession modeling was the use of population dynamics. For example, Hulst (1979) proposed three types of succession models. The first was a succession model that had a response but no competition—that is, it was a question of the effect of the vegetation community response to changes in environmental conditions. Succession can be described by a standard differential equation:

$$dN_i/dt = r_i N_i (k_i - N_i)/k_i$$

where $N_i$ is the number of individuals of $i$th type, $r_i$ is the rate of exponential growth, and $k_i$ is maximum population density.

The second type of succession model was based on the effect of suppression of one type by another due to intrusion. In this case the coefficient of competition can be used:

$$\alpha_{ij} = \frac{\partial[dN_i/dt]}{\partial N_j} \Bigg/ \frac{\partial[dN_i/dt]}{\partial N_i},$$

and the model itself can be written as:

$$dN_i/dt = r_i N_i \left( k_i - \sum_j \alpha_{ij} N_i \right) \Big/ k_i$$

The third type of succession model was alternative in character. If population density is a function of the size of previous vegetation species in the first two models, then in the third type maximum population density $k_i$ depends on time.

Organic matter production by the vegetation community depends on physiological processes in a plant's organisms, resources of solar radiation, moisture, elements of mineral nutrition, temperature, and many other internal and external factors. When modeling production processes one usually selects photosynthesis, respiration, matter motion, transpiration, cell division, etc. There exist many correlations between these factors and plant production. The formulas of Lieth (1985) are most often used. For example, to evaluate the annual production of plants $R_p$ ($\mathrm{g\,cm^{-2}\,yr^{-1}}$) as a function of annual mean temperature $T$ (°C) and annual mean precipitation $r$ (mm) the following evaluation can be used:

$$R_p = 3{,}000 \min\{(1 + \exp(1.315 - 0.119T))^{-1}, (1 - \exp(-0.000664r))\}$$

In a more general case, the vegetation biomass density $P$ over a given territory can be modeled by the balance equation

$$dP/dt = \min\{\rho_P, R_{PP}\} - M_P - T_P$$

where $\rho_P$ is the maximum biomass density ($B/D$)-coefficient of the considered type of vegetation, $R_{PP}$ is vegetation production with smaller $\rho_P$ current values of the $B/D$-coefficient, $M_P$ and $T_P$ are the values of dying off and expenditure on respiration per unit time, respectively.

The application of parametric descriptions of productivity in beogeocenotic models is an important supplement to the monitoring of vegetation communities. Unfortunately, existing databases and knowledge do not contain any synthesized information about large numbers of empirical correlations, such as the dependences of productivity and other functional components of vegetation communities.

The problem of model description of competition between individual plants is especially difficult in numerical biogeocenology. The mutual impact of plants is mainly described by so-called competition indices. A free-growing plant is one whose functioning is independent of resources and determined only by species characteristics and external parameters. Plants interact when they affect the same resources and when they mutually penetrate free-growth territories. Cenopopulations are bound by more complicated and diverse relationships, and as a result they differentiate niches and habitats. According to Mirkin (1986) who introduced five types of strategies for plant functioning in the environment ($K$ violents, $S$ ecotopic patients, $S_k$ phytocenotic patients, $R$ true explerents (ruderal—growing on waste ground or among rubbish), and $R_k$ false or phytocenotic explerents), a systematic approach to model description of competition and survival is possible. All these strategies are inherent to different degrees to individual species, and therefore survival model synthesis is determined by a concrete set of requirements of the plant to parameters of the niche. A plant's niche is the totality of the space it occupies, resources it uses, and other factors like the presence of pollinators, phytophagans, etc.

Each plant or in some cases their totality $N$ interacts with the environment $H$ (including other plants), exchanging some resources $V_N$ for the volume of resources $W_N$ obtained. The structure $|N|$ of plants or their community and behavior $\underline{N}$ are in equilibrium with the structure $|H|$ and behavior $\underline{H}$ of the environment, which is described by a minimax criterion.

The goals $N_S$ and $H_S$ of the vegetation continuum $N$ and the environment $H$, respectively, can be opposite, partially coinciding, or indifferent. Due to development of an optimal strategy, plants of different species adjust themselves to the levels of $(V, W)$-exchange possible in a given territory. Thus, due to differentiation of niches, there exist multi-level vegetation communities (e.g., in a forest). The factor of niche differentiation is diverse and covers such peculiarities of different species of plants as location of roots at different depths, different time of blossoming, needs of species for different minerals, etc. Generally, a plant's niche can change in time due to the dynamics of the plant itself and shifting of neighbors' niches. This diversity in possible situations needs to be considered when constructing models of vegetation community functioning.

A characteristic vegetation community model is the discrete model of Galitsky (1985). Let us spend some time looking at it. An equation that describes a change in plant biomass $B$ is written as:

$$dB/dt = k(t)(\alpha B^k + (B/B_f)^\gamma B_f - \alpha B^k$$

where $k(t) = \min\{A/A_f(t), 1\}$ indicates the territory occupied by the plant, $A$ is the area available for plant growth, $A_f$ is the area needed for free growth, $B_f$ is the biomass of a free-growing plant, $\alpha$ is the expenditure on balance (i.e., on biomass support), and $\gamma$ is the indicator of territorial deficit.

The basis of Galitsky's (1985) model is that the biomass of each plant—located in the cell of the common matrix it occupies—changes independently of other plants in the community until either the plant itself or any adjacent ones die off due to terrestrial deficit. At the moment of dying off, adjacent plants interact by distributing themselves throughout the cleared territory, and after a change in the $A$ value, the model equation holds again. As a result of the dying-off of individual plants the initial mosaic changes locally, gradually covering more and more territory. The dying-off criterion is simple: a plant is considered dead if its biomass falls below some threshold level $B_{\text{dead}}$ ($B_{\text{dead}}(t) = \varepsilon B_f(t)$, $0 < \varepsilon < 1$).

## 3.6 IMPROVEMENT OF GLOBAL MODEL STRUCTURE

GMNSS synthesis is based on its definition as a self-organizing and self-structuring system. Coordinated functioning of its elements in time and space is provided by the process of natural evolution. The anthropogenic constituent in this process threatens to break this integrity. Attempts to parametrize at a formal level the process of co-evolution of nature and humans as biospheric elements have led to a search for a single description of all processes in the NSS which would combine the efforts of various spheres of knowledge in studies of the environment. This synergetic element forms the basis of many studies on global modeling.

Global modeling refers to recognized, though not always accepted, spheres of knowledge, and many works have been recently written (Andreev, 2003; Danilov-Danilyan *et al.*, 2005; Demirchian, 2001; Kondratyev, 1999b) about its intrusion into non-traditional spheres of human life. Despite the lack of understanding of the role of models in NSS studies, it should be acknowledged that models make it possible to avoid the unpredicted consequences of large-scale transformation of the environment, being able to synthesize NSS trends on the basis of the environmental fragments. Further improvement of global models already developed is impossible without overcoming the duality of the anthropogenic component in current ecodynamics. Humans, like other elements of the environment, form their behavior based on response to the environment which they themselves are changing. Therefore, parametrization of social processes has to be one of the main priorities of further development of the global model.

Adding units to the global model describing the course of various processes in the NSS requires development of a flexible technology of coordination of their spatial and temporal scales, which, in turn, necessitates creating problem-oriented databases. As a result, the structure of a global model should permit development in such a way that there is no change in the hierarchy of units formed. This is possible by introducing mechanisms of adaptation of the model structure to changing descriptions of bond configurations in the NSS. Figure 3.3 shows such a structure in which the role of the synthesizing mechanism is played by the basic data line which coordinates the inputs and outputs of individual units and controls their compatibility both with each other and with the database.

Measurements and studies of global processes in the environment planned by the GCP program are aimed at overcoming difficulties which appear in global modeling and at adapting accumulated data and knowledge to the structures of global models currently available. A way of overcoming these difficulties involves GIMS technology, which makes it possible to use the global model in the regime of its constant adaptation to current data of global monitoring of the environment, providing thereby a means of parametrization of the spatial and temporal variability of biospheric and geospheric processes combined with the procedure of measurement planning.

Figure 3.4 demonstrates the global model built in to the GIMS structure through the DAT unit which controls the interface between other units of the model and the database updated as a result of the monitoring regime. The CON unit adjusts the information domain of the global model and its internal bonds to conditions of the simulation experiment and ensures control of the coordination of information fluxes among the model's units. The REP unit compiles report documents of the simulation experiment and provides the bonds between the global model and other information systems. Other units of the global model are divided into two classes which differ in their modeling technologies. Evolutionary modeling technology units parametrize

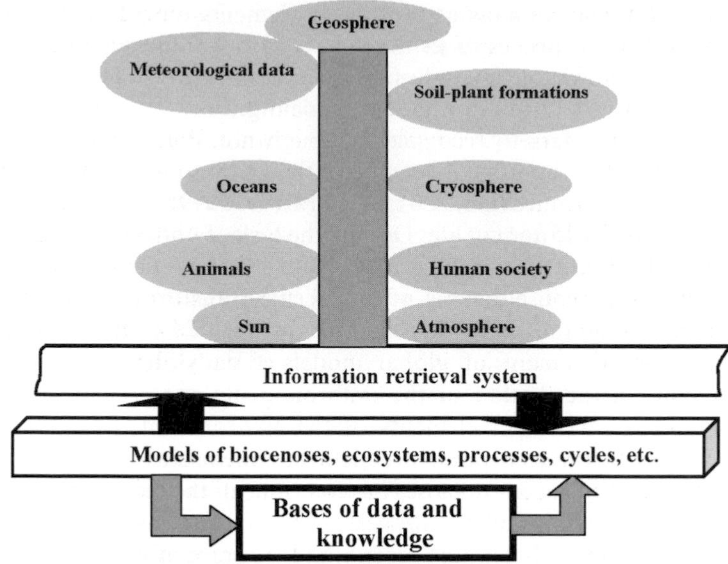


**Figure 3.3.** Organization of the global model of NSS functioning.

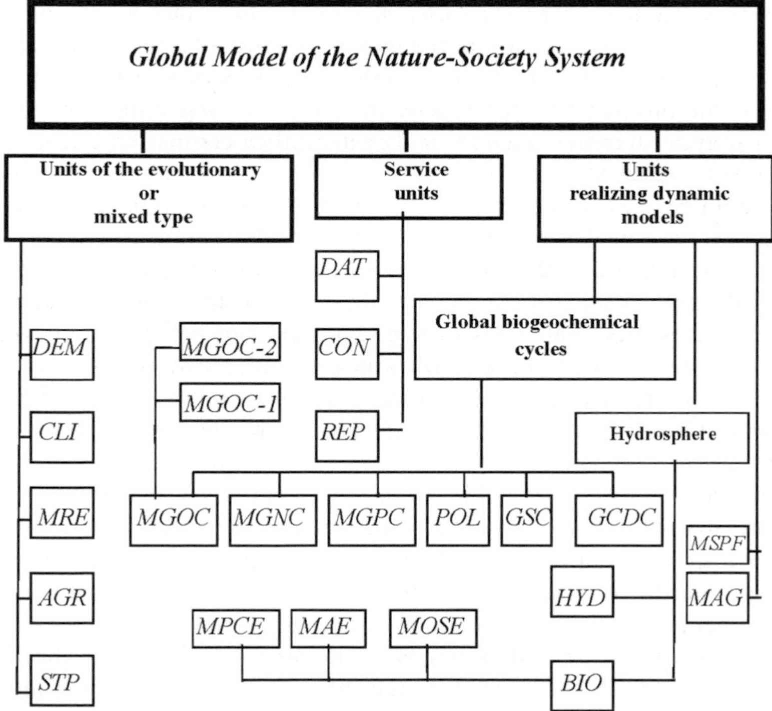


**Figure 3.4.** Block-scheme of the global NSS model.

demographic (DEM) and climatic (CLI) processes, creating scenarios of mineral resource control (MRE) and development of scientific–technical progress (STP), as well as modeling agricultural production (AGR). Dynamic model units are based on realization of balance relationships. The units that parametrize global biogeochemical cycles cover all the phase space of the NSS shown in Figure 3.1. Here are models of the global cycle of carbon (GCC), sulfur (GSC), phosphorus (MGPC), nitrogen (MGNC), oxygen and ozone (MGOC), and general pollutants (POL). To describe all the processes in the hydrosphere, the global model structure contains units which describe how hydrodynamic processes (HYD) and the water ecosystem function in different climate zones (BIO). The global model structure also contains models of the Okhotsk Sea ecosystem (MOSE), Peruvian current (MPCE), and equatorial upwelling (MAE). To broaden the adaptive capabilities of the global model, the structure of many units foresees various versions of parametric descriptions of the processes at work in the NSS.

Figure 3.1 shows the basic elements considered in the GMNSS version described here. As seen from Figure 3.4, concrete realization of each GMNSS unit is determined by what is known about the processes reflected in this unit. Thus, the units responsible for modeling biogeochemical and biogeocenotic processes can be described by differential equations.

The processes of photosynthesis, dying-off, and respiration of plants in land ecosystems can be parametrized by taking bonds between the external and internal system of the vegetation community into account—for instance, temperature dependences of photosynthesis and evapotranspiration of plants, plant–atmosphere gas exchange, impact of solar energy on exchange and growth processes, relationships between vegetation and soil processes, interaction of vegetation cover with the hydrological cycle. Climatic and anthropogenic processes can be parametrized by combining the equations of motion and balance with the evolutionary model. The GMNSS could be improved by broadening the functions of units which parametrize the biogeochemical cycles of substances and elements. This broadening should be directed at updating geochemical models, since the mixing and transformation of environmental pollutants in the geosphere plays the principal role in the formation of conditions for existence of all living beings, from protozoa to humans. It is especially important to improve and apply hydrogeological models in the GMNSS that describe the motion of groundwater and pollutants (heavy metals, radionuclides, organic compounds). The motion of underground water flows depends both on soil characteristics (density, porosity, temperature, etc.) and on external factors such as climate change.

Natural systems, both untouched and anthropogenically changed, are very difficult to describe parametrically. Therefore, detailed elaboration of the numerous processes responsible for the motion and distribution of pollutants is needed when preparing a geochemical model.

Among these processes, the most important are the following:

- chemical reactions which transform the mass of pollutants, converting them into different states;

- transfer of chemical elements through solution and scattering within geological structures;
- biological decomposition and transformation of dangerous chemical elements into harmless forms;
- processes of sedimentation and solution can increase and decrease the porosity and penetrability of the medium and thereby change the velocities of water fluxes; and
- heat transfer can change the liquid flux characteristics and direction of chemical reactions.

There are a number of geochemical models that address these processes (Zhu and Anderson, 2002):

- Solubility models, which usually answer the following three questions:
  (1) What are the concentrations of active ions and molecules in the solution?
  (2) What is the concentration of various minerals in the system and, hence, the course of reactions needed to achieve equilibrium?
  (3) What is a stable distribution of elements in a stable solution?
- Course-of-reaction models, which calculate the succession of states involved in equilibrium, including a step-by-step transformation of masses between phases within the system, with input or output of elements depending on temperature and pressure.
- Balance-of-mass-transformation models, which parametrize the processes involved in water solution.
- Bound-mass-transition models, which describe the totalities of interdependent processes of motion and division of pollutants between different phases.

Of course, extension of GMNSS units due to inclusion of these types of models needs further study, but should increase the accuracy of parametrization of both biogeochemical cycles and biocenotic processes. It is in this way that methods of global ecodynamics are being developed that foresee consideration of the interactive mechanisms of processes in the NSS and biological adaptability under conditions of abrupt changes in the environment. One such study was planned as part of the GCP (Goudie, 2001; Kondratyev and Krapivin, 2004; Lomborg, 2004; Steffen and Tyson, 2001).

As already mentioned, interest in the problem of carbon dioxide and other minor gas components of the atmosphere has arisen as a result of the concern of scientists, economists, and politicians about the possible consequences of expected climate change. Since fossil fuel burning and deforestation are the most significant sources of anthropogenic increase of $CO_2$ and $CH_4$ concentrations in the atmosphere, these gases are undergoing the most thorough study. Both these gases absorb IR radiation from the Earth's surface, and therefore affect the planetary climate. In the late 20th century, politicians suggested an unfortunate way of reducing this impact, formulated in the Kyoto Protocol. In fact, a real and objective solution of this problem consists in a deep study of the biogeochemical cycles of all gases connected with the carbon cycle. This approach is the focus of the GCP: the complex analysis of available data on the carbon cycle in the environment and the accomplishment of needed measurements. The project combines the work of three programs: IGBP, IHDP, and WCRP. The carbon cycle is considered the central cycle of the Earth system connected with climate, water cycle, cycles of biogenic salts, and biomass production due to photosynthesis on land and in the oceans. The human factor in the carbon cycle is determined by factors of food production and formation of habitat conditions.

The GCP foresees studies of three themes: (1) models and variability; (2) processes, control, and interactions; and (3) future carbon cycle dynamics. This structure for global carbon cycle studies should create databases and knowledge which will make it possible to specify all the elements involved in Figure 1.4 and to increase thereby the accuracy of estimates of carbon flux distribution between biospheric reservoirs.

The main idea of the scheme in Figure 1.4 is that the basic components of the global carbon cycle are its major reservoirs in the atmosphere (mainly $CO_2$), oceans (various layers of water masses and bottom sediments), land biosphere (vegetation, litter, and soil), and fossil fuel carbon (activated during the last 200 years due to human activity). The needed degree of detail in description of these reservoirs and the carbon fluxes between them can only be found by modeling the global carbon cycle by gradually increasing its complexity and related knowledge accumulation.

The current idea of carbon flux variability is based on using global scale space and ground observations of concentrations of atmospheric $CO_2$ and other gases, modeling the dynamics of the atmosphere, oceans, and biogeochemical processes and on using mass balance principles. As a result, certain stereotypes have been formed of global ideas about carbon balance in the biosphere:

- About half the $CO_2$ emitted to the atmosphere from fuel burning is absorbed by land and ocean sinks of carbon.
- Most fossil fuel consumption takes place in the Northern Hemisphere.
- Atmospheric $CO_2$ concentration increases considerably due to a change of land cover in the tropics.
- The amplitude of variability in atmospheric $CO_2$ concentration is commensurate with carbon emissions due to fossil fuel burning.
- The interannual variability of carbon fluxes between the atmosphere and land biocenoses exceeds that between the atmosphere and oceans.
- The complete pattern of ocean sources and sinks of atmospheric $CO_2$ is known: tropical waters emit $CO_2$, and high-latitude water bodies assimilate it.
- Maximum interannual variations of $CO_2$ flux at the atmosphere–ocean boundary are observed in the equatorial zone of the Pacific Ocean.

As a result, the following questions arise:

- According to what law do carbon sources and sinks change in time?
- What principles should form the basis for an inventory of carbon fluxes in the environment?

- What is the contribution of human activity, including fuel burning and land use, to the distribution of sources and sinks of carbon?
- What are the social aspects of the impact on the carbon cycle?
- What are the mechanisms underlying the global scale manifestations of regional and local processes of changes in carbon fluxes?

The GCP foresees a search for answers to these and other questions by choosing top-priority directions of study:

- to determine the role of oceans in the control of seasonal and decadal changes in carbon fluxes in the atmosphere–ocean system;
- to assess the role of the land biosphere in the formation of interannual variability of atmospheric $CO_2$ concentration;
- to determine the meridional distribution of carbon sinks in the Northern Hemisphere, with emphasis on the role of North America, Eurasia, Asia, and Europe;
- to assess the importance for the carbon balance of its fluxes between the oceans and land and carbon supplies in inland water bodies;
- to determine the role of depositions and horizontal transfer of matter in the coastal zone and open waters in the formation of the structure of carbon sources and sinks;
- to understand the spatial pattern and amplitudes of carbon sources and sinks in southern latitudes of the World Ocean;
- to determine the role of land tropical regions in the formation of carbon fluxes in the environment with emphasis on the processes of land cover changes;
- to understand differences in the use of fossil fuels in industrial sectors and individual regions of specific countries;
- to explain the principal forces involved in and the consequences of different ways of regional development of reservoirs and fluxes of carbon;
- to analyze the social and regional patterns of vulnerability and adaptability to changes in the carbon cycle;
- to determine the role of biological pumps and the processes of solubility in global and regional carbon cycles;
- to understand the differences and resolve contradictions in measurements of the rates and patterns of changes in land use and land cover transformation; and
- to develop methods of extrapolating the results of studies of the local and regional processes of carbon cycle to global scales.

The carbon fluxes between biospheric reservoirs are known to be controlled by numerous processes that are inadequately known, the most significant of which include:

(i) physical processes in the atmosphere, oceans, and land hydrosphere;
(ii) biological and ecophysiological processes in the oceans and on land; and
(iii) processes of the use of fossil fuel carbon by humans.

To understand these and other processes of carbon flux formation in the dynamic environment, new paradigms and views of the totality of natural and anthropogenic events are needed. It is necessary of course to have here a problem-oriented database to explain cause-and-effect connectivity in the NSS.

We already have a database that explains in detail many processes in this system, outlining the mechanisms underlying their appearance and recommending strategies for data collection and modeling:

- Direct $CO_2$ exchange at the atmosphere–ocean boundary is regulated by the physical processes of vertically mixing water layers above the thermocline with deep waters and the additional impacts of biological processes which pump carbon out of the water and mix it with deposits.
- There are key biological mechanisms of carbon flux formation in the ocean. In general, these mechanisms are connected with the functioning of the carbonate system; put more simply, they involve the vital functions of autotrophic and heterotrophic organisms, processes of dead organic matter decomposition, and respiration of living organisms, as well as nutrient circulation and photosynthesis.
- It has been established that World Ocean response to the increasing partial pressure of $CO_2$ in the atmosphere is non-linear both in space and in time.
- It is known that the interaction between light, water, nutrients, and the roughness regime follows the non-linear laws described by numerical models.
- Many significant feedbacks from the global carbon cycle have been well parametrized: physiological responses of plants to changes of temperature, air and soil humidity, aerodynamics of exchange processes between the Earth's surface and the atmosphere, dynamic and thermodynamic processes within the atmospheric boundary layer.
- When the concentrations of atmospheric $CO_2$ reach a certain level and under non-limiting conditions from light, water, and nutrients, photosynthetic assimilation by land plants becomes physiologically saturated.
- At regional and continental scales, $CO_2$ exchange between land ecosystems and the atmosphere is regulated by several factors:
    (1) extreme climatic phenomena, such as drought, large shifts in seasonal temperatures, changes in solar radiation flux;
    (2) changes in the frequency of fires, cloudless skies, and other large-scale fluctuations in the environment which lead to marked short-term losses of carbon from many biospheric reservoirs;
    (3) changes in biome boundaries and their types (e.g., substitution of evergreen by broad-leaved forests, forested areas by grass cover, or meadows by forests); and
    (4) biodiversity losses and intrusion of exotic species to territories of stable vegetation communities.

All studies foreseen by GCP plans have been aimed at obtaining reliable forecasts of the consequences of human activity in both the near future and over the long term.

For this purpose, it is necessary to develop a method to assess how soil–plant formations and World Ocean areas assimilate $CO_2$ from the atmosphere. Observations show that in the 500,000 years before 1860 the climate system had functioned in a relatively narrow range of changes in temperature and in $CO_2$ and $CH_4$ concentrations (during the last 100,000 years $CO_2$ concentration in the atmosphere has varied from 180 to 280 ppmv). The last century has substantially broadened the spectrum of processes involved in substance and energy transformation in the environment due to numerous new types of influence on biospheric elements. The human factor in these processes has brought forth an uncertainty of the future dynamics of these elements, which necessitated the working out of scenarios of the likely development of economics, science, and politics. The level of uncertainty here is so high that real forecasts are impossible. According to some estimates, the level of carbon emission by 2100 could reach 30 GtC/yr, which, at current rates of anthropogenic impacts on the environment, could lead to atmospheric $CO_2$ concentration reaching 950 ppm. These estimates though based on expert opinion come from fairly primitive models. Therefore, the GCP foresees accumulation of a vast base of knowledge, formation and testing of scenarios, development of models, and analysis of current trends as top-priority directions for studies of global environment change. A set of inductive considerations and reasons can be used as the "building blocks" for scenario synthesis:

- the concentration of greenhouse gases will increase for some time;
- the current and expected increase in GHG concentration—especially $CO_2$—will probably be followed by global warming and other climate changes;
- apart from physical factors like ocean circulation future assimilation of $CO_2$ by the oceans from the atmosphere will grow as atmospheric $CO_2$ concentration increases;
- the response of the ocean thermocline and circulation systems to climate change (especially temperature increase and changes in the hydrological cycle) will probably be non-linear and will be followed by increased stratification of the upper layer of the oceans, which could reduce deep-layer ventilation;
- increased stratification of the upper layer of the ocean will change the provision of nutritious salts to surface layers which will lead to a change in the role of phytoplankton in carbon assimilation and cause large-scale fluctuations in atmosphere–ocean interactions, especially in the Arctic Basin, the Okhotsk Sea, and other Northern Hemishere seas;
- there is a discrepancy in estimating changes in the land sinks of carbon in light of emerging trends for land biosphere transformation;
- careless and ill-considered anthropogenic changes to the Earth's surface will lead in the near future to a change in climate and the natural carbon cycle;
- instabilities and numerous states of equilibrium could be connected with biophysical and biogeochemical systems in which climatic and hydrological processes take place in addition to carbon and nutrient cycles—these phenomena could happen due to non-linear feedbacks in the re-distribution of energy and substances between the atmosphere, land, oceans, and ice;
- even without dynamic events—such as instabilities and increase in equilibrium

states—an increase in atmospheric $CO_2$ concentration between 1860 and 2100 will cause certain shifts in the climate–carbon system, changing the earlier distribution levels of $CO_2$ to regions;

- anthropogenic changes to the carbon cycle interact with the biophysical aspects of formation of this cycle, and hence could lead to variations in the global climate system following known laws;
- humankind also responds to threats to its well-being from undesirable climate change, though these responses are very uncertain (e.g., the Kyoto Protocol); and
- the potential responses of human society to undesirable climate change may take geographical, social, economic, and natural history directions:
  (1) changes in $CO_2$ emissions due to a transition from traditional to non-traditional types of fuels or an advance in technology;
  (2) improvement of the processes of production and use of energy, and a change in life style;
  (3) transition to renewable energy sources (bioenergy, solar energy, wind);
  (4) increase in the role of forests to assimilate excess carbon;
  (5) increasing the carbon sink from the biosphere, as a result of engineering, to geological formations and ocean depths; and
  (5) development of new technologies in agriculture leading to GHG emission reduction.

Taking all of this into consideration requires a universal complex approach to study the consequences of alternative scenarios for civilization development. Clearly, the search for safe stable functioning of the NSS is impossible without modeling the various scenarios; hence the need for a global model of how this system functions. Such a model and how it could be used were proposed in Krapivin and Kondratyev (2002) and Kondratyev and Krapivin (2005c). A substantial aspect of GMNSS development is a detailed description of how the Arctic and Antarctic systems function as interactive components in the global climate system. Arctic environment variability in the context of global change was studied by Bobylev *et al.* (2003). It is evident that the Arctic is a unique and important environment in the Earth system. The role of the Arctic Ocean and sea ice in climate change has been studied incompletely. Existing observational data cannot explain many of the processes in the Arctic environment. The importance of such observational data for climate models is indisputable. Nevertheless, the GMNSS needs a formal description of the interactions between the Arctic Basin and other regions of the globe. Such an effort was made by Overpeck *et al.* (2005) who declared that the Arctic Basin is moving towards a new state that falls outside the realm of glacial–interglacial fluctuations that have prevailed during recent Earth history. Present rate of sea ice loss, deglaciation of Greenland, and destabilization of permafrost in some areas point up the existence of variations in the Arctic cryosphere not observed in the recent past. But, Overpeck *et al.* (2005) said little about global warming. They proposed considering the Arctic system as the interactions between the nine key components (hubs) listed in Table 3.2. The interactions between these components can be unidirectional or bidirectional, strong or week, and positive or negative. Table 3.2 represents the main interactions

**Table 3.2.** Symbolic characteristic of interactions between essential components of the present Arctic system. A "+" represents a positive interaction. A "−" represents a negative interaction (Overpeck *et al.*, 2005).

| Component (hub) | 1 | 2 | 3 | 4 | 5 | 6 | 7 | 8 | 9 |
|---|---|---|---|---|---|---|---|---|---|
| 1. Sea ice | | + | | | − | − | − | − | + |
| 2. Thermohaline circulation | − | | + | − | | | | | |
| 3. Precipitation minus evaporation | + | − | | + | | | + | + | − |
| 4. Terrestrial ice | | + | | | + | + | | | |
| 5. Population | | | | | | | | − | |
| 6. Economic production | | | | | + | | | | |
| 7. Marine primary production | | | | | + | | | | |
| 8. Terrestrial biomass | − | | | | | | | | + |
| 9. Permafrost | | | | | | + | | − | |

between the components of the Arctic environment. This conceptual representation allows the GMNSS to be further updated by expanding the functions of its different blocks. It is clear that different interactions need additional investigation before their parametrization.

Arctic sea ice is the main indicator of global climate change. Sea ice data sets come from various sources (McCarthy *et al.*, 2001). Polar ice cover involves many parameters—such as area, thickness, and density. Existing results of observational and modeling analyses can be used as another means of updating the GMNSS. The Arctic's sea ice is a major driver of the global weather system. It covers an average of 14 million $km^2$ at its greatest extent in February and about half this at its smallest extent in the northern autumn (Chapman and Walsh, 1993). It acts as a natural refrigerator for the planet due to the reflection of solar energy away from the Earth. The Arctic's sea ice is home to a wide variety of wildlife, especially mammals, fish, and sea birds.

Detailed data on the extent of Arctic sea ice has been available from ground observations since 1953, and from satellite observations since 1972. These data point up two key factors controlling changes in ice distribution in the Arctic. The first factor is the intense warming that has occurred over the past 30 years in Siberia, Alaska, and Western Canada despite a weaker cooling trend in western Greenland and the Labrador Sea. The second factor is the major change in Arctic air circulation patterns, which leads to relatively warm spring and summer air masses, as well as to the melting of some sea ice and driving much of what remains away from the shore towards the central ice pack. A major consequence of these changes is a lengthening of the warm season during which sea ice can melt. Observation data have shown that the melt season since 1979 has become longer by 8% per year. Although this warming is likely down to natural forces, the question about the role of greenhouse gases in this process remains unanswered. Unfortunately, current science is unable to separate these reasons reliably. Many other questions arise concerning the correlation between

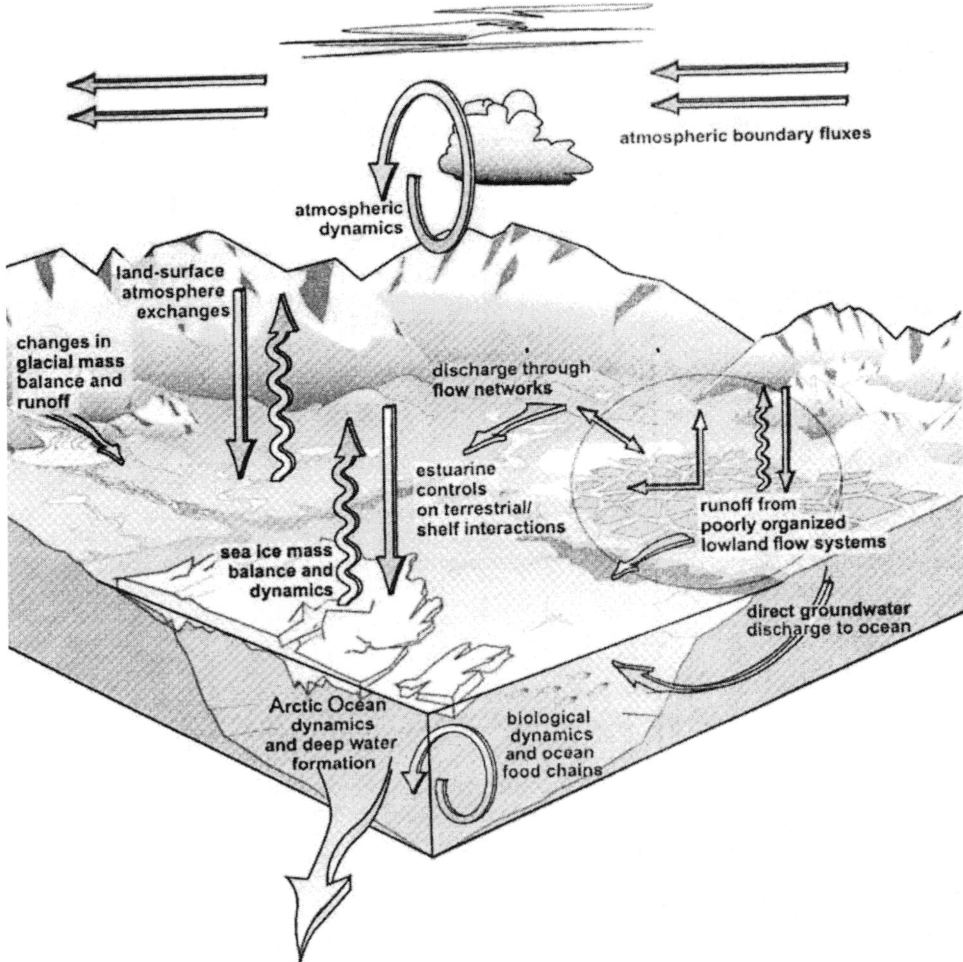


**Figure 3.5.** Key components of Arctic hydrology (*http://www.grida.no/climate/ipcc_tar/wg2/608.htm*).

the ice sheet and climate change, two of which are fundamental: What is the impact of changing ice sheets on global sea level rise? Is it possible to predict changes in ice sheet volume and hence changes in global sea level?

Glaciers and ice sheets are key mechanisms in Arctic Basin dynamics. They modulate Arctic Ocean level by storing water deposited as snow on the surface and discharging water back into the ocean through melting and via icebergs. At the present time satellite observations give operative data about the topography and properties of the ice sheet in polar regions. However, very little information is available about variations in the properties of the subglacial bed where there exist

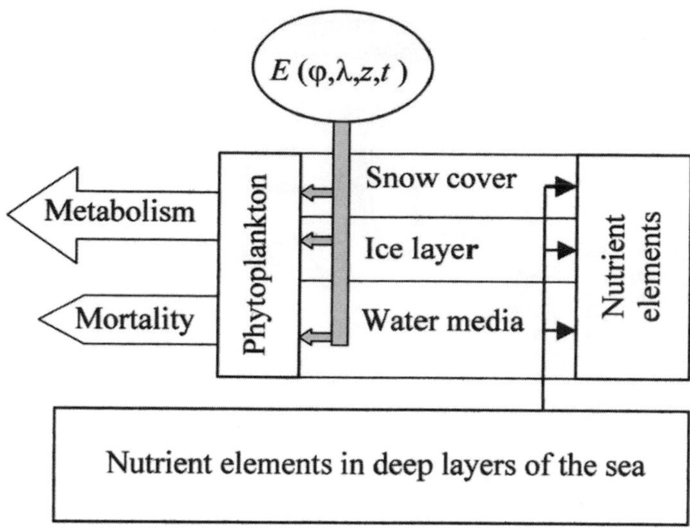


**Figure 3.6.** The conceptual structure of the model of the organic matter cycle in an ice-covered sea (Kondratyev *et al.*, 2004).

changes that control many processes in the Arctic Basin—in particular, Arctic hydrology.

The hydrology regime of the Arctic depends on the thermally sensitive cryosphere and its controlling influence on the water cycle. The major hydrological processes and related aquatic ecosystems of the Arctic Basin are affected by snow and ice processes, as well as the major Arctic rivers that originate in more temperate latitudes and pass through cold regions before reaching Arctic seas. The precipitation regime also greatly impacts the hydrology of the Arctic Basin. This is the reason the climate block of the GMNSS considers the patterns of snow and ice field distribution, as well as river and wetland structure. Figure 3.5 describes key linkages among land, ocean, and atmosphere in the Arctic hydrological cycle. These linkages need to be taken into account in the GMNSS to increase its precision for future assessment of global ecodynamics.

A significant Arctic environment component is biological productivity that is irregular in its spatial distribution. The ecozone permanently covered by ice has low biological productivity. In this area seals and polar bears are the main large animals. Figure 3.6 shows the formation of biological productivity in Arctic seas. As in all complex ecosystems, there are many relationships between species. The principal food source of the main groups of animals in the Arctic seas is photosynthesis, which varies widely throughout the year. Algae, phytoplankton, and macroalgae are primary producers that grow both in the water environment and in the ice ergocline (Legendre and Legendre, 1998). Of these, phytoplankton produce most of the edible material that is consumed by zooplankton and some fish. Small invertebrates are consumed by large pelagic vertebrates—fish, birds, and mammals. The Arctic food web is the regulator of processes in the *atmosphere–ice–water* system and controls the gas and energy exchange in formation of the $CO_2$ biogeochemical cycle.

The role of the Arctic environment in global ecodynamics is reflected via such important yet inadequately studied components of northern latitudes as the permafrost.

# 4

# Global change and geoinformation monitoring

## 4.1 INFORMATION PROVISION FOR GLOBAL ECOLOGICAL STUDIES

At present, many countries do not have an information system that would provide information support for environmental studies. Studies of natural–anthropogenic systems undertaken by different national and international organizations and departments are uncoordinated and the important results obtained are not always available for the mass user. Moreover, there are no single standards for presentation and storage of ecological information or for development of the technology to process it. Therefore, most experts dealing with environmental problems have scant access to new databases, and in most cases ecological information remains uncoordinated or unavailable. As a result of this lack of coordination, many scientists have no access to international ecological information services.

Development of a national information service aimed at using highly effective technologies and one that interacts with world information structures of natural monitoring is possible within the framework of the International Centre of Dataware for Ecological Studies (ICDES). This body is tasked with collecting, categorizing, and processing consistently (i.e., using a single standard) the data of ground and satellite observations of natural and anthropogenic systems on Earth, and with ensuring easy access to this information. The goals and tasks of ICDES include the following elements:

- preparation of information sets on the distribution of ecological data at world and regional centers of data collection on the state of environmental systems;
- synthesis of thematic databases aimed at supporting top-priority national and international programs of environmental studies;
- coordination of databases at scientific organizations in all countries;
- compiling catalogs of satellite and ground observational data and providing users with easy access to them;

- provision and coordination of the input from departmental and national thematic databases to international information networks, analysis of database completeness, as well as preparation and propagation of information systems on a regional level to increase the efficiency of assessments of the state of natural–anthropogenic systems from various spatial scales up to globally;
- development of standards for data presentation in geoinformation monitoring systems, and coordination of studies tasked with creation of new information technologies of database control in order to broaden the capabilities of computer technology for dialog systems;
- dataware provision of networks that facilitate the on-line interaction of ICDES with users on the state of developments in hardware/software support; and
- convening scientific conferences on the problems of dataware provision for ecological studies.

Let us now describe some of these elements.

### 4.1.1 Key problems of geoinformation monitoring

Synthesis of a structure of information support for ecological studies to solve the problems of environmental diagnostics depends on the priorities given in these studies. Unfortunately, there remains an unsolved basic problem of geoinformation monitoring: selection of top-priority directions for studies to take and determination of its needed infrastructure. The continuing practically independent development of technical and theoretical means of geoinformation monitoring hinders the solution of this problem. This was first formalized at a global level in Arsky *et al.* (2000), Kondratyev (1999a), and Marchuk and Kondratyev (1992). It was only after these studies that it became possible to formulate the provisions which determine information structure and the need for information levels in geoinformation monitoring. In fact, there are several levels of user demand for information support. These levels determine the hardware and software structure of the monitoring system. In its state of readiness, monitoring data can be divided into five levels: measurement results, corrected measurements, model coefficients, identifiers, and knowledge. Identifiers are semantic information structures which determine the distribution of the environmental elements for a given spatial resolution and which can be used to control information fluxes in the determination of and search for objects *a priori* specified by the information user. These may be concrete natural–anthropogenic formations, phenomena, and processes. Moreover, the network of identifiers serves to form scenarios and subsequently models for their assessment. The ICDES information service system should provide access to level 6 of the database (i.e., knowledge) through the user's interface and should permit the user to make copies and modify the data. The structure of the information field of the geoinformation monitoring system for each level forms several branches in its directions and objects of studies as well as in its spatial scales. Key top-priority problems of global monitoring were analyzed in Marchuk and Kondratyev (1992). In this work a set of spatial scales was developed to help guide the designers of monitoring systems. However, these solutions were not absolutely objective, being based as they were on solution of the problems facing observation planning. One of ICDES' functions is solution of this problem.

### 4.1.2 Integration of information about the environment

Geographic Information Systems (GIS) are the most developed means of dataware provision for ecological studies. They give a wide spectrum of operations from separate series of data, connecting and mixing them by a set of criteria and providing the processes of information exchange between specialists in different sciences. Fragmented data generated by the sources of different physical structures in various formats are here synthesized into a single information field. As a result, data from ground systems—both stationary and moving—of environmental observations can be combined as well as data from aircraft and satellite remote-sensing. Of importance here is the possibility of mapping environmental elements by reference to geographic coordinates and by selecting specific formations. This is absolutely necessary in the search for and identification of extreme situations like floods, dust storms, landslides, volcanic eruptions, etc. (Sims and Gamon, 2003; Timoshevskii *et al.*, 2003; Wright, 2003; Zarco-Tejada *et al.*, 2003; Zhan *et al.*, 2002)

GIS technology is based on the methods and algorithms necessary for formalized transformation of geographic information. In accordance with a discrete set of spatial scales of space digitization, a series of schematic maps $\{\Xi_i\}$ can be found, each carrying information on the structure and subject content of sections of the environment at a given scale. The respective software accomplishes the procedure of information integration from different sections into level $S$ of mapping, prescribed by the user of this information: that is, it gives $\Xi = \bigcup_{i \in S} \Xi_i$. GIMS technology is the GIS that have been developed toward realization of the formula $GIMS = GIS + model$. In other words, GIS functions spring to life as a result of introducing a new coordinate grid: the temporal scale. As a result, the user obtains a tool to forecast and, hence, based on level 6 of the database, can dynamically integrate ecological information. Provision of this transition is another of ICDES' functions.

### 4.1.3 Systematic approach to information provision in extreme situation identification

One important function of the geoinformation monitoring system is to detect and identify the extreme situations that can happen in the environment. Much experience in solving this problem using radiophysical, acoustical, optical, and other methods suggests that for a successful result it is necessary to synthesize a wide spectrum of multi-channel measurements and to use a hierarchical structure in searching and processing of information from different levels of the monitoring system: satellite/flying laboratory/floating laboratory/ground or a stationary observation point on the surface of the water. The problem of detection and identification of abnormal phenomena in the environment at each level can be solved using respective technologies, but its prompt solution can only be provided by all levels jointly. The satellite level provides a periodical survey of environmental elements and transmits a primary signal about the possible appearance of an abnormal formation in the next level. The passage of the signal through all levels gives an answer to whether environmental parameters have really changed at some level or whether it is a false signal.

The observation process at each level is divided into four stages: (1) barrier search; (2) search within the area; (3) tracing; and (4) contact retrieval. In the first case, an extreme situation in a confined area can be detected due to passage across its boundary. The search within the area uses detection instruments distributed throughout the area and operating within a certain monitoring regime. This monitoring regime is possible when expecting floods, landslides, snow avalanches, volcanic eruptions, etc. The third and fourth regimes of the monitoring help to realize the procedure of decision making about an approaching natural disaster.

Hence, the problem of dataware provision for ecological monitoring in a search for extreme situations is a system problem. Each level of the monitoring system needs a certain set of information structures. Therefore, detailed study of this set is yet another of ICDES' functions. Thus, when accomplishing a multi-stage hierarchical search of anomalies determined by some set of criteria, the monitoring system forms an information field that can be used by nature protection services and other organizations. The ICDES should develop the necessary documentation and legal principles for this procedure.

### 4.1.4 Ecological monitoring

An important constituent of geoinformation monitoring is ecological monitoring. This serves as a source of information about the processes in ecosystems and gives initial information for ecological models. As Spellenberg (2005) noted, ecological monitoring, though too expensive for many countries, should cover the whole planet. Rich countries and international organizations must provide the financial support. This necessity is dictated by the following reasons:

- as the processes involved in many ecosystems have been studied inadequately, organization of the way in which they are monitored could broaden ecological knowledge and consequently increase the reliability of forecasts for the coming decades;
- development of effective strategies for ecosystem control is impossible without systematic observation;
- present anthropogenic disturbance to ecosystems can lead to their change in the future, which can only be detected and recorded under organized adaptive monitoring;
- long-term observations of ecosystems will make it possible to predict the onset of negative consequences of anthropogenically based changes; and
- ecological monitoring will make it possible to identify and systematize changes taking place in biological communities.

To optimize ecological monitoring it is clear that it is necessary to determine the key characteristics, observations of which should be considered paramount to increasing the reliability of estimating ecosystem dynamics. These characteristics include:

- cyclic changes;
- length of delay of ecosystem response to external forcings;
- indicators of the stability and direction of development of community structures; and
- sensitivity of ecosystems to changes.

Of course, when discussing global processes it is necessary to understand that it is impossible to organize ecological monitoring everywhere. Therefore, priority regions should be chosen and their role in the global processes should be indicated as well as the frequency of observations; experts normally choose the following regions:

- regions where powerful anthropogenic impacts on the environment take place as a consequence of controlled land use;
- regions where, despite anthropogenic processes being little developed, there is a danger of changes in biological communities due to human activity connected with the production of new plant species; and
- regions where ecological monitoring is not developed, but knowledge of the processes of environmental degradation is important for prediction of global ecodynamics.

Among the regions that arouse the special interest of experts is the Antarctic where huge masses of ice are concentrated and which is surrounded by some of the most biologically rich water areas of the World Ocean. This region, being far removed from densely populated territories, seems to be the least subjected to direct pollution. However, the global processes of pollutant propagation and climate change inevitably influence Antarctic ecosystems. During the last 200 years the zones between 40°S and the Antarctic have been subjected to the growing fishery of plankton, fish, and whales. Therefore, prior knowledge about changes in the Antarctic environment is important for prognostic assessment of the consequences of anthropogenic process development in other remote territories (Chapman and Walsh, 1993).

The conceptual aspects behind selecting regions where ecological monitoring should be developed (up to a certain level of knowledge) are connected with the search for constructive mechanisms of assessing the reliability of this knowledge and its significance for understanding global ecodynamic processes. Such a mechanism was proposed in Kondratyev *et al.* (2002, 2003a, 2006b).

The problem of gaining operational multi-purpose control of the physical, biological, and chemical properties of the environment in zones where the branching networks of anthropogenic pollutant propagation are concentrated and intensive changes in other external conditions are observed is the subject of studies for many nature protection programs at national and international scales. The complexity of this problem is determined by the spatial extent of territories, the presence of numerous external factors affecting natural processes, and mainly by the nonstationary functional characteristics of the environment. To assess the state of an ecosystem, it is necessary to take into account the interaction of numerous physical,

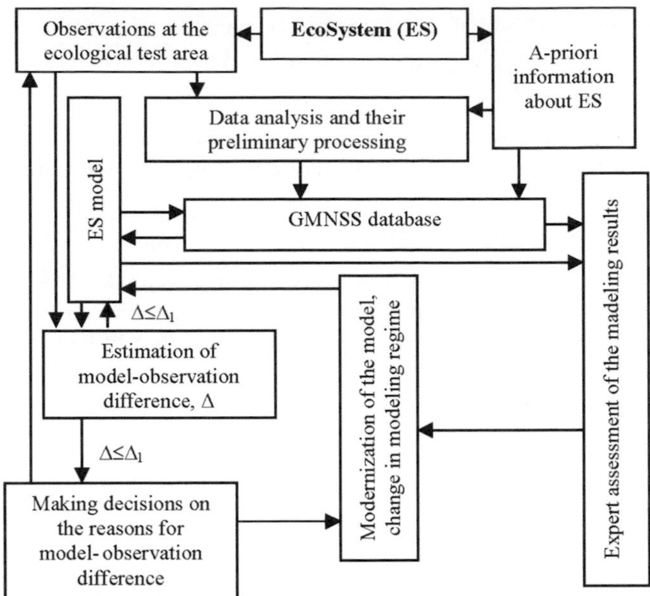


**Figure 4.1.** The organization of ecological monitoring using an adaptive modeling regime. Notation: $\Delta$ = integral or subject estimate of difference between modeling and observational data; $\Delta_1$ = permissible level of difference for $\Delta$ estimation.

chemical, and biological processes, which requires the development and application of systematic methods and new computer technologies. All this puts a strict demand on which method of simulation system synthesis should be used to integrate the indicated processes into a single structure. We choose GIMS technology because it provides an adaptive regime of field measurements as the basis for the suggested method of synthesis of the simulation system for ecological monitoring. Figure 4.1 demonstrates an alternative use of modeling and measuring regimes in ecological monitoring.

GIMS technology simplifies the organization needed for regular ecological monitoring and makes it possible—due to the coordinated use of observational means and numerical models—to optimize economic costs. A good example of an information system that applies GIMS technology is shown in Figure 4.2. This system is used by ecological monitoring services for expert comparison of episodic assessments of individual environmental elements with modeling results. This comparison could result either in correction of some components of the ecosystem model or in additional measurements. On the whole, realization of the scheme in Figure 4.2 is possible with regular satellite observations of areas of different types of the Earth's surface, temperature, air moisture content, wind speed and direction, water body salinity, etc. Of course, optimization of ecological monitoring requires additional studies in order to develop the requisite set of algorithms of observational data-processing as well as creation of criteria to assess the state of ecosystems. One such indicator of the state of ecosystems is the biological complexity of the territory of the ecosystem.

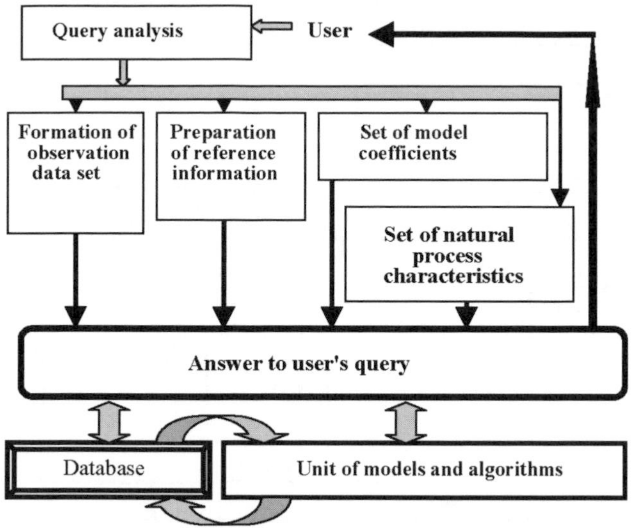


**Figure 4.2.** The dialog regime of algorithmic support for ecological experiments.

## 4.2 OPEN SYSTEMS IN GLOBAL ECOINFORMATICS

The information infrastructure of a standard geoinformation monitoring system (GIMS) should be adapted in such a way that its resources are automatically adjusted to the subject sphere. As a rule, known GIMSs do not have this function since open system technology is almost never used when synthesizing the systems involved in environmental control. Progress here is hindered by the absence of ways of setting up a computing environment for GIMS which governs information fluxes between the data-processing system and the base of knowledge and establishes standardized interfaces between a GIMS and its users.

Some progress in providing the functions of open systems to a GIMS has been achieved recently by using GIS technology. With the help of this technology, the problems of scaling and joining multi-discipline databases and their accessibility from the Internet have been solved. However, the thematic and subject referencing of databases has not been standardized, which creates serious obstacles to their practical use and especially to determination of the database structure. These difficulties are overcome using GIMS technology. GIMS technology proposes a mechanism for combining GIS technology with simulation modeling, which makes it possible to project software with a broad set of functions for problem orientation. These achievements depend on creation of a software package for formalized presentation of environmental objects. However, studies here are in their formative stages. Of course, here one should analyze the many branches in expert and learning system theory involved in solving problems concerning environmental control and establish interactions between many computer projection technologies—such as CASE (Computer Aided Software Engineering) technology—by developing the open system theory of environmental monitoring (Jones, 1990).

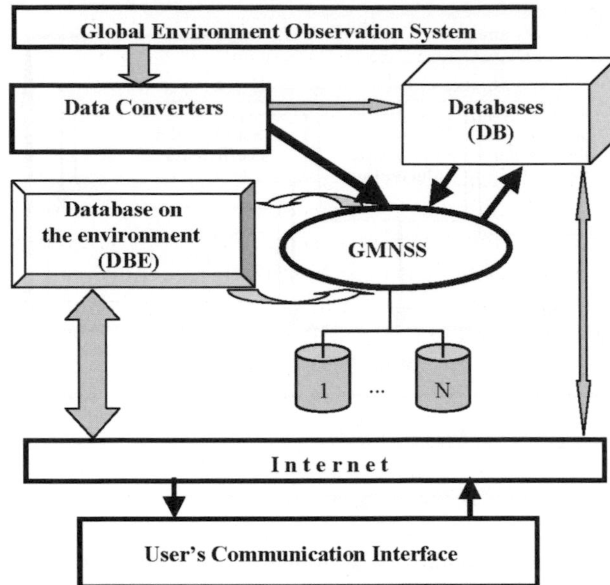


**Figure 4.3.** GIMS technology used for synthesis of the global system of environmental control using standard means of telecommunication.

Systems of ecological and geophysical monitoring that do not depend on spatial scales are being developed with almost no centralized control regarding standardization. The investigators and authors of such systems are faced with serious difficulties when using the results of such systems. Recently achieved progress in standardization of the global information infrastructure has been the catalyst for similar studies in geoinformation monitoring—mainly in development of the global GIMS. Information fluxes in the global GIMS are shown in Figure 4.3.

The Internet holds many global and national databases. Use of these data sources on the environment is based on a multitude of different standards. But, until information channels acquire the rank of intrasystem provision, the standardization problem will remain. There is no single DBE level which would agree with either other levels of the GMNSS or DB. Therefore, creation of interfaces within the GMNSS is one of the top priorities.

GMNSS synthesis cannot be accomplished without there being standardizing interfaces between units DB and DBE. Many models of biogeochemical, biogeocenotic, climatic, and socio-economic processes in the environment have been developed and used that do not correlate with each other and, moreover, do not take into account the applied aspects of other users. This drawback with a large part of DBE can be removed by introducing into GMNSS an interface between the user and database. For instance, to parametrize the process of evapotranspiration, there are no fewer than 10 models whose inputs are different in principle (Nitu *et al.*, 2000b). From DBE analysis, the GMNSS interface should either choose one of them or ask the user for additional data.

A global GIMS, due to its specific character, should be calculated over the long term and allow for evolutionary development. It is necessary here to meet the requirements of the the functional reliability of the user interface facility. Substitution or addition of individual GIMS units should not affect the structure of its internal and external information fluxes.

Environmental monitoring concerns science-intensive experiments. It requires creation of a system of conjugated sub-systems of observation and data-processing. Bringing about such a global system is only possible by coordinating spatial and temporal scales. The Tiller scale is an example of such coordination (Kondratyev *et al.*, 2003a). Within each unit of this scale, using GIS technology, cartograms are developed that show the general plan of a territory, the river network structure, dislocation of settlements and other anthropogenic objects, inventory schemes of the use of land resources, boundaries of geographic zones and structures, topographic imagery of the territory, and prognostic scenarios of territorial changes. Spatio-temporal coordination of environmental elements should cover all possible combinations of correlations between spatial sizes and temporal scales. This stage of GIMS synthesis is labour-intensive and crucial. The following standardization algorithm is proposed for its accomplishment.

The entire territory of the Earth's surface $\Omega = \{(\varphi, \lambda)\}$ is portrayed as a geographical grid with steps $\Delta\varphi$ in latitude $\varphi$ and $\Delta\lambda$ in longitude $\lambda$. The vertical structure of the atmosphere, soil, and hydrosphere is presented as a totality of layers of thickness $\Delta z$, where the axis $z$ is perpendicular to the surface of a cell $\Omega_{ij} \subseteq \Omega$, and has its origin at the World Ocean level ($z = 0$). Thus, the environment $V$ is divided into volume compartments $V_{ijk} \subseteq V$. The procedure of digitization can continue within each compartment $V_{ijk}$ with no allowance at the $l$th level where the value $l$ is determined either expertly or proceeds from other criteria. For instance, in the U.S. National Arctic Program started in 1992, $l = 5$ and each level of digitization of the Arctic region is referenced to a certain spatio-temporal classification of objects and processes in the environment. At the present level of science development, the grounds for the choice of this scale raise doubts. Nevertheless, there is no other algorithm to replace the expert one. Therefore, for this procedure to reach a certain level of formalization it is suggested that GMNSS synthesis should be based on such a hierarchy of compartments and be provided with some converters—from data and knowledge bases to a parametric space of the GMNSS. In other words, we propose a multitude of semantic structures $\{A_s\}$ whose elements will perform the functions of identifiers of objects and processes in the environment. As a result, the GMNSS will have as input only these structures and they alone will connect the data and knowledge bases with individual GMNSS units. Figure 4.4 explains the procedure involved in the user–GMNSS interaction. All functions of data conversion, knowledge exception, and its adaptation to user demands are performed by the control unit of the GMNSS numerical experiment. Adjustment of the GMNSS subject unit is made using the built-in editor of semantic structures $\{A_s\}$, access to which is realized in several ways due to hierarchical dialog units and pictograms. By programming GMNSS units one can generate the workspace of the scenario chosen by the user.

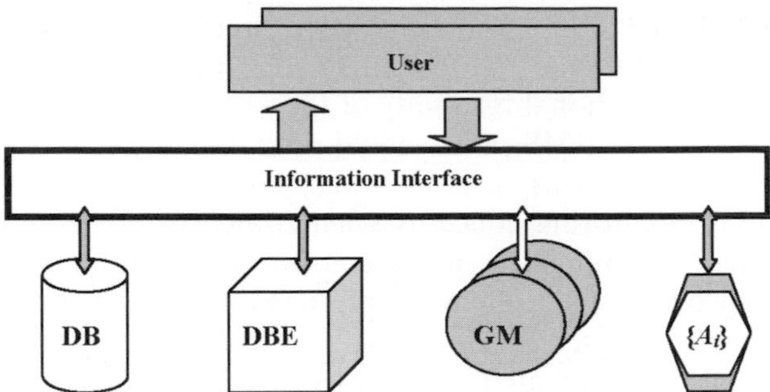


**Figure 4.4.** GIMS resource control.

A set of basic identifiers provides a degree of universality of GMNSS inputs and allows the user to synthesize scenarios for inadequately parametrized natural and anthropogenic processes. In general, $A_s = \|a_{i_1,i_2,\ldots,i_s}\|$, where the element $a_{i_1,i_2,\ldots,i_s}$ identifies an object, process, phenomenon, event, or other section of the environment through a parametric description of its image using a combination of symbols or any attributes of such symbols, including color shades. The parameter $s$ indicates the dimension of the information pattern understood in a broad sense. Experience shows that most structures $A_s$ have four dimensions: $i_1$ latitude, $i_2$ longitude, $i_3$ height or depth, and $i_4$ time. The level of discreteness is provided by the hierarchy of spatio-temporal scales in DB, and if these scales are inconsistent with user demands the algorithms of spatio-temporal interpolation are used.

Note, in conclusion, that development of the software environment of open systems for data and knowledge bases in global geoinformation monitoring, and adjustment of this environment to simulation experiment conditions have only been possible as a result of generation of the many interface functions that optimize the use of resources of the whole GMNSS system and govern the input/output structure. The ultimate goal of such studies is to create the operational environment to control information fluxes in the global infrastructure of hierarchical systems of environmental monitoring.

## 4.3 TECHNOLOGIES AND SYSTEMS OF GLOBAL MONITORING

### 4.3.1 Space-borne information systems

A current system of ecological monitoring is the space-borne observatory Aqua launched by NASA within the Earth Observing System (EOS) program on 4 May 2002 to near-Earth orbit (705 km) with an orbital time of 98.8 minutes. The Aqua instrumentation is characterized in Table 4.1. The system mainly measures water components in the surface layer of the atmosphere and on the Earth's surface, separating the liquid, vapor, and solid phases of water; it also delivers information about the climate system. Six basic measuring sub-systems enable one to obtain data,

**Table 4.1.** Equipment on board the space observatory Aqua (Parkinson, 2003).

| Measuring system | Characteristics of the measuring system |
|---|---|
| Atmospheric Infrared Sounder (AIRS) | AIRS has 2,382 high-resolution channels, 2,378 channels measuring IR radiation in the range 3.7 to 15.4 μm, other channels cover visible and near-IR regions (0.4–0.94 μm) |
| Advanced Microwave Sounding Unit (AMSU) | AMSU is a 15-channel gauge to measure: upper-atmosphere temperature; radiation in the range 50 to 60 GHz at frequencies 23.8, 34.4, and 89 GHz; water vapor; and precipitation. Spatial resolution 40, 45 km |
| Humidity Sounder for Brazil (HSB) | HSB is a 4-channel microwave device measuring humidity (183.31 GHz) and radiation (150 GHz). Horizontal resolution 13.5 km |
| Clouds and the Earth's Radiant Energy System (CERES) | CERES is a broad-band 3-channel scanning radiometer. One channel measures reflected solar radiation in the range 0.3–5.0 μm. Two other channels (0.3–100 μm and 8–12 μm) measure reflected and emitted radiant energy at the top of the atmosphere |
| Moderate Resolution Imaging Spectroradiometer (MODIS) | MODIS is a 36-channel scanning radiometer in the visible and IR ranges (0.4–14.5 μm), aimed at obtaining biological and physical information about the atmosphere–land system |
| Advanced Microwave Scanning Radiometer for EOS (AMSR-E) | AMSR-E is a 12-channel scanning passive radiometer to record land cover radiation at frequencies 6.9, 10.7, 18.7, 23.8, 36.5, and 89.0 GHz with regard to horizontal and vertical polarization of the signal. Antenna diameter is 1.6 m, scanning period is 1.5 s |

the use of which, for instance, in the global model can lead to assessment and forecast of the NSS state.

Let us briefly consider the characteristics and capabilities of the Aqua observatory (Parkinson, 2002, 2003). Its AIRS/AMSU/HSB sub-system is aimed at reconstructing the profiles of air temperature and humidity up to 40 km. Horizontal resolution varies between 1.7 and 13.5 km depending on the channel used. Moreover, this sub-system can diagnose solar radiation fluxes in clouds and estimate the precipitation rate. Also, the sub-system gives information on atmospheric minor gas components, land and water surface temperatures, heights of the tropopause and stratosphere, some characteristics of the cloud layer, as well as long-wave and short-wave radiation fluxes. Clearly such sub-system data need respective algorithms to process them. Creation of such algorithms can broaden considerably the significance of information obtained from the sub-system.

The CERES sub-system has been tested on the TRMM (the Tropical Rainfall Measuring Mission) and Terra satellites launched by NASA in 1997 and 1999, respectively. Scanning is made at fixed and rotating azimuth planes. This facilitates assessment, for instance, of the greenhouse effect of water vapor in the upper troposphere and calculation of many important characteristics of clouds and aerosols.

There are many problems in assessing the causes of global climate change that cannot be solved for lack of data about energy processes in the "atmosphere–ocean" and "atmosphere–land" systems. MODIS provides information about the different characteristics of these systems. In particular, for the oceans, measurements are made of primary production, photosynthetically active radiation, organic matter concentration, surface layer temperature, sea ice condition and its albedo, and chlorophyll concentration. For land, pure primary production is determined as well as the type of Earth cover, change in vegetation index, surface temperature and its radiance, and state of snow cover and its albedo. In the atmosphere, measurements are made of the optical thickness and microphysical properties of cloud cover, optical characteristics and size distribution of aerosols, ozone content, total supplies of precipitated water, and the profiles of temperature and water vapor. Spatial resolution constitutes from 250 m to 1 km depending on the problem to be solved.

The MODIS sub-system was first carried by Terra. The Aqua satellite carries a modified version of MODIS–Terra: two lines 11.28 μm and 11.77–12.27 μm have been included increasing the feasibility of environmental temperature measurements.

The AMSR-E sub-system has been developed by the Japanese National Agency for Aeronautics (NASDA) and is a successor to SMMR (Scanning Multichannel Microwave Radiometer), SSM/I (Special Sensor Microwave/Imager), MOS (Microwave Scanning Radiometer), and TMI (TRMM Microwave Imager). Having such a powerful information sub-system on board Aqua will make it possible to solve problems of precipitation rate assessment, calculation of integral water content in the air column over a given territory, measurements of water surface temperature and wind speed over the sea, mapping sea ice solidity and its temperature, mapping soil humidity, and calculation of snow cover depth by estimating its water content. The spatial resolution varies from 5 to 56 km.

All the information flows from Aqua are transmitted to and recorded at both the Alaska Ground Station and the Svalbard Ground Station. Then the data are transmitted to the Goddard Centre for Satellite Data Processing. In parallel, and in almost real time, the AMSR-E data are transmitted to NASDA and the Japanese Data Centre for Fishery, where they are used for numerical weather prediction and assessment of fish supplies in the industrial zones of the World Ocean.

The means of space-borne Earth observation developed in recent decades cover practically all significant spheres of NSS functioning. It should be noted here that these means, unfortunately, have been developed without correlation with global modeling results, which in many cases has led to the information redundancy of satellite systems. Nevertheless, satellite systems do now deliver data on the following parameters:

- interactions in the Sun–Earth system (physical mechanisms that control the motion of matter between biospheric and geospheric formations);
- atmospheric dynamics (physical and chemical processes, meteorological parameters, hydrological fluxes);

- dynamics of the oceans and coastal regions (circulation, wind roughness, water temperature and color, surface layer productivity, photosynthesis, transformation of inorganic matter to organic, phytoplankton, chlorophyll, marine nutrition chains, fishing, structure of ocean populations);
- processes in the lithosphere (parameters of the Earth's rotation, tectonic activity, dynamics of continents and glaciers, soil moisture, topographic structures, resources);
- functioning of the biosphere (characteristics of atmosphere–biosphere interaction, global structure of soil–plant formations, biomass distribution, agricultural structures, forest areas, snow-covered areas, structure of rivers and other water systems, precipitation, food supplies, pollution); and
- climate system dynamics (Earth radiation budget, minor gas components of the atmosphere, greenhouse gases, temperature, cloudiness, long-term climate trends).

Remote-sensing data are the main source of operational information on the systems that control the global ecological, biogeochemical, hydrophysical, epidemiological, geophysical, and even demographic situation on Earth. The remote geoinformation monitoring of objects and processes in the environment currently provides huge volumes of data which can be used to help solve many problems in NSS control. Unfortunately though, the problem of coordinating the requirements of global databases with the structure of satellite monitoring remains unresolved. The development of methods to process remote-sensing data lags behind the progress in technical equipment of satellite monitoring systems. It is hoped that GIMS technology will enable such a lag to be overcome and to balance the algorithmic and technical aspects of remote sensing (Petzold and Goward, 1988; Schmugge, 1990; Werner and Newton, 2005).

The remote-sensing systems of the Earth listed in Table 4.2 are only a small part of a large spectrum of satellite systems with wide-ranging characteristics designed to collect data on the state of various land and sea elements as well as on atmospheric parameters. The Aqua and Terra satellite observatories are EOS elements of such systems and have the purpose of keeping an eye on the planet's health. They measure many of the characteristics of land systems, oceans, and the atmosphere.

Table 4.2 contains information about some systems of space observation. These systems reflect a certain history of development of remote-sensing technology and data-processing. Of course, the technology of sensing in the optical region is the most highly developed. However, in recent years, technologies involved in remote-sensing the microwave spectral region have been intensively developed (Table 4.3). Passive sensing of the environment using microwave radiometers gives effective results in assessing thermal characteristics and determining the structure of land cover. Active radiometry gives information about the physical structure and electric properties of environmental objects. The combined use of instruments in the optical and microwave regions will make it possible to increase the efficiency of remote monitoring systems due to mutually overcoming the drawbacks inherent in each region. There is also the added advantage of using polarization effects in measuring systems

**Table 4.2.** Some environmental observation systems and their equipment.

| System | Characteristic of the system |
|---|---|
| ADEOS | Improved satellite Earth-observing system equipped with an advanced radiometer of visible and near-IR range (AVNIR), ocean color and temperature scanner (OCTS) |
| ERS-1, 2 | European satellites for remote sounding of the environment used within the ESA program |
| EOS | Earth-observing system equipped with clouds and ERB sensors, altimeter, acoustic atmospheric lidar, laser wind meter |
| UARS | NASA satellite launched in 1999 to study the upper atmosphere |
| MOS | Sea-observing satellites of the Japanese National Space Agency equipped with VNIRs and microwave scanning radiometer |
| Envisat | ESA satellite to observe the environment |
| ERBS | NASA satellite to study the Earth radiation budget |
| GMS | Geostationary meteorological satellite |
| GOES | Geostationary satellite to observe the environment |
| GOMS | Russian geostationary satellite to observe the environment |
| JERS | Japan/NASA satellite to study Earth resources |
| Aqua (NASA-EOS) | Equipped with six sensors (see Table 4.1) |
| Landsat 4, 5 | Equipped with thematic mapper in the visible and near-IR regions, launched in 1982 and 1984. Spatial resolution 30 m |
| Landsat 7 | Equipped with advanced thematic mapper, launched by NASA in 1999. Spatial resolution 15 m |
| Spot 1–3 | Satellites of the French Space Agency equipped with multi-spectral and pan-chromatic sensors with resolution 20 m and 10 m, respectively |
| Ikonos | Partially functioning satellite launched in 1999, equipped with multi-spectral and pan-chromatic sensors with resolution 4 m and 1 m, respectively |
| Terra | Launched by NASA in 1999, carries ASTER radiometer with resolution 15 m |

(Chen *et al.*, 2003; Freeman, 2005; Karam *et al.*, 1992; Kirdiashev *et al.*, 1979; Richardson and Le Drew, 2006; Sasaki *et al.*, 1987, 1988; Shutko, 1986).

Synthesizing space-borne systems of environmental monitoring is not only a complicated scientific–technical problem but also is rather expensive. Therefore, a search for effective solutions to this problem requires application of a systematic

**Table 4.3.** Brief characteristics of some remote-sensing devices (Kramer, 1995).

| Device (sensor, radiometer, measuring system) | Used frequencies | Resolution, sensitivity |
|---|---|---|
| • Advanced Airborne Hyperspectral Imaging Spectrometer (AAHIS) | 440–880 nm | 3 nm |
| • Airborne Imaging Spectrometer (AIS) | 900–2,400 nm | 9.3 nm |
| • Compact Airborne Spectrographic Imager (CASI) | 418–926 nm | 2.9 nm |
| • Compact High Resolution Imaging Spectrograph Sensor (CHRISS) | 430–860 nm | 11 nm |
| • Airborne Ocean Color Imager Spectrometer (AOCI) | 443, 520, 550, 670<br>750<br>11,500 nm | 20 nm<br>100 nm<br>2,000 nm |
| • Wide-angle High-Resolution Line-imager (WHiRL) | 595 nm | 20 nm |
| • Airborne Imaging Microwave Radiometer (AIMR) | 37 and 90 GHz | 3 dB |
| • Airborne Multichannel Microwave Radiometer (AMMR) | 10, 18.7, 21, 37, and 92 GHz | 0.5 K |
| • Advanced Microwave Precipitation Radiometer (AMPR) | 10.7, 19.35, 37.1, and 85.5 GHz | 0.2–0.5°C |
| • Airborne Water Substance Radiometer (AWSR) | 23.87, 31.65 GHz | 0.1 K |
| • CRL Radar/Radiometer | 9.86, 34.21 GHz | 0.5 K |
| • Dornier SAR (DO-SAR) | 5.3, 9.6, 35 GHz | 1 m |
| • Electromagnetic Institute Radiometer (EMIRAD) | 5, 17, 34 GHz | 1 K |
| • Electromagnetic Institute SAR (EMISAR) | 5.3 GHz | 2 × 2 m |
| • Electronically Steered Thinned Array Radiometer (ESTAR) | 1.4 GHz | +(3–4)° |
| • Microwave Radiometer "Radius" | 0.7, 1.425, 5.475, and 15.2 GHz | 0.5–1.5 K |

approach to analysis of information fluxes needed for reliable prognostic estimates of the trends of global ecodynamics. This requires, on the one hand, schematic solutions in measurement planning and, on the other hand, development of adaptive technologies of processing monitoring data. A prospective approach to the problem of measurement planning is the radio-occultation method developed by Yakovlev (2001). The idea of this approach consists in the simultaneous use of two satellites, one carrying a signal-emitting system, the other a receiver. Signal propagation along the satellite-to-satellite route leads to a dynamic change in its characteristics, which makes it possible to retrieve atmospheric parameters along this route. Since this route is constantly changing as a result of the motion of satellites, information can be continuously obtained about the influence of atmospheric effects at different heights on the propagation of microwaves in the chosen region, and on the basis of this information, by solving inverse problems, it is possible to estimate density, pressure, temperature, water content, and other parameters of the atmosphere. Most informative is the system when microwaves propagate along the stationary satellite-to-satellite route. For this configuration of the monitoring system, effective theoretical models have been developed of the dependence of refraction attenuation, refraction angle, frequency, and absorption of radio beams by the atmosphere. The transition from the binary structure of the system of microwave propagation through the atmosphere to many satellites will make it possible to optimize the global system of environmental monitoring.

### 4.3.2 General conception of geoinformation monitoring

It is clear from our discussion that the notion of climate means the totality of environmental phenomena necessary for the formation of human habitats; this notion can be expressed in its characteristics, such as temperature, precipitation, humidity, composition and motion of the atmosphere, dynamics of the oceans, etc. The climate system has three dimensions: space, time, and human perception. Each is historically connected with all the processes in the NSS; therefore, development of a global system consisting of models and technical means is urgent for obtaining reliable estimates of the NSS state and especially for prediction of its future trends. A complicated and rather contradictory element of such a system is the global climate model, since its synthesis requires solution of the problem of parametrization of the interaction between NSS components. The complexity of this problem is determined by the multi-dimensionality of the phase space of the climate system. Models already developed divide this space into the atmosphere, land, and ocean. Depending on spatio-temporal division and assumed simplifications and approximations, models of different types appear. This is connected both with the presence of developed conceptual and theoretical approaches and with available empirical information. Nevertheless, the following four types are the most important components of the climate system present in the models: radiation, dynamics, processes at the boundaries, and spatio-temporal resolution. Depending on orientation, the models are divided into energy–balance, radiative–convective, statistical–dynamical, and models of atmospheric general circulation.

As a result of the global character of environmental problems, their solution requires joint efforts from the international community, which has been recently manifested in international and national programs of global change studies and the development of space-borne systems of geoinformation monitoring. Special attention to these problems has been paid in the U.S.A. For example, since 1995 the U.S. National Science Foundation has financed the MMIA (Methods and Models for Integrated Assessment) program, which is studying the role of large-scale technogenic and socio-political processes in the dynamics of physical, biological, and human systems. Since the beginning of the 21st century, the National Science Fund has been supporting more than 20 research programs, connected in various ways with the global problem of environmental change—among them are such programs as GGD (Greenhouse Gas Dynamics), U.S. GLOBEC (U.S. Global Ocean Ecosystems Dynamics), and EROC (Ecological Rates of Change). Special attention is paid to studies of the high-latitude environment and Arctic systems.

Important aspects of studies into global change of the environment include technical equipment of the system of data collection, organization of the way in which they work, and reducing monitoring data to compatible formats. Meteosat-Wefax is one such format (Kramer, 1995). It is used by a system of geostationary satellites (Meteosat-5 and 7, INSAT, FY-2B Wefax, GMS, GOES–West, GOES–East) that provide a flow of 400 images of the Earth's surface per day with spatial resolution from 2.5 to 10 km. Among other widely used formats of image transmission APT, HRPT, and PDUS merit a mention. Polar-orbiting satellites APT give, on average, 12 images of the Earth's surface per day at frequencies of 136.5 MHz (NOAA-12 and 15) and 137.62 MHz (NOAA-14 and 17). The HRPT satellites have a resolution of 1.1 km in five spectral regions (two visible and three IR) at a rate of about 12 images per day. The PDUS satellite is used in the Meteosat series mainly to map Europe at 2.5-km resolution.

This goes to show that the means of remote-sensing have been sufficiently developed to provide the GMNSS with real-time data for prompt assessment of the environmental state and prediction of its key characteristics. However, the role of available data from remote-sensing in studies of global environmental change is confined mainly to revealing and recording partial phenomena without analyzing their global role (Chen *et al.*, 2000; Del Frate *et al.*, 2003; Dong *et al.*, 2003; Levesque and King, 2003; Timoshevskii *et al.*, 2003).

Thus, the rapid accumulation of knowledge, gigantic scientific–technical progress, and global human impact on nature pose the problem of comprehensive assessment of the state of the environment and analysis of possibilities of long-range forecast of the changes taking place. Scientific studies of anthropogenic impacts on the environment have led to the conclusion that solution of the problem of objective control of natural phenomena dynamics and human impact on them is only possible by creating a single system of geoinformation monitoring based on the GMNSS. Within international and national programs on NSS studies and their capacious databases on environmental parameters, information has been accumulated on the dynamics of natural and anthropogenic processes at various scales, and models of biogeochemical, biogeocenotic, climatic, and demographic processes have been developed. The technical base of global geoecoinformation monitoring includes effective means of collection, recording, accumulating, and processing observational data from space-borne, flying, ground, and floating platforms.

However, despite considerable progress in many spheres of environmental monitoring, the main problem remains unsolved: substantiation of an optimal combination of all technical means, creation of an effective and economical structure of monitoring, and preparation of reliable methods for prognostic assessment of environmental dynamics under conditions of anthropogenic forcing. Recent studies testify to the possibility of developing a global model that may be of use (in the adaptive regime of its application) in substantiating recommendations for the structure of monitoring and formation of database requirements (Kondratyev *et al.*, 2002).

Monitoring systems should be based regionally and include a bank of models of biogeochemical, biogeocenotic, and hydrodynamic processes in the biosphere taking into account their spatial heterogeneity. They should have the purpose of combining existing regional databases and explain how the systems of environmental observations work. By cooperating with specialists who already possess developed climatic and socio-economic models, the NSS model is now ready for development. With this model one would be able to assess the completeness and quality of available databases and give recommendations on further development of their structure.

The next stage in modifying the model is to study the interaction of the NSS with processes in the adjacent space and to actively participate in creating a global model that takes the processes in the magnetosphere and the impact of cosmic space into account.

The ultimate aim is to create a system able to predict the development of natural processes and to assess the long-term consequences of different scale impacts on the environment. The sphere of use of this scheme covers various problems of environmental protection on regional and global scales: ecological examination of forecast change in the structure of Earth's cover, hydrological regimes, and atmospheric composition.

Construction of the GMNSS is connected with consideration of the following important problems:

- Systematization of data on regional and global changes and discovering the constituents of biospheric processes and the structure of biospheric levels.
- Development of a conceptual model of the biosphere as a component of the global geoecoinformation monitoring system.
- Inventory and analysis of existing banks of ecological data and selecting the structure they should take.
- Classification of spatio-temporal characteristics and cause-and-effect feedbacks in the biosphere to develop the level of coordination between spatial and temporal scales of ecological processes.
- Creation of banks of models of standard ecological systems, as well as biogeochemical, biogeocenotic, hydrological, and climatic processes.
- Study of the processes involved in the biosphere–climate interaction.
- Search for laws to explain the impact of solar activity on biospheric systems at different spatial scales.
- Construction of models of regional and global biotechnological fields, and hence development of algorithms synthesizing the industrial unit in the global model.
- Formation of a bank of scenarios of the co-evolutionary development of the biosphere and human society.
- Development of demographic models.

- Parametrization of the factors involved in scientific–technical progress in the sphere of biospheric resources.
- Modeling the future for urban building and use of the Earth's resources.
- Search for new information technologies for global modeling that could bring about a reduction in data and knowledge base requirements.
- Linkage of the computer hardware using the principles of evolutionary technology.
- Further development of the concept of ecological monitoring and creation of a theoretical ecoinformatics base.
- Development of methods and criteria of assessment of the stability of regional and global natural processes.
- Analysis of biospheric and climatic structure stability.
- Synthesis of the NSS model and development of the computer means to accomplish computational experiments to assess the consequences of different scenarios of anthropogenic activity.

Analysis of the results of recent studies on these problems shows that new methods need to be developed for successful global modeling and for systematic analysis of natural processes. In addition, methods of data-processing aimed at synthesizing a balanced criteria to be used for information selection and consideration of the hierarchy of cause-and-effect feedbacks in the NSS also need to be developed. The global model should be based on a set of regional scale models in which the region is considered as a natural sub-system interacting through biospheric, climatic, and socio-economic connections with the global NSS. Such a model would describe the interaction and functioning of all processes in the sub-system at different levels of the spatio-temporal hierarchy, would cover both the natural and anthropogenic processes characteristic of a standard region, and would initially be based on the available information. The model structure is aimed at its use being adaptive.

Combining a system of environmental information collection, a model of the way in which a regional geoecosystem functions, a system of computer cartography, and a means of artificial intelligence should result in a geoinformation monitoring system of a standard region (GIMS region) that could be used to:

- assess the impact of global change on the regional environment;
- assess the role of regional environment changes in changes of the climate and biosphere on Earth;
- assess the ecological state of the atmosphere, hydrosphere, and soil–plant formations in the region;
- form and renew information banks about ecological, climatic, demographic, and economic parameters of the region;
- promptly map the landscape;
- forecast the ecological consequences of the effect of anthropogenic forcings on the environment in the region and adjacent territories;
- typify the Earth's land and sea cover, natural disasters, populated areas, surface contamination of landscapes, hydrological systems, and forested areas; and
- assess the level of population safety in a given region.

Substantiation of a GIMS region means selecting the components of the biosphere, climate, and social environment that are characteristic of a given level of the spatial hierarchy. The sequence of actions needed to develop a GIMS region project is based on experience of GIS development and construction of numerical models of the functioning of large natural–economic systems. This experience determines the expedience of the following structures for a GIMS region:

- sub-system of data collection and express-analysis;
- sub-system of primary processing and accumulation of data;
- sub-system of computer mapping;
- sub-system of assessing the state of the atmosphere;
- sub-system of assessing the state of soil–plant cover;
- sub-system of assessing the level of ecological safety and risk for human health in the region;
- sub-system of identifying the causes of damage to ecological and currently unpolluted situations; and
- sub-system of intellectual support.

The unification and specialization are characteristic features of the evolutionary software as a consequence of the minimum of a priori information, effective mechanisms of adaptation, and the modular principle of realization. With an orientation toward up-to-date personal computer engineering and a diverse range of application problems, these specific features have made it possible to work out evolutionary computation technologies in which, with the aid of active dialogue with the user, a set of software modules is realized as well as the adjustment of the evolutionary facility to the specificity of the problem being solved. Regardless of the field of application, the evolutionary technology or artificial neural networks with software support is characterized by adaptability, flexibility, dynamism, and self-correction, while the main distinction consists in the high effectiveness and adjustability under the conditions of the maximum informative uncertainty, including that of an irremovable character. Actual evolutionary modeling technology gives a new structure of GMNSS.

Analysis of the state of databases and the way in which they are updated shows that current monitoring systems have no self-contained network. Based on the experience of using aerospace monitoring centers in many countries, it is suggested that a standard system of autonomous multi-purpose and specialized ground and aircraft centers be employed to assess the current state of the elements of natural and anthropogenic landscapes, prediction of ecological and unpolluted situations of populated areas, environmental diagnostics, anomalous phenomena, natural disasters, and technogenic catastrophes. A network of such centers coordinated with the NSS would be able to resolve the following problems:

- thematic mapping the results of ground and aerospace measurements and surveys made using contact and remote methods in the optical, IR, and microwave regions;

- current estimates and forecast of ecological and unpolluted states of dwelling areas; industrial and fishing zones; sources of drinking water and rivers; agricultural, forest, and hunting areas; places of accumulation of garbage and industrial waste; accumulation of mineralized and contaminated water as potential sources of extreme situations;
- territorial–seasonal inventory typifying quasi-homogeneous and anomalous sites and the sites of floristic background and soils, water bodies, and the atmosphere; geophysical fields of different nature and mapping their local parameters; identification of the sources of Earth cover, air, and water basin pollution; their impact on populated localities and environmental elements;
- operational examination and highly accurate fixing of coordinates on small sites and objects that have non-ordinary reflective and radiative characteristics; detection of anomalous phenomena, natural disasters; dynamics of the consequences of such events in populated localities; and environmental parameters;
- identifying targets for search-and-rescue emergencies on land and at sea; collection of expensive metal remains of rockets; shipwrecks in rivers and seas; assessment of fire risk to taiga tracts and peat bogs; oil and gas extracting, mining, and other enterprises; identification of sites with an increased content of methane in the surface atmosphere that are potential sources of emergency situation at hydrocarbon pipelines; and
- combining the results of aircraft observations with space-derived images and ground measurements to retrieve an objective pattern of ecological and currently unpolluted situations and to substantiate ways of rationally using the Earth's resources, agricultural, forest and industrial potentials, water basins and rivers, and transport and irrigation systems.

Small, medium-sized, and unmanned flying vehicles are used as autonomous aircraft centers. They are equipped with radar, radiometric, TV, laser, and photographic means of observation as well as with systems of referencing observation results to characteristic points of the surface and local coordinates.

At present, it is possible to make a single system from regional models of the most important natural processes, including processes in the magnetosphere. Available knowledge and accumulated data make it possible to realize the local and regional levels of the spatial hierarchy. Use of such a system will enable one to formulate the database requirements needed to provide a more detailed spatial division of natural systems and processes, even down to consideration of the role of individual elements of ecosystems.

An ecological monitoring system has to include three basic directions:

(1) *Ecosociology*—study of the human impact on nature and interactions in the NSS.
(2) *Ecoinformatics*—recording, storage, transmission, analysis, synthesis, modeling, and presentation of information about the environmental state.
(3) *Ecotechnology*—creation of technologies and technical means that have a minimal impact on the environment and humans.

Ecosociology entails:

- analysis of social structures affecting the environment, their motives, goals, methods, and technical means;
- analysis of the necessity and adequacy of anthropogenic impacts of a social structure on the environment;
- analysis of optimal allocation of an industrial and agrarian infrastructure and habitation in the environment; and
- working out eco-protection measures that take into account socio-regional, socio-national, and socio-global interests.

Ecoinformatics entails obtaining diagnostic, prognostic, and practical information about the state of the environment and any change in it as a result of anthropogenic and natural forcing. Such information should reflect the following:

- quantitative estimation of the state of the environment and natural resources;
- quantitative estimation of anthropogenic impacts and detection of harmful and excessive production;
- current state and maintenance of regional (in some cases, district) simulation models of interactions in the "production–environment", "agriculture–environment", and "dwelling–environment" systems;
- current state and maintenance of an ecomodel of the region;
- current state used by such an ecomodel in national and international cooperation;
- current state that makes it possible, based on a simulation ecomodel, to construct prognostic schemes of the impact of harmful and excessive production on the regional environment;
- current state used in prognostic modeling of environmental change in the development of resources; and
- current state used, on the basis of simulation and prognostic models, when working out recommendations on either development or reduction of production and house-building in the region.

An ecoinformation monitoring system is a three-level scheme based as it is on satellite, ground, and aircraft observations. An enlarged functional scheme of the system consists of the following basic components:

- complex observation system (data collection) from space, aircraft, and ground facilitating accomplishment of regular *in situ* and remote measurements of environmental parameters;
- sub-system of data-receiving/transmitting from space and ground observation means using satellite, radar, wire, and fiber-optic communication channels;
- sub-system of primary data-processing from different observation systems;
- sub-system of data accumulation and systematization (databases); and
- group of sub-systems to solve applied (thematic) problems of monitoring.

Task-oriented and information–technical systems of satellite observations refer to the following classes:

(a) Systems of non-operational detailed observations from space-borne platforms using data from multi-zonal and topographic surveys.
(b) Systems of operational observations which are divided into sub-classes:
    (i) *systems of planned surveys* of problems in the global operative control of the state of the atmosphere and the Earth's surface at a small spatial resolution of continuous (quasi-continuous) observations using space-borne platforms on geostationary orbits and regular observations using space vehicles in either highly elliptic or circular near-polar orbits; and
    (ii) *systems of detailed observations* which ensure data for the chosen objects or regions on the Earth's surface are obtained at a high or medium spatial resolution of continuous (in highly elliptic orbits) and regular (in medium to high circular orbits) functioning.

At present, for most mid-orbital space vehicle projects, both sub-problems have been resolved.

For specific aircraft sensing of regions with a complicated ecological situation, an aircraft scanner is needed that is able to carry experimental instrumentation and function both when piloted and unmanned.

A regional ecomodel and its operational support require a highly productive information-calculating network that is based on a distributed network of the current means of computer techniques with the respective software.

The regional ecological monitoring system should become the nub of ecological examination of the problem of territorial control. It will provide every part of the region with access to information about the environmental state by means of available sites or pages at some global network (e.g., the Internet). Based on this, one should be able to make competent decisions on many vital problems in both social and personal spheres.

Ecotechnology entails:

(c) Assessment of real and potential danger as a result of manufacture or production of goods connected with emissions of substances harmful both for the environment and for humans.
(d) Assessment of the technological processes and detection of the stages of these processes at which these substances are emitted.
(e) Correction of technological processes under the preservation of production quality and reduction of harmful impact on the environment and humans.
(f) Development of instrumentation for corrected technological processes.
(g) Instruction of personnel when working under new conditions and with new machines.
(h) Constant control of technological discipline observance.
(i) Qualified examination of technological processes and respective instrumentation.

For the manufacture or production of goods to have a minimal impact on the environment and human organism depends on the use of highly intellectual computer technologies that are capable of controlling all processes of production and accomplishing optimal control of technological equipment.

All these directions of ecological activity can be concentrated in a single organizational structure: a regional center of ecological monitoring which is an element of the global GIMS.

Such a regional center of ecological monitoring should consist of three structural units corresponding to the basic directions of ecological activity: ecosociology, ecoinformatics, and ecotechnology. Each of them, in correspondence with its goals, can be divided into several sections:

(1) Ecosociology:
- analysis and prediction in the nature–society system;
- nature protection measures; and
- decision-making and interaction with the powers that be.

(2) Ecoinformatics:
- reception and analysis of space-derived information;
- aero-survey by manned and unmanned aircraft;
- ground observations using a network of sensors and ground stations;
- control of human health;
- modeling and maintenance of the regional ecomodel; and
- information-calculating center.

(3) Ecotechnology:
- ecoanalysis of industrial enterprises;
- ecoanalysis of coal-, gas-, and oil-extracting industry;
- control of technology observance; and
- examination of technological and engineering solutions.

### 4.3.3 Evolutionary technology of modeling

Ecoinformatics stimulates broadening the role of computing facilities and numerical modeling in non-traditional spheres of their application, such as ecology, biophysics, and medicine. Model experiments have become traditional, and the term "numerical experiment" is now used in many serious studies, even experiments with global ecological models. However, in all these studies the presence of a more or less adequate model is assumed a priori. For concrete use of the model as a series of numerical experiments a universal computer is needed. As a rule, insurmountable difficulties appear here connected with limited memory and speed. As a result, an investigator is constantly faced with the insoluble contradiction between the desire to increase the accuracy of the model and computer capability. It is clear that development of a model that fully represents the real object is ultimately impossible. On the one hand, complete consideration of all parameters of the object leads to the problem of multidimensionality on a massive scale. On the other hand, simple models taking into account a small number of parameters are just not good enough to represent complicated objects and processes of interest to the investigator. Moreover, in such spheres as physics of the ocean, geophysics, ecology, medicine, sociology, etc., creation of a satisfactory model is principally impossible due to the practical impossibility of obtaining complete information. It is also impossible to determine the type of model, and hence the use of currently developed modeling technologies turns out to be unpromising. In particular, attempts at a complex approach to synthesizing global models have shown that traditional methods of formalizing processes in the environment and society cannot give reliable estimates and forecasts of global change. Nevertheless, a search for new effective ways of synthesizing the global system of control of the environmental state capable of giving reliable estimates of the consequences of anthropogenic activity has led to evolutionary-type systems (Bukatova and Makrusev, 2004). As a result, a new theory of modeling called evolutionary informatics has appeared (Bukatova and Rogozhnikov, 2002; Krapivin *et al.*, 1997).

We shall use the term "model", though its interpretation does have other meanings. We shall deal with the description of objects unpredictably changing in time that currently make it impossible to avoid information uncertainty at any juncture. Hence, the model, in a broad sense, should provide continuous adaptation to the varying behavior and structure of the observed object.

So, let a real object $A$ have an unknown algorithm of the way in which it functions, and only some prehistory $\{Y\}^n$ of final length $n$. It is necessary to simulate the functioning of $A$ with the help of models created in a system operating in real time. It is clear that a succession of increasingly adequate models $\{A_k\}$ should be formed whose reliability is tested by prehistory. The studies of many authors have resulted in a new approach to creating learning systems based on modeling the mechanisms of evolution. Evolutionary modeling, on the whole, can be presented as a hierarchic two-level procedure. At the first level, there are two alternating processes called conditionally the process of structural adaptation and the process of application. At the $k$th stage of adaptation, in the process of functioning of the structural adaptation algorithm, a succession of models $\{A_{s,i}\}$, $i = 1, \ldots, M_s$ is synthesized, a memory of which is formed from $K$ of the most effective models $\langle A_s^1, \ldots, A_s^k \rangle$. At the $s$th stage of application (following the $k$th stage of adaptation) the system uses models stored in memory, continuing to choose the most effective ones.

The principal scheme of the $i$th step of the $k$th stage of model adaptation means that a real object is presented by prehistory $\{Y\}^n$, and on this basis a structural adaptive synthesis of models (structural adaptation) is accomplished. The $i$th stage starts by choosing a "mutant" $A_{ki}^j$, which is either $A_{ki}^j$ or the model used already at the previous $(i-1)$th step of the $k$th stage of adaptation. From the structure $|A_{ki}^j|$, with the help of the regime of random changes, the structure $|A_{ki}|$ is synthesized. The model $A_{ki}$ is introduced into the unit of evolutionary selection where, in accordance with the criterion $F$, it is either memorized or forgotten.

The procedure involved in the evolutionary selection of models provides practically unlimited time for the system to function under conditions of unavoidable information uncertainty. As a rule, apart from prehistory $\{Y\}^n$ not meeting the requirements of statistical homogeneity, in this case one has no other information.

It is clear that under these conditions one has to maximize the use of all available information—in particular, information about the functioning of the stages of adaptation and use. The algorithm of parametric adaptation is used to adapt first-level parameters: characteristic number of each change mode, multiplicity of the use of change mode, size of the list of change modes, probabilistic distribution of the list of change modes, probabilistic distribution of $K$-models $A_k^j$, memory volume at the stage of adaptation, length of prehistory, etc. (Bukatova and Rogozhnikov, 2002; Bukatova and Makrusev, 2004).

One possible way of increasing the efficiency of evolutionary software tools is to develop specially designed processors for evolutionary systems of information processing. These would include an enlarged functional basis, large-unit (multiprocessor) construction with a homogenous elemental base, distributed intelligence at the system level, processors, elements, variability of functioning and structure (composition and connections) of synthesized models, hierarchical distribution of all levels of control.

Analysis of the algorithmic aspects of specially designed evolutionary processors makes it possible to formulate some basic requirements and the problems likely to be met and resolved when developing the evolutionary hardware. Algorithm-type problems appear here that are inherent at the system, processor (unit), and elemental levels of realization: maximum parallelizing, enlarged primitive operations, provision of maximum tenability, formation of distributed memory and control, selection of models, synchronization and control, input–output of information.

Maximum parallelizing can be accomplished by setting up parallel hardware for computational processing including information exchange between individual processors, accomplishment of primitive operations by micro-processors (operations on structural changes and assessment of synthesized models), etc. For instance, for hardware implementations of perceptron-type systems, parallelizing increases the speed of the system by two to three orders of magnitude, and making operations in basic blocks concurrent adds to this a further one to two orders. Since parallelizing the functioning of synthesized models in each micro-processor increases the amount of time saved, such time reserves can be built up.

The problem of the synthesis of a network with a variable structure arises with the creation of special hardware with a flexible structure based on the new principles of information processing. Recent advances in microelectronics can help to solve this problem (Elinson and Sukhanov, 1984). Computer hardware of evolutionary type is based on the mobile structure of microprocessors managed by the control hardware synthesized from optical super large integral schemes.

Maximum tenability provides adaptability to changes in input signal characteristics, dynamics of processing conditions (change in multi-processing), a priori prescribed information, and other factors involved in the evolutionary computational process. In full measure, this tenability is provided by current correction of parameters, transition to other computational procedures, and varying the composition and structure of connections between basic elements. All this increases the probability of achieving the necessary state for a specially designed processor and ensures its reliability.

Parallelizing the evolutionary systems of processing at different levels requires distribution at the levels of accumulation and storage of information then used for control and management of respective processes, procedures, and primitive operations. In this way, a substantial shortcoming is overcome in the successive (von Neumann) architecture of traditional computing devices in which processing and storage of information are separated, which leads to loss of time in transmission of data.

In general, the multi-level structure of distributed memory is hierarchical, when accumulated information about the functioning of elements, for instance, is used to assess the functioning of a specially designed processor.

The control procedure in traditional computing devices is centralized and carried out in a successive mode that is time-consuming, especially in multi-processor systems. But level-distributed control can substantially increase the system's speed due to parallelizing the processes of correction (learning). It is clear that distributed memory and control interact, and their hardware implementations follow the same principles. In particular, distributed memory makes it possible to individually carry out integral corrections at any level of the system.

The synchronization and control of primitive operations and procedures within a set of processors as well as standard computational processes between the totalities of processes and other units of a specially designed processor is an algorithmic and circuit-technical problem. As part of multi-processor realization of an evolutionary computational process, asynchronous interaction of parallel fragments of the algorithm is expedient and, in general, an asynchronous circuit technique is necessary. Systolic (pulsatory) processors can be analogous to this approach (Elinson and Sukhanov, 1984). One means of hardware implementation of information input–output is the use of optical integrated circuits.

## 4.4 COORDINATION OF THE GLOBAL MODEL WITH DATABASES

Development of a geoinformation monitoring system needs resolution of some problems in the organization of measurement data flow. An important problem is synthesis and control by those databases responsible for the reliability of prognosis of the state of the natural object under study. Solution of this problem was given top priority in several national and international programs on study of the environment (Kondratyev, 1999a, 2000).

Collection and accumulation of data in the monitoring regime needs regular processing for the use of GIMS technology to be meaningful. In this case, it is necessary—on the basis of noise-loaded information that is fragmentary in time and space—to form spatial images of natural objects. In addition to these operations, it is necessary to assess the model parameters and, on the basis of assessment of discrepancies between measured and prognostic states of the investigated system, to either correct the models or change the monitoring regime. The existing systems of database control do not resolve the problem of adaptation of environmental models to the database.

Problems in the sphere of environmental studies have been determined by many international projects investigating the functioning of natural and anthropogenic systems. The IGBP was set up to study the interaction of these systems and create a hierarchical NSS model (Kondratyev, 1999a). Experience from this identified the data series necessary to describe the global system:

- *Geological data* (ebb and flow of tides, geochemical processes, geographic and geomagnetic surveys, gravity data, hydrological processes).
- *Marine system data* (sea surface temperature, World Ocean geophysics, seawater salinity, wave processes in the World Ocean, living biomass and dead organic matter in the World Ocean).
- *Ecological data* (species of living organisms, soil–plant formation over land, ecological zoning of the Earth's surface, phytomass and total organic matter, microbiological processes in the biosphere).
- *Hydrological data* (precipitation, evapotranspiration, soil moisture, river run-off, hydrological network over land).
- *Atmospheric data* (temperature and atmospheric pressure, chemistry of the atmosphere, wind speed, cloudiness, air humidity).
- *Associated data* (political boundaries, coastline contours, topographic maps, socio-economic structures).

One of the most important problems in dealing with databases is the lack of correlation between their structure and models. To meet this requirement, the database structure should have two special features. First, the presentation of data in models should be separated from their physical presentation—that is, transformation of one presentation into another should be carried out by data control software tools. This separation is called the physical independence of data and models. Second, the presentation of data in a partial model should be protected from changes in the global logical structure and from changes in the data requirements of other models. To achieve this independence, it is necessary to separate the presentation of data in each model from the total logical presentation and to provide the possibility of adding new fragments of data without modifying the model that uses them. This separation is called the logical independence of data. It can be realized by data standardization and introduction of a multitude of semantic structure identifiers $\{\Xi_i\}$, which establish a single correspondence between the model and the requested information field. Standardization can be accomplished by determining the level of data readiness:

(1) Data of direct measurements representing phenomena at the moment of measurement.
(2) Processed data corrected for exclusion of possible distortions in the process of measurement.
(3) Estimates of parameters of the model of the process under study obtained using corrected data.
(4) Cartographic presentation of data referred to a concrete spatial scale.

The identifier indicates the path from one level of data readiness to another through interfaces with models. Interface paths are determined by the type of model. For instance, for models of atmospheric Gauss-type pollution, the database should contain information about the type of source, landscape topography, height of the source, wind speed and direction, etc. However, depending on the technology parametrizing the process of pollutant scattering in the atmosphere, ramifications can occur in the database. For example, if it is connected with classification of synoptic situations, the choice of model is simplified. Otherwise, a standard version of the model is used. In other words, a multitude of identifiers should connect the bank of models and the database in a universal way, permitting the models to be modified without other changes. For example, the identifier of soil–plant formations is used when choosing biogeochemical models. This identifier is shown in Table 4.4. Using this identifier one can easily describe the global distribution of soil–plant formations by a 2-D matrix $\Xi_1 = \|a^1_{ij}\|$, whose elements $a^1_{IJ}$ correspond to pixels $\Omega_{ij}$ of geographical digitization of the Earth's surface. Equating $a^1_{IJ}$ with one of the symbols in the right-hand column of Table 4.4 makes it possible to establish a single correspondence between database and the type of biogeocenotic unit of the global model that should function in territory $\Omega_{ij}$. As a result, comparing the water surface in the territory with the point symbol gives the shape of $\Xi_1$ shown in Figure 4.5.

Identifying database fragments by their levels of readiness and bases of knowledge ensures the independence of the model structure and the structures of databases. Their symbolic coordination creates a new, fifth level of database, with the structure of every NSS element in pixel $\Omega_{ij}$ taking the form of a set of catalogs schematically shown in Figure 4.6. As a result, the global model, based on this interface with other databases, provides the user with direct access to the NSS structures needed for study. The unit of database control analyzes the user's request to choose spatial scales $\Delta\varphi$ and $\Delta\lambda$ and, based on some basic scales (Table 4.5), adapts the whole simulation system to the user's scenario using algorithms of spatio-temporal interpolation. The user can then study a confined space using capabilities of the GMNSS. In this case, the user answers the system's request and forms the structure of the studied space by correcting the content of respective thematic identifiers. The most important are built in to the control unit and, depending on the actions of the user of the global model, necessary corrections to the symbolic presentation of elements of the process under study are made.

Depending on the level of data readiness, a hierarchy of the problems under study can be considered. The first level is used to reveal the local features of natural processes. The second level provides the input information for data-processing systems. The third level makes it possible to form spatial images. The fourth level creates the conditions for introduction of GIS technology and organization of monitoring systems with forecast capabilities. The fifth level ensures the use of GIMS technology.

These levels of data readiness need to become part of the respective structure of databases. The fourth level of data readiness is the largest and is adjusted for use in geoecosystem numerical models. Sub-levels can appear at the fourth level to meet the requirements of a given type of model. For instance, evolutionary modeling is

**Table 4.4.** Identifier of the types of soil–plant formations following the classification of Bazilevich and Rodin (1967).

| Type of soil–plant formation | Identifier |
|---|---|
| Arctic deserts and tundras | *A* |
| Alpine deserts | *B* |
| Tundras | *C* |
| Mid-taiga forests | *D* |
| Pampas and grass savannahs | *E* |
| Northern taiga forests | *F* |
| Southern taiga forests | *F* |
| Sub-tropical deserts | *G* |
| Sub-tropical and tropical grass–tree thickets of the tugai type | *I* |
| Tropical savannahs | *J* |
| Salt marshes | *K* |
| Forest tundra | *L* |
| Mountain tundra | *M* |
| Tropical xerophytic open woodlands | *N* |
| Aspen–birch sub-taiga forests | *O* |
| Sub-tropical broad-leaved and coniferous forests | *P* |
| Alpine and sub-alpine meadows | *Q* |
| Broad-leaved coniferous forests | *R* |
| Sub-boreal and saltwort deserts | *S* |
| Tropical deserts | *T* |
| Xerophytic open woodlands and shrubs | *U* |
| Dry steppes | *V* |
| Moderately arid and arid (including mountain) steppes | *W* |
| Forest–steppes (meadow steppes) | *X* |
| Variably humid deciduous tropical forests | *Y* |
| Humid evergreen tropical forests | *Z* |
| Broad-leaved forests | + |
| Sub-tropical semi-deserts | & |
| Sub-boreal and wormwood deserts | @ |
| Mangrove forests | # |
| Lack of vegetation | * |

characterized by the presence of prehistory in the form of fragments about the real state of the phenomenon under study. As a result, a structure should be formed at the fourth level with data accumulated from another level over a certain time period. Evolutionary modeling removes restrictions on the time invariance of data series that are usually imposed by other methods of modeling.

The prevailing types of models using bases of knowledge from the data domain of geophysics and ecology are oriented toward the third level of data readiness. At this level the database should contain estimates of model coefficients as functions of space coordinates, time, and types of objects under study.

```
  W    150   120    90    60    30    0     30    60    90   120   150    E
N ....................****. **********........................................ N
  ....................AAA..**********......**...............A................
  .............AAA. AAAA.....*********..............C......AAAAA.....A........
  ....CCCC.....CCA. A. AAAC...*****.........LLL.....C. CCCCCMMMMCAAAAAAAA.....
  MC. FFFMFMFFFFCCCCCAA..MM.***...L......DDDFF. FFFFCLLLFFFMFFFFFMAAAMMMMMMM
  ....LLFFFFCCCCCCCC...C.....L..........RG. DDDDDDDDDFFFFDDDDDDDDDMMFFMCCM.
  .....L...FFFFGGGFF...CFC...........+...R. RGGGGGGGDGGGGGGGGGDDDDDDM..L....
  ...L......+RGGGGGDDF. FFFF..........++. ++RRRRXXXXGXXXXGGGGGGDDGDDD...L....
50...........RRXWXXGGDDDDD. D...........+++++XXWWVVVVVVVGGGGGGGGGDDD........50
  ............RXWWX++. +++G............+++++XXWWV. S. @@QQSGVVVVGXRR..........
  ............BBBXXX++++..............++..P++..WQ. KS&QQQSSSSSSXX++..+.......
  ............RBUWXP+PP.............PP....P. WWWW&K&&QQSSSSSQXX. +..P........
  .............@UUUPPPP..............&&&&.....&&WUUUUUQBBQQ+++...P.........
30.............@@@UP..P.............HHHHHHHHHHHHHHHTHHHJPQQQQPPPP............30
  ..............@@@.................HHHHHHHHH. HH. TTTHNNYZZPPPP.............
  ...............U@...J............TTTTTTTTTTTTTTT...NNI. YYYI..............
  ................YYZ...J..........TTTTTTTTTTT. TT....JN..YYY...............
  ..................YZ.............JJJJJJTJJJNTT.....NN...YY..Y............
10....................Z. JJ..........JJJJJJJJINNTT.....N....Y...............10
  ....................ZJJZ..........YJZYJJJJJJT..........Z................
  ....................ZZZZJ............ZYYYYJNN..........Z..Z............
  ...................QZZZZJZZ..........YYZYJJJ............Z.Z............
  ...................WZZZZJJJJJ.........JYYNNJ...................ZZ.......
10....................ZZZZJJJJ..........JNNNN......................Z......10
  ....................WWZZZJJY..........JNNNN..Z...............NN........
  .....................WWNZYJY..........JJJNJ. J...............NNNNN.......
  .....................TNNYYY...........TJJJJ. N............JJTTJJJ......
  .....................TNNYP............HHJE...............JJTTTTNN.....
30......................UUEY..............HEE................HHTTTJJ......30
  .....................UUEI..............E..................U...UUP......
  .....................UEE......................................P.
  ....................+VV.................................P....+..
  ....................+V..........................................
50....................+E.........................................50
  .....................+...........................................
S ................................................................ S
  W    150   120    90    60    30    0     30    60    90   120   150    E
```

**Figure 4.5.** Representation of the types of soil–plant formations by pixels in the GMNSS spatial structure.

**Table 4.5.** Scales of presentation of cartographic information characteristic of developed monitoring systems.

| Scale | 1:1250 | 1:2500 | 1:100,001 | 1:25,000 | 1:50,000 | 1:250,000 | 1:625,000 |
|---|---|---|---|---|---|---|---|
| Spatial resolution (km) | $0.5 \times 0.5$ | $1 \times 1$ | $5 \times 5$ | $10 \times 20$ | $40 \times 40$ | $250 \times 200$ | $500 \times 500$ |

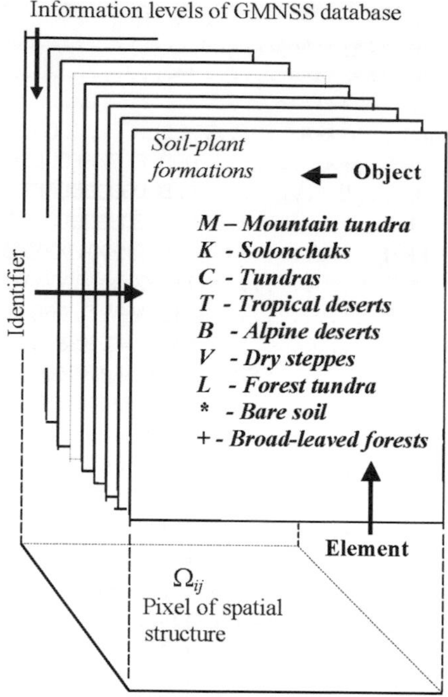


**Figure 4.6.** Information levels of the global database and their cartographic identification in the GMNSS.

## 4.5 THE GIMS-BASED EFFECT IN MONITORING REGIME CONTROL

There are many parameters describing the environmental conditions on Earth. Among them are soil moisture and moisture-related parameters such as the depth of a shallow water table and contours of wetlands and marshy areas. Knowledge about these parameters and conditions is very important for agricultural needs, water management, and land reclamation, for measuring and forecasting trends in regional to global hydrological regimes, and for obtaining reliable information about water conservation estimates.

In principle, such information may be obtained by using on-site measurements and remote-sensing and by getting access to a priori knowledge-based data in GIS databases. But the problem that arises here consists of solving the following:

- What kind of instruments are to be used for conducting so-called ground-truth and remote-sensing measurements?
- What is the cost to be paid for on-site and remote-sensing information?
- What kind of balance should there be between the information content of on-site and remote-sensing and the cost of these types of observations?
- What kind of mathematical models can be used both for the interpolation of data

and their extrapolation in terms of time and space with the goals of reducing the frequency and thus the cost of observations and increasing the reliability of forecasting the environmental behavior of the observed items?

These and other problems can be solved by using a monitoring system based on combining the functions of environmental data acquisition, control of data archives, data analysis, and forecasting the characteristics of the most important processes in the environment. In other words, this unification forms the new information technology called GIMS technology. The term "GeoInformational Monitoring System (GIMS)" is used as a parameter in the formula: $GIMS = GIS + Model$. The relationship between the GIMS and GIS is shown in Figure 4.7. Evidently, the GIMS is a superset of the systems shown in Figure 4.8. There are two views of the GIMS. In the first view the term "GIMS" is synonymous with "GIS". In the second view the definition of the GIMS expands on the GIS.

GIMS technology allows construction of the optimal structure for a monitoring procedure directed at the assessment and prediction of environmental sub-systems.

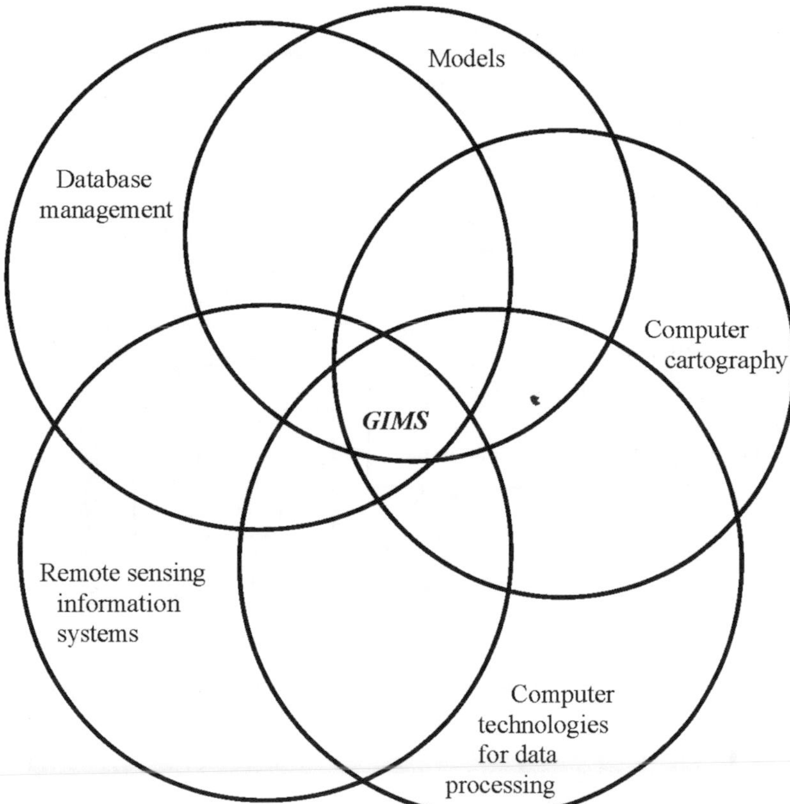


**Figure 4.7.** GIMS technology as the intersection of many scientific disciplines.

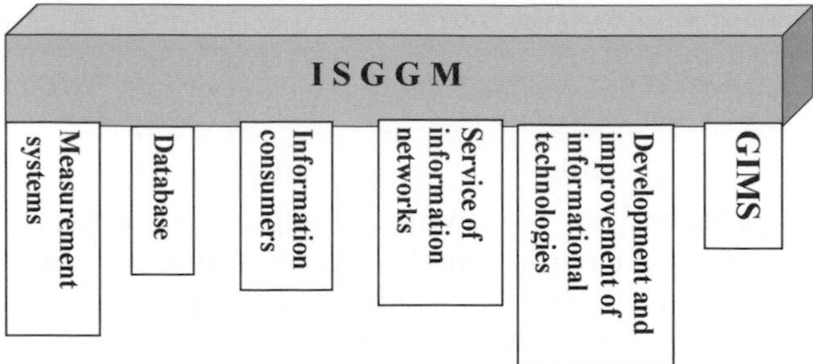


**Figure 4.8.** Integrated System for Global Geoinformation Monitoring and its structure.

This can be realized by means of an adaptive process that gives the possibility of combining the model structure and observations. Schematically, this process is represented by Figure 4.9. In reality, a search for the optimal combination between model complexity and observations can only be achieved by synthesizing models that have different structures and parametrical bases and by describing the various schemes for the realization of observations. To demonstrate the effectiveness of this process Table 4.6 contains the comparative results characterizing possible consequences when

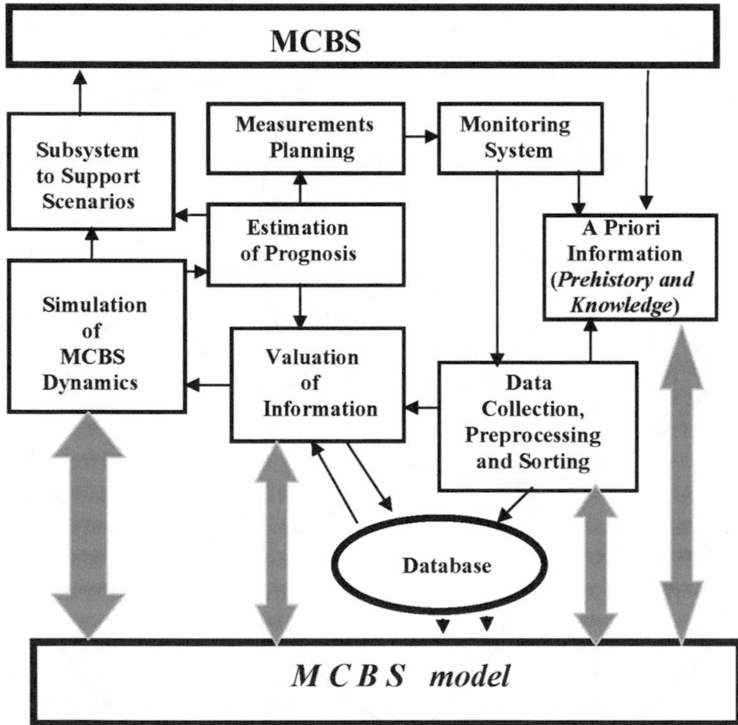


**Figure 4.9.** Conceptual diagram of using the MCBS model in the adaptive regime of geoecoinformation monitoring.

**Table 4.6.** Comparison of on-site measurements and results of simulation experiments using different model modifications related to the vertical distribution of Aral Sea water temperature. Designation: SMAHF = Simulation Model for the Aral Hydrophysical Fields, SCT = SMAHF block simulating the process of convective mixing in seawater, MWD = SMAHF block describing water density, and MSF = SMAHF block parametrizing current flows in the Aral Sea.

| Depth (m) | Real water temperature (September 2005) (°C) | Results of simulation experiments | | | |
|---|---|---|---|---|---|
| | | SMAHF | SMAHF without SCT | SMAHF without SCT and MWD | SMAHF without SCT, MWD, and MSF |
| 0 | 18.3 | 18.4 | 19.0 | 19.2 | 19.5 |
| 5 | 18.3 | 18.4 | 18.7 | 18.8 | 19.1 |
| 10 | 18.1 | 18.0 | 18.1 | 18.3 | 19.0 |
| 15 | 18.0 | 18.0 | 17.4 | 16.9 | 17.3 |
| Average error (%) | | 0.55 | 3.12 | 3.72 | 4.95 |

different models are used. In this case the Simulation Model for the Aral Hydrophysical Fields (SMAHF) is used to reconstruct the spatial distribution of water temperature (Bortnik *et al.*, 1994). The results of the simulation experiment given in the third column of Table 4.6 deviate minimally from observed values. The fourth column contains temperatures evaluated by removing from SMAHF the simulation process of convective mixing in seawater. The modeling results given in the fifth column correspond to the SMAHF version, but excludes the description of water density. Finally, the results in the sixth column show the importance of taking current flows in the sea into consideration. This example shows the importance of model description of hydrophysical and hydrological processes taking into account detailed parametrization of basic components that play significant roles in water regime formation. It is clear that the main problem in synthesizing simulation models to describe hydrophysical and hydrological processes is in achieving a balance between the modeling procedures influencing the seawater regime. Nevertheless, even in this case GIMS technology guarantees reliable results.

Let us consider another example in which environmental parameters change slowly over time. The task is to reconstruct the soil moisture distribution in a restricted area by combined use of model and remote observations. Figure 4.10 shows the monitoring scene when soil moisture is measured using L-range microwave radiometers on board a flying laboratory. When the model of regional water balance described by Kondratyev *et al.* (2004) is used, GIMS technology allows the optimal distance between parallel flying routes to be calculated. It is seen that such precision for soil moisture mapping depends on the distance between the routes taken by the aircraft. The curves in Figure 4.11 allow the monitoring regime to be optimized to give such precision. Such a procedure can be used in any region where operative

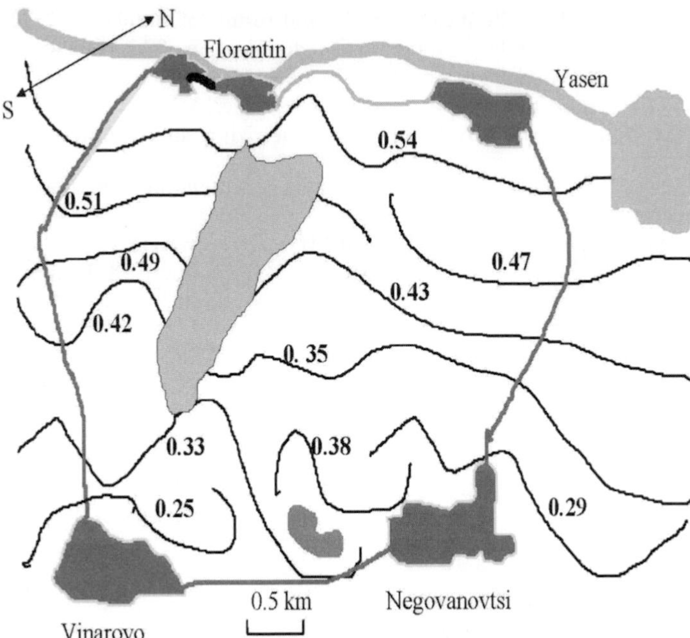


**Figure 4.10.** GIMS technology application to the task of microwave-mapping the soil moisture in Bulgaria at the beginning of May. The moisture values in $\mathrm{g\,cm^{-3}}$ are shown on the curves. Designations: 1 grassland, 2 reservoir. Direction of movement of an in-flight laboratory is shown by the arrows.

knowledge of soil moisture content is needed to take a decision about the regime of irrigation system functioning.

## 4.6 THE GIMS-BASED METHOD OF VEGETATIVE MICROWAVE MONITORING

### 4.6.1 Introduction

Solution of the majority of applied problems within agrometeorology, forestry, animal husbandry, and other areas of human activity regarding the protection of nature is difficult for the reason that effective methods of control of the *soil–plant formation* (SPF) system are insufficiently developed. In recent years, the global carbon cycle problem has acquired special significance because of the greenhouse effect. Knowledge of the state of the SPF gives a full rpresentation of the spatial distribution of carbon sinks and sources on the Earth's surface.

As is well known, of the remote-sensing techniques the most effective for observations of SPF environmental parameters is microwave radiometry. However, these observations are a function of different environmental conditions that mainly depend on SPF type. This is the reason it is necessary to develop data-processing methods for

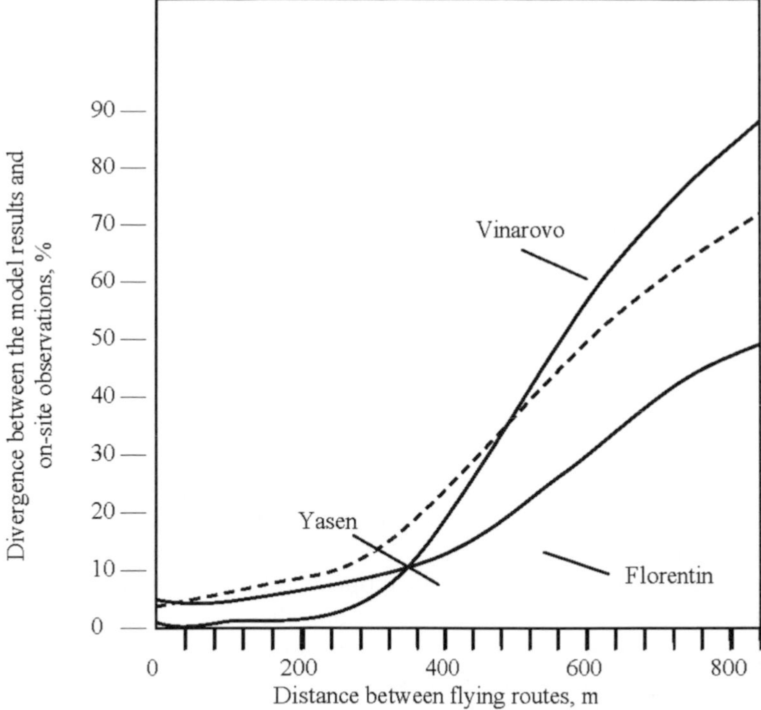


**Figure 4.11.** Dependence of precision of soil moisture content on distance between flying routes in the experiments in Bulgaria (see Figure 4.10).

microwave monitoring that allow reconstruction of SPF characteristics in light of vegetation types and that provide the possibility of synthesizing their spatial distribution.

As noted by Chukhlantsev *et al.* (2003), the problem of microwave remote-sensing of vegetation cover requires studying how electromagnetic waves (EMWs) attenuate within the vegetation layer. Solution of this problem is made possible by combining experimental and theoretical studies. Vegetation cover is commonly characterized by varied geometry and additional parameters. Therefore, knowledge of the radiative characteristics of the SPF as functions of time and spatial coordinates can be acquired by combining on-site measurements and models. The general aspects of such an approach have been considered by many authors (DeWitt and Nutter, 1989; Del Frate *et al.*, 2003; Levesque and King, 2003). However, these investigations were mainly restricted to investigation of models describing the dependence of the vegetation medium on environmental properties, as well as the correlation between the morphological and biometrical properties of vegetation and its radiative characteristics.

One prospective approach to solution of the problems arising here is GIMS technology. Combination of an environmental acquisition system, a model of how a typical geoecosystem functions, a computer cartography system, and a means of artificial intelligence will result in creating a geoinformation monitoring system for any typical natural element, a system that is capable of fulfilling many tasks arising in microwave radiometry of global vegetation cover. A GIMS-based approach, in the framework of EMW attenuation by vegetation canopies, allows a knowledge base to be set up that establishes the relationships between experiments, algorithms, and models. The links between these areas are adaptive and represent the optimal strategy for experimental design and model structure. The goal is to explain and assess the application of the GIMS method to the tasks of reconstructing the spatial and temporal distribution of SPF radiative characteristics.

### 4.6.2 Vegetation media as the object under study of electromagnetic wave attenuation

Many investigators (Levesque and King, 2003; Zarco-Tejada *et al.*, 2003; Zhan *et al.*, 2002; Chukhlantsev and Shutko, 1988; Chukhlantsev *et al.*, 2003, 2004; Del Frate *et al.*, 2003; Engman and Chauhan, 1995; Dong *et al.*, 2003; Ferrazzoli and Guerriero, 1996) support an effective technology for microwave monitoring of vegetation covers. But greater accuracy in this technology is required for the reason that the various vegetation covers, especially vegetation canopies, are too dynamic and complicated for experimental study or model parametrization. There are two considerations here. One is connected with the creation of a highly productive technology to estimate land cover characteristics. The other is connected with the climate change problem as manifested in the greenhouse effect. In these cases GIMS technology recommends a balanced scheme for solution of the problems in land cover monitoring. As follows from Armand *et al.* (1997), combination of experimental and theoretical studies of attenuation of microwave radiation by vegetation cover is required. It is actually possible to synthesize the experimental dependence between attenuation and a restricted set of vegetation parameters. Estimation of how microwave radiation is attenuated by vegetation cover in real time is only possible by applying microwave models and interpolation algorithms.

From the point of view of microwave remote-sensing, knowledge of the water content of vegetation is an important in synthesizing a model of the extent to which microwave radiation is attenuated by vegetation. This knowledge can be acquired by making use of the fact that water absorption features dominate in the spectral reflectance of vegetation in the near-infrared spectrum (Sims and Gamon, 2003). Using such indexes as the normalized difference vegetation index (NDVI), the plant water index (PWI), the leaf area index (LAI), the simple ratio vegetation index (SRVI), and the canopy structure index (CSI) it is possible to determine the canopy structure and photosynthetic tissue morphologies. The correlation between leaf water content and the leaf level reflectance in the near-infrared has been successfully studied by many authors who linked leaf and canopy models to study the effects of leaf structure, dry matter content, LAI, and canopy geometry. As a result, it is possible to parametrize the forest structure including the canopy and crown closure, stem density, tree height, crown size, and other forest features (Shutko, 1986; Zarco-Tejada *et al.*, 2003).

Quantitative information about SPF properties may be obtained by means of different remote-sensing techniques based on passive microwave data (Basharinov *et al.*, 1979; Camillo *et al.*, 1986). This approach is particularly effective at estimating the hydraulic parameters of soil–plant formations. Knowledge of these parameters allows the water balance of the territory occupied by vegetation of a given type to be parametrized and the growth of plants to be modeled. Under this approach a comparison of remote data registered at different EMW ranges may provide highly precise estimates of the water content in the various layers of the SPF (canopy, trunks, stalks, and soil). Moreover, by combining data analysis and remote-sensing the parameters of ground vegetation cover and its role in flows of sensible and latent heat from the surface to the atmosphere can be estimated.

Thus, a set of vegetation-related parameters can be generated at specific time and space resolutions as the basis for calculation of EMW attenuation by the vegetation layer. Available remote monitoring data can be used as input information for algorithms and models to synthesize the spatio-temporal distribution of attenuation effects.

### 4.6.3 Links between experiments, algorithms, and models

Field observations are the initial stage in determining the important properties of vegetation cover. Many problems exist in providing model descriptions of biological and physical processes operating at different time scales. The tasks of describing EMW attenuation by the vegetation cover has time scales ranging from less than a second to hours or days and has various spatial scales too. Consequently, parametrical description of EMW attenuation by the vegetation cover demands an enormous number of experimental observations, the planning of which depends on the model type and the vegetation component under investigation. It is here that effective synthesis of final results at given space–time scales can be realized by application of GIMS technology. This allows one to solve the problem of scaling the *soil–vegetation–atmosphere* model and to describe the radiative transfer processes from individual plants or small plots to large-scale biomes.

As mentioned above, satellite data can be used to calculate many parameters of vegetation cover, including unstressed stomatal conductance, photosynthetic capacity, fraction of photosynthetically active radiation absorbed by the green portion of the vegetation canopy, canopy reflectance, transpiration, and other important environmental characteristics. These data and the results of field experiments allow model experiments to be generated (Krapivin and Phillips, 2001a).

The GIMS-based method uses a priori information to start the model experiment. Under this procedure, a set of algorithms operate over the whole data set and determine the input parameters of the model used. The model structure and its coefficients change according to fluctuations in the difference between experimental and model estimations of the EMW attenuation effect. Usually, the central part of the model describes the fundamental processes within the SPF, such as radiation balance, water circulation (evapotranspiration, root water uptake), photosynthesis, and mortality.

### 4.6.4 Microwave model of vegetation cover

#### *4.6.4.1 Two-level model of vegetation cover*

The *atmosphere–vegetation–soil* system (AVSS) is a subject that receives much consideration (Figure 4.12). Microwave emission in the AVSS is a result of the combination of absorption and diffraction processes within the media and repeated reflection at their boundaries. A most thorough study of microwave emission models (MEMs) for possible AVSS configurations has been undertaken by many authors (Chukhlantsev and Shutko, 1988; Chukhlantsev, 2006). Usually, the AVSS emissivity capability is described as:

$$e = 1 - |R|^2 \tag{4.1}$$

where

$$R = \frac{R_{21}\exp[-2i\chi h] + R_{32}}{R_{21}R_{32} + \exp[-2i\chi h]}; \qquad \chi = \chi_1 - i\chi_2 = \frac{2\pi}{\lambda}[\varepsilon_2 - \sin^2\theta]^{1/2};$$

where $\theta$ is the angle between the direction of electromagnetic wave propagation in the atmosphere and the *atmosphere/vegetation* interface; $R_{21}$ and $R_{32}$ are the complex Fresnel coefficients for the *atmosphere/vegetation* and *vegetation/soil* interfaces, respectively; and $h$ is vegetation cover height.

The Fresnel coefficients for the interface between the $i$th and $k$th medium are defined by the expression:

$$\hat{R}_{ik} = |R_{ik}|\exp(-iq_{ik}) \tag{4.2}$$

where $q_{ik}$ is the phase shift.

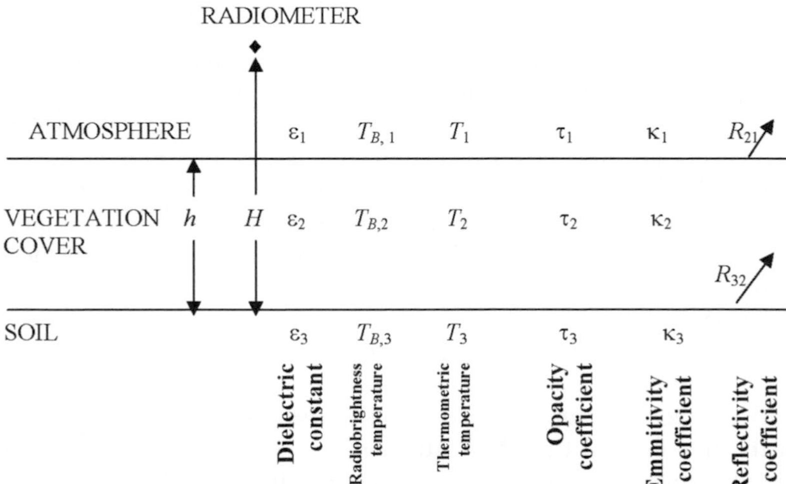


**Figure 4.12.** Diagram of model-based calculation of the shielding influence of vegetation cover in microwave monitoring.

From (4.1) and (4.2) one gets the following representation for emissivity:

$$e = \frac{r - \hat{R}_{32}^2 \exp(-4\chi_2 h) - \hat{R}_{21}^2 - 2\kappa_1 \sin q_{21} \sin \xi}{r + \kappa_1 \cos(q_{21} + \xi)}$$

where

$$\xi = q_{32} + 2\chi_1 h;$$

$$\kappa_1 = 2\hat{R}_{21}\hat{R}_{32} \exp(-2\chi_2 h);$$

$$\hat{R}_{ik} = |R_{ik}|;$$

$$r = 1 + \hat{R}_{21}^2 \hat{R}_{32}^2 \exp(-4\chi_2 h);$$

$$\hat{R}_{ik} = \left\{ \frac{(\gamma_i - \gamma_k)^2 + (\delta_i - \delta_k)^2}{(\gamma_i + \gamma_k)^2 + (\delta_i + \delta_k)^2} \right\}^{1/2}; \tag{4.3}$$

$$tg q_{ik} = \frac{2(\delta_i \delta_k - \gamma_i \gamma_k)}{\gamma_i^2 - \gamma_k^2 + \delta_i^2 - \delta_k^2}; \tag{4.4}$$

$$\gamma_k = \begin{cases} \dfrac{A_k}{a_k} & \text{for horizontal polarization;} \\[2ex] \dfrac{A_k \varepsilon_k'' - \beta_k \varepsilon_k''}{(\varepsilon_k')^2 + (\varepsilon_k'')^2} & \text{for vertical polarization;} \end{cases}$$

$$\delta_k = \begin{cases} \dfrac{\beta_k}{a_k} & \text{for horizontal polarization;} \\[2ex] \dfrac{A_k \varepsilon_k'' - \beta_k \varepsilon_k'}{(\varepsilon_k')^2 + (\varepsilon_k'')^2} & \text{for vertical polarization;} \end{cases}$$

$A_k = \sqrt{(a_k + b_k)/2}$; $\beta_k = \sqrt{(a_k - b_k)/2}$; $a_k = \sqrt{b_k^2 + (\varepsilon_k'')^2}$; $b_k = \sqrt{(\varepsilon_k')^2 - \sin\theta}$; $\varepsilon_2 = \varepsilon_2' - i\varepsilon_2''$ is the dielectric permittivity of vegetation cover; $\varepsilon_3 = \varepsilon_3' - i\varepsilon_3''$ is the dielectric permittivity of soil; and $\varepsilon_1 = \varepsilon_1' - i\varepsilon_1''$ is the dielectric permittivity of air with $(\varepsilon_1' = 1, \varepsilon_1'' = 0)$.

Following Chukhlantsev and Shutko (1988), the Fresnel coefficients are determined as functions of the vegetation cover parameters. The dielectric constants of the vegetation and soil are input parameters to formulae (4.3) and (4.4). Considering the vegetation and soil as double-component mixtures of dry matter and water, the dielectric constants can be determined by the following expressions:

$$\sqrt{|\varepsilon_2|} = \rho_P \sqrt{\varepsilon_B} + (1 - \rho_P)\sqrt{\varepsilon_W}; \qquad \sqrt{|\varepsilon_3|} = \rho_B \sqrt{\varepsilon_B} + (1 - \rho_B)\sqrt{\varepsilon_W}; \tag{4.5}$$

where $\rho_P$ and $\rho_B$ are the relative volumetric concentrations of water in plants and soil, respectively; and $\varepsilon_B$ and $\varepsilon_W$ are the dielectric constants of dry soil and water, respectively.

As was shown empirically by Engman and Chauhan (1995), $\varepsilon_3' > \varepsilon_3''$ and both the real and imaginary parts of the soil dielectric constant are increasing functions of volumetric moisture content. These functions can be approximated by the following formulae:

$$\left.\begin{aligned} \varepsilon_3' &\approx c_1 + (c_2\rho_B + c_3\rho_B^2)\exp(c_4 - c_5/\lambda), \\ \varepsilon_3'' &\approx (d_1\rho_B + d_2\rho_B^2)\exp(d_3 - d_4/\lambda), \end{aligned}\right\} \tag{4.6}$$

where $\lambda$ is the wavelength (cm), and $c_i$ and $d_i$ are constants depending on the soil type. For example, soil that consists of 30.6% sand, 55.9% silt, and 13.5% clay is characterized by the dependencies (4.6) with $c_1 = 2.35$; $c_2 = 52.4$; $c_3 = 31.1$; $c_4 = 0.057$; $c_5 = 1.22$; $d_1 = 7.1$; $d_2 = 46.9$; $d_3 = 0.0097$; and $d_4 = 1.84$.

Thus, taking into account the dependencies (4.6) it becomes possible to optimize the microwave range for passive remote-sensing. Of course, there exist a set of unsolved problems relating to the model corrections necessary to consider surface roughness and other obstacles distorting the brightness temperature $T_B$ and a set of special features arising in inverse task solution. A basis for future model refinements is the correlation between $T_B$, atmospheric transmissivity for a radiometer at height $H$ above the soil, smooth surface reflectivity $R$, and thermometric temperatures of the SPF $T_{s-v}$ and atmosphere $T_a$. This correlation can be expressed by the Schmugge–Shutko formula (Schmugge, 1990; Shutko, 1986):

$$T_B = t(H)[RT_{\text{sky}} + (1 - R)T_{s-v}] + T_a \tag{4.7}$$

where $T_{\text{sky}}$ is the contribution from the reflected sky brightness. Typical remote-sensing applications use microwave wavelengths $\lambda \geq 1$ cm and in this case atmospheric transmission approaches 99% and $T_a + T_{\text{sky}} \leq 5°\text{K}$ (Engman and Chauhan, 1995).

A more precise correlation than (4.7) can be achieved by considering the influence of surface roughness on soil reflectivity:

$$R' = R\exp(-g \cdot \cos^2\omega),$$

where $\omega$ is the angle of incidence, and $g$ is the roughness parameter ($g = 4\sigma^2k^2$, with $\sigma$ being the root mean square height variation of the soil surface and $k = 2\pi/\lambda$).

Formulae (4.5) and (4.6) can be simplified by correlating soil water content $\rho_B$ and vegetation water content $\rho_P$ (DeWitt and Nutter, 1989):

$$\rho_B = 78.9 - 78.4[1 - R'\exp(0.22\rho_P)]$$

For uniform media, the expression (4.7) is often represented in the following way:

$$T_B = \frac{(1 - \hat{R}_{21}^2)\{[T_3(1 - \hat{R}_{32}^2)]e^{-\tau_2} + T_2(1 - e^{-\tau_2})(1 + \hat{R}_{32}^2 e^{-\tau_2})\}}{1 - \hat{R}_{21}^2\hat{R}_{32}^2\exp(-2\tau_2)},$$

where the opacity coefficient $\tau_2$ can be evaluated by means of the following formula (Kirdiashev *et al.*, 1979):

$$\tau_2 = \frac{4\pi Q}{\nu\rho_1\lambda I_m(\sqrt{\varepsilon_2})},$$

where $\nu$ is vegetation volume density and $\rho_1$ is the specific density of the wet biomass.

The coefficient $\hat{R}_{21}$ is estimated by the following expression:

$$\hat{R}_{21} \approx \alpha_2[1 - \exp(-\tau_2)].$$

There is a number of simpler and more complex models of the AVSS. For example, the model describing the forest medium, as sketched in Figure 4.12, was developed by Ferrazzoli and Guerriero (1996) and is based on the Rayleigh–Gans approximation of the electromagnetic properties of coniferous leaves (needles). Vegetation cover is represented as a two-layered structure. The first layer is the canopy composed of leaves and branches, represented by disks and cylinders, respectively. The second layer of the vegetation cover consists of trunks described by near-vertical cylinders. Such a model structure allows microwave forest emission to be simulated.

Models for layered vegetation have been studied by many authors (Chukhlantsev, 2006; Ferrazzoli and Guerriero, 1996; Karam *et al.*, 1992). The main mathematical task is solution of the radiative transfer equation. Different models consider the wide-frequency range of both deciduous and coniferous forests and account for the distributions of branch size and leaf orientation. The basic conclusion of these studies indicates that detailed knowledge of the forest structure is important for model adequacy. For example, taiga forests and forest tundra have such important components as lichens that are characterized by lowered photosynthesis and can influence the attenuation index (Petzold and Goward, 1988). The tasks arising here have yet to be studied.

#### 4.6.4.2 Analytical model of vegetation cover

To understand the dependence on environmental parameters of the radiobrightness temperature registered by a radiometer at height $H$ above soil covered by plants, it is useful to consider some simple parametrical descriptions of this dependence. A simplified microwave model can be written as the following integral:

$$T_B(H) = e_s T_s \exp\left[-\int_0^H \gamma(x)\,dx\right] + \int_0^H T(y)\gamma(y)\exp\left[-\int_0^y \gamma(x)\,dx\right]dy + \zeta(H) \quad (4.8)$$

where $\gamma$ is the absorption coefficient, $e_s$ is the soil emissivity coefficient, $T_s$ is soil temperature, $T$ is temperature of the atmosphere, and $\zeta$ is model precision. There exist different simplifications of equation (4.8). For example, often various approximations for $T(y)$ and $\gamma(y)$ are used:

$$T(y) = \begin{cases} T_2, & 0 \le y \le h; \\ T_1, & h < y \le H, \end{cases} \qquad \text{or } T(y) = T_0 + T_1 y + \cdots + T_n y^n$$

and $\gamma(y) = \gamma_0 + \gamma_1 y$ or $\gamma(y) = \text{const}$ (Basharinov *et al.*, 1979).

For the satellite monitoring case it can be shown that:

$$T_B = e_s T_s p_{\text{sat}} + \sum_{l=1}^{n-1} D_l(0) \quad (4.9)$$

where $p_{sat}$ is atmospheric transmittance between soil surface and satellite,

$$D_1 = T(h), D_2 = \frac{1}{\gamma(h)} \cdot \frac{dT(h)}{dh}, \ldots, D_l = \frac{1}{\gamma(h)} \cdot \frac{dD_{l-1}(h)}{dh}, \ldots.$$

By cconsidering different analytical representations for the vertical profile $T(h)$, formula (4.9) can calculate brightness temperature $T_B$ with an error equal to $D_n(0)$.

A simpler microwave model than (4.8) for the canopy–surface brightness temperature can be given by (Sasaki *et al.*, 1987, 1988):

$$T_B(\lambda, \theta) = e_2 T_2 + \hat{R}_{21}(\theta) T_{B,\text{sky}}(\theta_1, \theta_2) \tag{4.10}$$

where $\theta$ is the angle of incidence, and $T_{B,\text{sky}}$ is the sky brightness temperature incident on the canopy level from direction $(\theta_1, \theta_2)$, with zenith and azimuth angles $\theta_1$ and $\theta_2$, respectively.

Model (4.10) simplifies the analysis of the influence of canopy temperature and wind speed, for example, on variations in brightness temperature. This analysis is possible with knowledge of canopy roughness as a function of wind speed.

### 4.6.5 The greenhouse effect and forest ecosystems

A principal aspect of anthropogenic impact on the environment is evaluation of the consequences of $CO_2$ emissions to the atmosphere. Published results estimating the greenhouse effect and excess $CO_2$ distribution in the biosphere vary widely and sometimes are contradictory or else they are too flatly stated. This is a natural consequence of all kinds of simplifications adopted in modeling the global $CO_2$ cycle. The GIMS makes it possible to create an effective monitoring system that allows estimation of the spatial distribution of carbon sinks and sources in real time.

Before we go any further, some problems need to be solved to assess the role of humankind's use of the Earth's surface. In particular, there is the problem of formalized description of the processes of change in the structure of land covers, such as afforestation, forest reconstruction, deforestation, and the associated carbon supplies. Understanding meteorological processes as functions of greenhouse gases is a key problem confronting humankind in the first decade of the third millennium. Solid knowledge of the meteorological phenomena at various spatio-temporal scales under conditions of varying supplies of $CO_2$ and other greenhouse gases is the only way of making correct and constructive decisions about global environmental protection.

The dynamics of surface ecosystems depends on interactions between biogeochemical cycles, which during the last decade of the 20th century suffered significant anthropogenic modification, especially to the cycles of carbon, nitrogen, and water. Surface ecosystems—in which carbon remains in the living biomass, decomposing organic matter, and the soil—play an important role in the global $CO_2$ cycle. Carbon exchanges between these reservoirs and the atmosphere take place through photosynthesis, respiration, decomposition, and burning. Human interference with this process is by changing the structure of vegetation cover, pollution of water basin surfaces and soil areas, as well as by direct emissions of $CO_2$ into the atmosphere.

The role of various ecosystems in the formation of carbon supplies to biospheric reservoirs determines the rate and direction of changes to regional meteorological situations and to global climate. The accuracy of assessment of the level of these changes depends on the reliability of the data on surface ecosystems (Elsner and Jagger, 2006; Filatov *et al.*, 2005; Guieu *et al.*, 2005; Mason, 2003; Melillo *et al.*, 2003; Odell, 2004; Parson and Fisher-Vanden, 1999; Schulze, 2000).

Existing environmental data show that knowledge of the rates and trends of carbon accumulation in surface ecosystems is uncertain. However, it is clear that surface ecosystems are important assimilators of excess $CO_2$. Understanding the details of such assimilation is only possible by modeling the process of plant growth; that is, considering the effect of nutrient elements in the soil and other biophysical factors on plant photosynthesis. Therefore, forest ecosystems and associated processes of natural afforestation, forest reconstruction, and deforestation should be studied in detail. In a forested area, the volume of the reservoir of $CO_2$ from the atmosphere is a function of the density of the forest canopy, and over a given period of time a change of this volume is determined by the level and character of the dynamic processes of transition from one type of forest to another. The causes of this transition can be natural, anthropogenic, or mixed. Biocenology tries to create a universal theory of such transitions, but so far there is only a qualitative description of the transitions observed.

# 5

# Decision-making risks in global ecodynamics

## 5.1 RISK CONTROL AND SUSTAINABLE DEVELOPMENT

Taking the definition of sustainable development given by Kobayashi (2005) as our basis, let us now consider trends in the strategy of human behavior toward NSS development, which depends heavily on the decisions taken. The degree of risk in making decisions at this stage of human development cannot be evaluated because results to be obtained in the immediate future and the consequences later on are unknown. In that sense, our current ineffective strategy of risk control regarding decisions made could lead to catastrophic consequences—in particular, the decisions taken by international organizations on nature use management, change in land cover, impacts on anthropogenic processes, etc. At a time of few methods for long-range forecast, an erroneous decision taken now can cause irreversible consequences with the possible development of undesirable processes in the environment. At the beginning of the 21st century, most key indicators of the habitats of living beings, especially those characterizing human safety, are nearing the critical state, and in some regions a life-threatening state (Ehrlich and Kennedy, 2005; Freeman, 2005; Gavin, 2001; Gillaspie, 2001; Gitelson *et al.*, 1997).

The appearance of environmental conditions of danger to people is mainly connected with the probability of a natural or technogenic catastrophe as a result of poor decisions made earlier. Catastrophes, in general, and natural disasters, in particular, threaten human life and therefore their prevention through prediction is a necessary element of those spheres of science dealing with the processes taking place in the NSS. Evaluation of the risk of occurrence of a natural disaster and subsequent damage requires interdisciplinary analysis of a vast amount of information about various aspects of NSS functioning. Ecological safety—taking a broad view—is closely connected with the socio-economic level of development of society over a given territory. For instance, according to Vladimirov *et al.* (2000)—discussing conditions in Russia—many indicators of society development are in a critical state.

Compared with Germany, Russia spends 50 times less on ecological safety. Other indicators, such as expenditure on nature protection and avoidance of ecological losses in Russia, deviate negatively from standard values by factors of 2.5 and 3.5, respectively. Moreover, most indicators of NSS development in Russia testify to a negative trend in the preparedness of Russian society to prevent natural disasters and meet their consequences. Whereas in many developed countries most natural disasters result in manageable levels of damage, in many regions of Russia they cause significant damage. For instance, the sudden snow fall on 13 October 2004 in Chelyabinsk resulting in a 40-cm snow layer brought urban life to a standstill for 24 hours.

Many natural phenomena require a formalized description for their prediction. Since all processes in the NSS are somehow interconnected, the desire to create a GMNSS with a broad set of functions is one way to resolve the global problem of predicting emergency situations in the environment. Statistical mean dependences created for description of correlations between the frequency of occurrence of natural disasters and their consequences in many cases are a sufficiently effective means to make strategic decisions on the prevention of natural disasters. Such a methodology can be seen in the widely known law of Richter–Gutenberg (Vladimirov *et al.*, 2000) which applies to the U.S. territory as follows:

$$\log N = \begin{cases} 1.65\text{–}1.35F & \text{for floods,} \\ 1.93\text{–}1.39F & \text{in case of tornadoes,} \\ 0.45\text{–}0.58F & \text{for hurricanes,} \\ -0.55\text{–}0.41F & \text{in case of earthquakes.} \end{cases}$$

where $F$ is the logarithm of the annual average number of victims for the last 100 years, and $N$ is the number of events.

Knowledge of the laws of occurrence of natural disasters, ability to predict catastrophic events, and the presence of mechanisms for warning of disasters do not provide complete protection for people and their homes from losses and destruction. It is necessary that these constituents be claimed by people who understand the risk and realize the danger. Vladimirov *et al.* (2000) introduced the notion of "safety culture" which they defined as the code of conduct, morals, and emotional response to cataclysms.

Risk as a measure of danger—evaluated by taking many factors into account—can serve as a guideline to resolve the problem of controling a set of circumstancess that can disturb human habitats and change conditions for the functioning of society. Risk includes the probability of an unfavorable event *and* the amount of resulting losses. In other words, risk reflects the measure of danger of a natural phenomenon including estimates of the levels of adversity for many different aspects. At the same time, risk also has a subjective constituent which can be measured with the help of formal methods of decision-making that take into account intuitive assessment of the situation and psychological norms of perception of the environment. Many problems are overcome when legislation that reflects the general principles of danger assessment is accepted (Glazovsky, 2005; Gorshkov, 1990; Hay, 2005; Hossay, 2005).

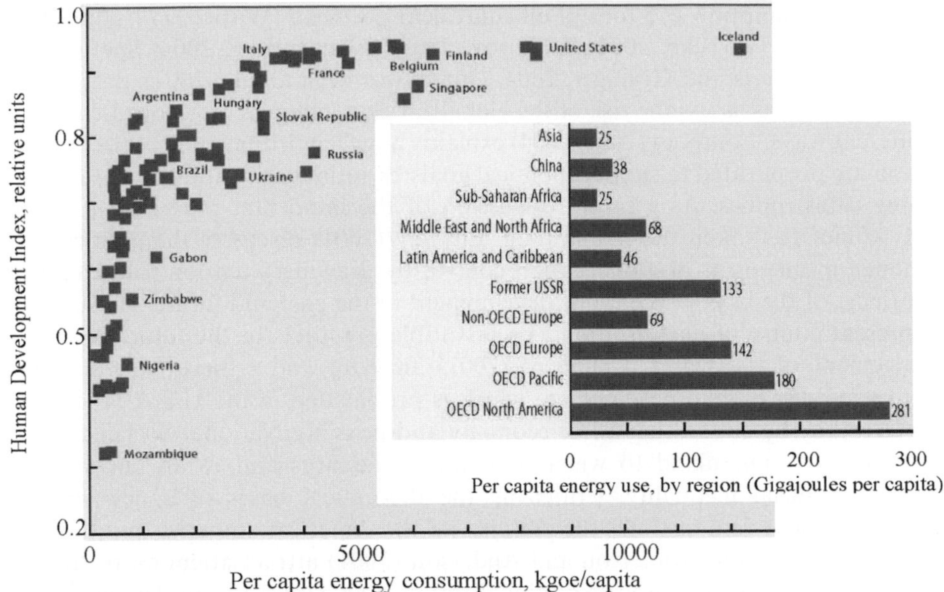

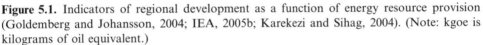

**Figure 5.1.** Indicators of regional development as a function of energy resource provision (Goldemberg and Johansson, 2004; IEA, 2005b; Karekezi and Sihag, 2004). (Note: kgoe is kilograms of oil equivalent.)

On the whole, the notion of risk covers areas characterizing the constituents of probability of an unfavorable change in conditions for human existence—that is, natural, economic, social, cultural, and religious constituents. Each of these areas is a complex of non-linear functions of a multitude of NSS parameters, prediction of whose changes is the subject of current global change science. Of importance here are the economic levers controling the processes of NSS development. The relationship between economy, natural resources, and energy in many respects determines the direction of global process development. In this sense, the main problem facing humankind consists in finding a compromise between the increasing needs of an ever enlarging consumption society and nature with its limited food and energy resources. Humankind appears to have reached the crossroads between the strategy of globalization and regional control. As can be seen in Figure 5.1, the level of risk of taking an unfavorable direction of development for many poor countries remains high as a result of the strategy of developed countries of appropriating excessive amounts of resources. Therefore, inclusion of a country or a region into the process of globalization is important from the viewpoint of evaluating the consequences. The index of globalization—which is used to evaluate this risk and which includes variables characterizing the level of economic integration, development of technologies, political aspects and openness of the society to information—does not foresee free transmission of excessive resources from one country to another. As of 2004, 62 countries—whose population constitutes 84% of global size—consume 96% of global GDP. The assistance to other countries is to be rendered to achieve the level of energy consumption at 5 tons of oil equivalent per head (Wisner *et al.*, 2005; Van den Bulcke and Verbeke, 2001; Sungurov, 2005; Spangenberg, 2004; Speth, 2004; Saiko, 2003; Purvis and Grainger, 2004; Quaddus and Siddique, 2004).

Many experts assess the risk of the globalization strategy to developed countries in different ways. Lomborg (2001, 2004) explains how "environmental myths" originate and are propagated to support political goals by intimidating the population with coming catastrophes. As a result, discussion of the important problems of global ecodynamics has taken the wrong path. Lindsey (2001) discusses the political and economic mechanisms of globalization control by drawing attention to the contradictoriness of the laws of economic development in the past and future. He supports the present course of globalization as a favorable prospect for the future economic arrangement of the world. Callinicos (2001) analyzes and criticizes the so-called "third way" for human development which is propagated in the U.S.A., the U.K., and Germany by advocating a free economy and possibly National Socialism. This way will lead humankind to wider economic differences and to an uncontrolled increase in social inequality. Criticizing the theoretical bases of a new market, Callinicos (2001) demonstrates the presence of the American imperial trend toward political globalization. Anderson and Anderson (2001) attract attention to the fact that there is more to the process of globalization than economy—no less important are culture, politics, and biology, since people live in a world of open systems and, depending on the interaction of these NSS components, many global processes can form. Tabb (2001) considers the process of globalization from the viewpoint of ethics whose coordination with economy is one factor preventing negative trends in the development of modern society. Trade wars, capital concentration, displacement of investment processes, and latent restriction of social rights are basic components of the present ideology of globalization (Kidron *et al.*, 2004; Kirchner, 2003; Kondratyev and Krapivin, 2006g; Kotliakov, 2000; Lovelock, 2003; Malinetsky *et al.*, 2003).

## 5.2 INDICATORS OF THE EFFICIENCY OF RISK CONTROL IN CASES OF NATURAL DISASTERS

Optimizing the risk to insurance from natural disasters becomes every year an ever more urgent problem, since economic losses are on the increase and are almost unpredictable. For instance, the insurance pay-out as a result of Hurricane Hugo in 1989 constituted $5 billion exceeding by more than 50% the maximum loss of any earlier natural catastrophe. However, 3 years later the insurance payment for losses from Hurricane Andrew exceeded fourfold those from Hurricane Hugo. Examples of unexpectedly large insurance losses are on the increase (Lehmiller *et al.*, 1997; Newbold, 2002; Rathford, 2006; Santos, 2005; Simpson and Ganstang, 2003).

The problem of risk control as a quantitative measure of the NSS property to ensure habitat safety in a given territory is a complicated mathematical problem of optimization which can be successfully solved using logical and probabilistic (LP) methods (Solozhentsev, 2004) based on construction, solution, and study of the LP—functions of risk. Boolean algebra combined with methods of numerical modeling of processes taking place in the NSS is the mathematical basis for these studies. The informative characteristic of such an approach to evaluating the risk of occurrence of a natural catastrophe is determined by the accuracy of prediction of development of natural–anthropogenic processes in a given territory and the reliability of produced scenarios at substituting individual elements in this forecast.

According to Solozhentsev (2004), to assess the efficiency of risk control one should use discrete multitudes of parameters $\{Z_j\}$ that affect the efficiency and the scale $Y$ of the efficiency indicator. In this case, introduction of the goal function $F = N_{jc} + \cdots + N_{kc}$, where $N_{jc}$ is the number of correctly identified events on the scale $Y$ in the class $j$, and maximization of $F$ makes it possible to determine the weights of the factors affecting efficiency. The complexity of this problem is determined by lack of reliable data on the statistical characteristics of these factors. The difficulties that crop up can be overcome using methods of assessing the level of self-organization and self-regulation of natural systems (Ivanov-Rostovtsev *et al.*, 2001) and synergetics (Chernavsky, 2004). In particular, the use of D-Self theory facilitates introducing a generalized characteristic of the NSS state in the form of a non-linear function $\xi_0 = \sqrt{\xi_i \xi_i^*}$, where $\xi_0$, $\xi_i$, and $\xi_i^*$ are NSS parameters meeting three axioms:

(i) *discreteness* of elements within and beyond the NSS with their interaction;
(ii) *hierarchy* of elements within and beyond the NSS; and
(iii) *contingency* between discrete elements of hierarchy of the structure within the system ($\xi_i$) and beyond the system ($\xi_i^*$) in the form of dependence $\xi_0 = \sqrt{\xi_i \xi_i^*}$.

Synthesis of the D-Self model as applied to the NSS is an individual problem and is not considered here. We only mention it because, according to Ivanov-Rostovtsev *et al.* (2001), the method of modeling the evolutionary dynamics of natural systems based on D-Self technology has some features in common with evolutionary modeling technology (Bukatova and Makrusev, 2004). In both cases, evolution of the natural system is considered as a discrete process of a change of its states in some parametric space. In the case considered, the economic indicators and social infrastructure of the territory are made into a separate level of hierarchy of parameters and considered as indicators of risk control efficiency. Other characteristics of NSS functioning are considered as external parameters.

The frequency and intensity of natural disasters are on the increase, which can be clearly seen in Table 5.1. Respectively, economic damage from natural disasters is also on the increase (Tables 5.2, 5.3; Figure 5.2). Therefore, detailed analysis of the relationships between parameters of natural disasters and regional NSS characteristics such as infrastructure, level of economic development, state of social order, level of education, and attitude to religion is an important stage of development for a regional strategy to meet the consequences of natural disasters. Losses from natural disasters between 1987 and 1997 reached $700 billion, and this means that—for poor countries—meeting the consequences of natural disasters in the years to come will become an insoluble problem. For instance, in 1998 total economic losses from natural disasters constituted $65.6 billion, 66% of which will have to be paid by developing countries. It is clear that economic mechanisms of support to regions

**Table 5.1.** Statistics of large-scale natural disasters (Ruck, 2002).

| Years | 1950–1959 | 1960–1969 | 1970–1979 | 1980–1989 | 1990–1999 | 1992–2002 | 2003 |
|---|---|---|---|---|---|---|---|
| Number of natural disasters | 20 | 27 | 47 | 63 | 91 | 70 | 107 |
| Economic losses ($ billion) | 42.1 | 75.5 | 138.4 | 213.9 | 659.9 | 550.9 | 64.6 |

**Table 5.2.** The ratio between real economic losses and insurance payments in the U.S.A. evaluated from the most substantial natural disasters.

| Decade | Number of events | Economic damage ($ billion) | Insurance payments ($ billion) |
|---|---|---|---|
| 1950–1959 | 12 | 30 | 3.7 |
| 1960–1969 | 15 | 41 | 7.4 |
| 1970–1979 | 17 | 45 | 7.8 |
| 1980–1989 | 13 | 50 | 18.0 |
| 1990–1999 | 35 | 219 | 82.0 |
| 1994–2003 | 39 | 161 | 66.0 |

**Table 5.3.** Insurance losses from large-scale natural disasters.

| Year | Damage from natural disasters ($ billion) |
|---|---|
| 1992 | 24.4 |
| 1993 | 10.0 |
| 1994 | 15.1 |
| 1995 | 14.0 |
| 1996 | 9.3 |
| 1997 | 4.5 |
| 1998 | 15.0 |
| 1999 | 22.0 |
| 2000 | 7.5 |
| 2002 | 10.0 |
| 2003 | 12.75 |
| 2004 | 45.74 |
| 2005 | 78.33 |

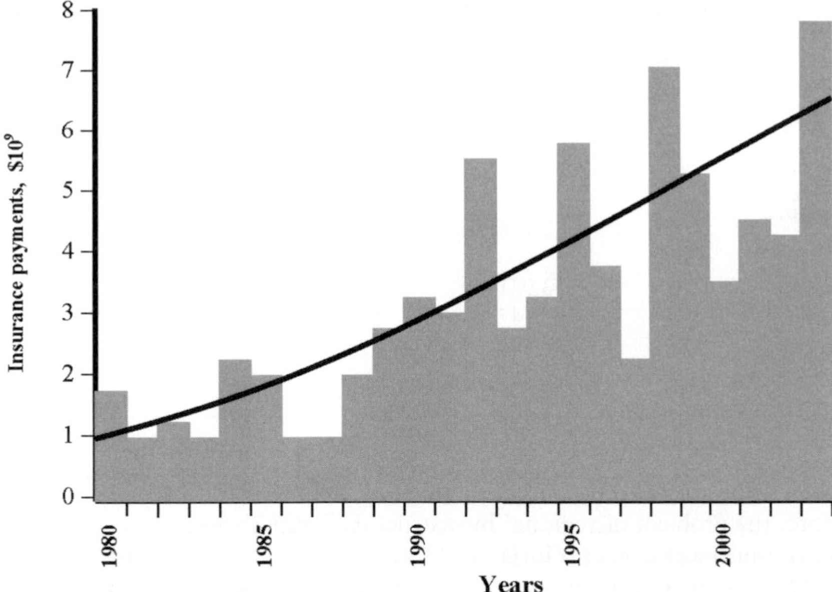

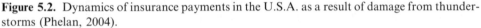

**Figure 5.2.** Dynamics of insurance payments in the U.S.A. as a result of damage from thunderstorms (Phelan, 2004).

damaged by natural disasters should be optimized by taking the real state of the NSS into account. From estimates of the World Bank, poverty is characterized by the following indicators:

- one-third of the population in developing countries are poor, 18% of them being beggars;
- about 50% of the world's poor and 50% of world's beggars live in South Asia;
- poverty prevails in rural localities; and
- basic means of subsistence for the poor is obtained from agriculture.

Development of infrastructure plays an important role in poverty reduction. However, it is infrastructure that is most damaged by natural disasters. For example, in Asia almost 70% of natural disasters are floods, each flood causing damage estimated recently at about $15 billion, with losses to agriculture contributing most. Therefore, the problem of risk reduction due to infrastructure transformation arises. In developing countries the solution of this problem is connected with optimizing processes of the use of land resources and working out ecological strategies. It is necessary to take into account the indicator of the growth of population density in developing countries. Quantitative and qualitative dynamism in urban development leads to an overproportional increase in the vulnerability of such territories. This trend of developing countries is provoked by either the economic interests or socio-political factors of a given region, when a large number of people are concentrated in a small area, which drastically increases the risk to human life in case of natural disasters. On the other

**Table 5.4.** Characteristics of natural disasters recorded in 2003.

| Region | Number of events | Number of victims | Economic losses | | |
|---|---|---|---|---|---|
| | | | Total | Insurance | Ratio of insurance payments to total losses |
| Africa | 57 | 2,778 | 5,158 | 0 | 0 |
| America | 206 | 946 | 21,969 | 13,274 | 0.6 |
| Asia | 245 | 53,921 | 18,230 | 600 | 0.03 |
| Australia/Oceania | 65 | 47 | 628 | 246 | 0.39 |
| Europe | 126 | 20,194 | 18,619 | 1,690 | 0.09 |
| *Globe* | *699* | *77,886* | *64,604* | *15,810* | *0.24* |

hand, as a defense against this situation protective constructions could easily be built. Therefore, the problem of regional infrastructure optimization is becoming with the passage of time more urgent (Gurjar and Leliveld, 2005). To some extent, solution of this problem favors putting into practice decisions of the World Summit on Sustainable Development (Johannesburg, South Africa) adopted at the 17th plenary meeting on 4 September 2002. This plan foresees mobilizing technical and financial assistance to developing countries in order to balance the economic, social, and ecological development of all regions of the world.

The ratio between economic losses in the case of a natural disaster and real insurance arrangements for subsequent financial investment to compensate for the consequences is one of the key problems in the formation of strategies of insurance companies. One successful attempt at formalizing the processes that arise is the use of models of catastrophes in assessment and control of the risk of extreme events (Grossi and Kunreuther, 2005). These authors concentrated on the risk of natural disasters and discussed the actual problem of controling terrorism risk. The goal of the study was to reduce losses from potential dangers. Pennsylvania University tested the proposed technology, carrying out several numerical experiments, which showed that a model approach to planning financial risk from natural disasters makes it possible to optimize the insurance of extreme events. The necessity of this can be seen from Tables 5.4 and 5.5, which demonstrate a very non-uniform regional distribution of the ratio of insurance payments to total losses from natural disasters. If, in addition, one takes into account the distribution of losses by the types of events (Tables 5.6 and 5.7), it becomes clear that model optimization can substantially raise the effects of insuring these events. Floods are among the most destructive and frequent of natural events. Analysis of the consequences of the many floods that have already occurred in the 21st century shows that they markedly change the social and economic development of regions. As for the real danger of floods for society, the European Economic Community decided to finance the RIBAMOD (River Basin Modelling) project, which had as its purpose the solution of the problem of hydrological risk control and covered the following important themes (Beven and Blazkova, 1998):

**Table 5.5.** Catastrophes in 2005 by region (Enz, 2006).

| Region | Number of events | % | Victims | % | Insured loss ($ million) | % |
|---|---|---|---|---|---|---|
| North America | 54 | 13.6 | 3,781 | 3.9 | 72,633 | 87.1 |
| Europe | 59 | 14.9 | 659 | 0.7 | 7,039 | 8.4 |
| Asia | 208 | 52.4 | 89,633 | 92.4 | 2,660 | 3.2 |
| South America | 21 | 5.3 | 943 | 1.0 | 47 | 0.1 |
| Africa | 41 | 10.3 | 1,851 | 1.9 | 49 | 0.1 |
| Oceania/Australia | 6 | 1.5 | 26 | 0.0 | 359 | 0.4 |
| Seas/space | 8 | 2.0 | 125 | 0.1 | 609 | 0.7 |
| *World total* | *397* | *100.0* | *97,018* | *100.0* | *83,396* | *100.0* |

**Table 5.6.** Distribution of natural disasters and respective losses in 2003.

| Type of event | Share of a given type of event (%) | Victims (%) | Total economic damage (%) | Insurance losses (%) |
|---|---|---|---|---|
| Earthquakes and volcanic eruptions | 12 | 61 | 10 | 1 |
| Hurricanes, tornadoes, storms | 43 | 2 | 39 | 76 |
| Floods | 28 | 5 | 21 | 8 |
| Other natural disasters | 17 | 32 | 30 | 15 |

**Table 5.7.** Distribution of natural disasters and losses in 2005 (Enz, 2006).

| Event | Number | Victims | Insured loss ($ million) |
|---|---|---|---|
| Floods | 61 | 5,017 | 3,464 |
| Storms | 48 | 4,354 | 73,512 |
| Earthquakes and tsunamis | 12 | 75,267 | 234 |
| Drought | 10 | 783 | 20 |
| Bush fires and heat waves | 12 | 2,549 | 623 |
| Cold and frost | 3 | — | 477 |
| Other natural catastrophes | 3 | 113 | — |
| *Natural catastrophes total* | *149 (37.5%)* | *88,083 (90.8%)* | *78,330 (93.9%)* |

- accumulation of databases on watershed basins known to initiate floods;
- assessment of risks from decisions on protection and reduction of the consequences of floods;
- development of technologies of planning measures on flood prevention and operative interference in flood control; and
- planning interactions between different groups of specialists when working on lessening flood damage.

Most countries that suffer floods have services for prevention of and protection from floods. The structure and equipment of these services are determined by regional special features of the causes of floods. Regardless of this however, there are general problems in assessing the risk to life, infrastructure, and economy from flooding. An effective way of solving these problems is to predict hydrological processes at regional and global levels. The efficiency of such predictions depends on the information content of nature-monitoring systems.

A natural disaster always leads to economic losses: in poor countries the population becomes even poorer, discussions of sustainable development take on life or death meanings, and problems of environmental protection and search for NSS equilibrium at a regional level become secondary (Freeman, 2000). Natural disasters destroy infrastructure, a key component of the economic growth of a region. For almost 18% of the global population poverty is critical; therefore, occurrence of a natural disaster in their habitat creates a complicated social situation, a way out of which is only possible with the assistance of developed countries. Therefore, a reduction of the risk of economic losses from natural disasters in developing regions is a top-priority problem for global ecodynamics (Kondratyev and Krapivin, 2005b; Kondratyev *et al.*, 2004).

## 5.3 SOCIAL AND HUMAN DIMENSIONS OF RISK

Human society has reached the point where the correlation between processes in the environment and social organization are so critical that the declaration of one man can drastically change global processes. For instance, Tony Blair declared in November 2005 at a meeting of 20 countries in London that the Kyoto Protocol should be reconsidered and that science will find the ways to resolve the problem of global warming. He also said that politically no country wants to sacrifice its economy to resolve the problem. Other ways needed to be found, ones that would not restrict economic growth or check the processes involved in raising living standards in the world. Here the statement of one man has been the catalyst for finding mechanisms of raising the living standards and poverty eradication at the same time as checking global warming.

According to Letz (2000), humankind has crossed the threshold of system adaptation making it too late to damp out deviations from acceptable values of NSS parameters and to preserve the habitat; therefore, immediate measures are needed on formation of a new view of the processes in the environment: especially, a technocratic paradigm determining the decisions and respective outlook of national leaders. Therefore, the socio-political constituent of risk is very important for evaluating and determining what it is that society needs to do to prevent damage from the potential consequences of decisions made. The most important problems to be resolved by governments are:

- reduction of the level of risk to human life as a result of public policy toward neutralizing negative processes in the environment;
- increase in the socio-political stability of society by setting up institutional and informative conditions to prepare the population for extreme circumstances;
- increase in the mobilization capabilities of society for preventive and adequate response to threats at the time of their development; and
- development of the legal basis and respective structures to meet the consequences of extreme events and disasters of natural and technogenic origin.

As Churin (2005) noted, social processes in modern society have no purpose-oriented control, and therefore it is difficult to evaluate the risk of decision-making. The interconnection between social and natural processes practically everywhere is determined by their interactivity. Therefore, the regulation of social processes inevitably affects the functioning of natural systems—hence, the necessity to develop legal mechanisms for transition to sustainable development.

## 5.4 ESTIMATION OF RISK IN THE MONITORING REGIME

### 5.4.1 Decision-making in risk assessment

A decision on the level of potential risk of possible change in the environment can be made by analyzing the prehistory of such events and using the methods of prediction of natural events. As a rule, risk assessment methods are based on statistical processing of data on parameters of the processes whose interaction can initiate undesirable changes in environmental characteristics. For instance, when assessing the risk to human health from environmental pollution, to make a serious decision a lot of information on the territory is needed:

- multi-year observations of the concentration of chemical elements, their allocation, and characteristics;
- data on the hydrology of the territory and synoptic characteristics;
- assessment of the ecological consequences of pollution and their impact on human health; and
- the state of protective constructions and the level of development and technical equipment of services for monitoring, preventing, and meeting extreme situations.

Just how efficient the systems of risk assessment are depends on the form and kind of applied procedure of decision-making (Potapov *et al.*, 2006). The most informative is when prediction procedures are combined with an environmental monitoring regime, which can foresee situations of decision-making in real time on the basis of information accumulation before the decision was made or from analysis of database fragments with no time reference. Statistical analysis of several events that follow the functioning of the monitoring system can be carried out by numerous methods whose applicability in each case is determined by the totality of probabilistic parameters that characterize the phenomenon under study. However, non-stationarity and parametric uncertainty in situations when each observation requires much effort and expense prompt a search for new methods of decision-making on the basis of observation data that are fragmentary both in time and in space.

With the development of alternative methods of making statistical decisions, the problem of finding objective estimates of the parameters of processes taking place in the environment has been substantiated anew. It is possible to consider and compare two approaches to this problem: the classical approach based on a priori restricted numbers of observations, and successive analysis based on the procedure of step-by-step decision-making. The development of computer technologies facilitates realizing both approaches in the form of a single system of making statistical decisions (Vald, 1960; Viscusi and Gayer, 2004; Wagner, 2005).

The classical way of making statistical decisions using the Neumann–Pearson method is based on $n$ measurements fixed beforehand and taken from a priori assumptions of the probable character of a set of observations $X = \{x_1, \dots, x_n\}$. Accepting hypotheses $H_0$ or $H_1$ is based on drawing the boundary of an optimal critical area $E_1$ in the form of a hyper-surface:

$$L_n = L_n(x_1, \dots, x_n) = f_{a_1}(x_1, \dots, x_n)/f_{a_0}(x_1, \dots, x_n) = C \tag{5.1}$$

where $f(x_1, \dots, x_n) = \prod_{i=1}^{n} f(x_i)$, $f_a(x)$ is the density of probability distribution for the variable $x$ with an unknown parameter $a$, and $C$ is the constant magnitude chosen when $E_1$ has a level of error of the first kind $\alpha$.

Relationship (5.1)—called the probability coefficient—serves as a key to the final choice between the hypotheses:

(1) if $L_n \le C$ then the hypothesis $H_0$ is accepted; and
(2) if $L_n > C$ then the hypothesis $H_1$ is accepted.

The content and semantic load of hypotheses $H_0$ and $H_1$ depend on a concrete problem. In real experiments, based on a sample $\{x_i\}$, an empirical and then continuous distribution $F(X) = f(x)$, but with an assumption of uncertainty of one or several values of the parameters. As a rule, based on one of the goodness-of-fit tests, a concrete form of distribution is chosen, and its parameters are evaluated from measured values $\{x_i\}$.

The Neumann–Pearson method and sequential analysis facilitate the construction of operative characteristics for decision-making without concretization of the type of density $f_a(x)$. Consider the case of a homogenous independent sample, when sample values $x_i$ $(i = 1, \ldots, n)$ are independent realizations of the same random variable $\xi$ with density $f_{a_0}(x)$ for hypothesis $H_0$ and density $f_{a_1}(x)$ for hypothesis $H_1$. The parameter $a$ of a true density $f_a(x)$ can differ from $a_0$ or $a_1$. It is shown that there are the following relationships for errors of the first or second kind ($\alpha$ and $\beta$):

$$\alpha \approx \exp[-0.5\{(E_{a_1}\xi - E_{a_0}\xi)(D_{a_0}\xi)^{-1/2}\}^2 n], \tag{5.2}$$

$$\beta \approx \exp[-0.5\{(E_{a_1}\xi - E_{a_0}\xi)(D_{a_1}\xi)^{-1/2}\}^2 n], \tag{5.3}$$

where

$$E_a\xi = \int_{-\infty}^{\infty} \ln[f_{a_1}(x)/f_{a_0}(x)]\, f_a(x)\, dx, \tag{5.4}$$

$$D_a\xi = \int_{-\infty}^{\infty} \{\ln[f_{a_1}(x)/f_{a_0}(x)]\}^2 f_a(x)\, dx - (E_a\xi)^2, \tag{5.5}$$

With a successive procedure, the operative characteristic is as follows:

$$L(a) \approx [A^{h(a)} - 1]/[A^{h(a)} - B^{h(a)}], \tag{5.6}$$

where $h(a)$ is the root of equation:

$$\int_{-\infty}^{\infty} [f_{a_1}(x)/f_{a_0}(x)]^{h(a)} f_a(x)\, dx = 1, \tag{5.7}$$

$A$ and $B$ are two thresholds for the likelihood coefficient $L_n(x)$, for which the following estimates are valid:

$$B \approx \beta/(1-\alpha), \qquad A \approx (1-\beta)/\alpha, \tag{5.8}$$

In accordance with this, as in the classical algorithm, $L(a_0) = 1 - \alpha, L(a_1) = \beta$. Hence, the average number of observations in sequential analysis can be evaluated:

$$\left.\begin{aligned} E_a v &= [(1-\alpha)\ln[\beta/(1-\alpha)] + \alpha \ln[(1-\beta)/\alpha]/E_{a_0}\xi \quad \text{for } a = a_0; \\ E_a v &= [\beta \ln[\beta/(1-\alpha)] + (1-\beta)\ln[(1-\beta)/\alpha]/E_{a_1}\xi \quad \text{for } a = a_1. \end{aligned}\right\} \tag{5.9}$$

With $a = a^*$, when $E_{a^*}\xi = 0$ and $E_{a^*}\beta^2 > 0$, we have:

$$E_{a^*}v \approx [-\ln[\beta/(1-\alpha)] \ln[(1-\beta)/\alpha]/E_{a^*}\xi^2 \tag{5.10}$$

According to (5.9) and (5.10), in a sequential procedure the number of observations for making a decision is a random variable $v$, whose average value $E_a v$ can be greater or smaller than $n$. To find out which $v$ value to use, it is necessary to know the distribution $P\{v = n\} = P_a$, for which the following expression is valid:

$$E_a v \cdot P_a(n) = W_c(y) = c^{1/2} y^{-3/2} (2\pi)^{-1/2} \exp[-0.5c(y + y^{-1} - 2)], \tag{5.11}$$

where

$$\begin{aligned}
0 &\leq y \leq n|E_a\xi| < \infty, \\
c &= K|E_a\xi|/D_a\xi = (E_a v)^2/D_a v > 0, \\
Dv &= KD_a\xi/(E_a\xi)^3, \\
Ev &= K/E\xi, \\
K &= \begin{cases} \ln A & \text{for } E_a\xi > 0, \\ \ln B & \text{for } E_a\xi < 0 \end{cases}
\end{aligned}$$

The distribution function (5.11) in Russian literature is called the Wald distribution:

$$W_c(x) = (c/2\pi)^{1/2} \int_0^x z^{-3/2} \exp[-0.5c(z + z^{-1} - 2)]\, dz \tag{5.12}$$

Universality of the Wald distribution follows from its duality with respect to the normal distribution (Basharinov and Fleishman, 1962):

$$W_c(x) = \Phi[(x-1)(c/x)^{1/2}] + \Phi[-(x+1)(c/x)^{1/2}] \exp\{2c\}, \tag{5.13}$$

where

$$\Phi(x) = 1/(2\pi) \int_{-\infty}^x \exp\{-t^2/2\}\, dt.$$

Note that if $|E_a\xi|$ and $|D_a\xi|$ are small compared with $\ln A$ and $\ln B$, then the distribution of a relative variable $v/E_a v$ determined from expression (5.11) will approximate the real distribution of this magnitude, even if $\xi$ is distributed and does not follow the normal law.

Theoretical constructions concerning the universality of distribution (5.12) are important for general assessment of the efficiency of the sequential procedure of decision-making. However, these constructions are not important in practical application of the Wald distribution. Therefore, synthesis of the system of automated decision-making as a GIMS unit has been accomplished without considering correlation (5.13). It has been done in connection with the fact that it often happens that the number of observations turns out to be small and the effect of asymptotic normality is not realized. The resulting situation is resolved by making a decision either following the procedure of the evolutionary algorithm (Bukatova and Rogozhnikov, 2002) or in accordance with the Mkrtchian algorithm (Mkrtchian, 1982).

### 5.4.2 Scheme of observation planning when using sequential data analysis

Making a decision on detection of some effect in the process of continuous environmental monitoring depends on the scheme of observation planning. The classical approach directs the observation system towards collection of a fixed volume of data with their subsequent processing in order to reveal certain effects or properties in the space under study. The method of sequential analysis does not separate these stages but alternates them. In other words, the monitoring of data-processing is made after

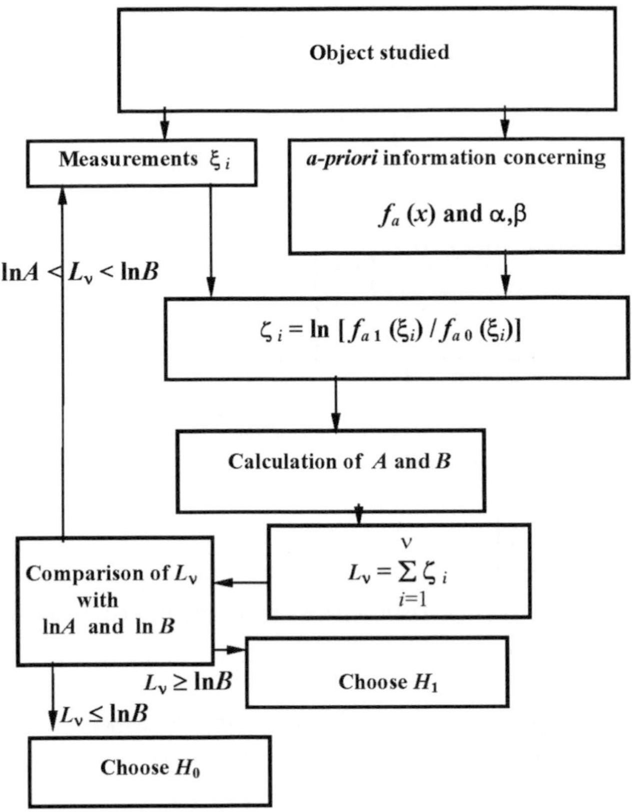


**Figure 5.3.** Experiment organization based on using the successive analysis procedure to choose between two competing hypotheses $H_0$ and $H_1$.

each measurement: this is shown in Figure 5.3 for the sequential procedure of decision-making and in Figure 5.4 for the classical procedure. It can be seen that the algorithmic load in the sequential procedure changes dynamically, whereas in the classical case methods of data-processing are used at the final stage. When forming the structure of an adequate system of automated decision-making, these approaches should be realized in the form of individual units, which should be chosen in conversational mode with the GIMS operator. The operator's decision can change in the dynamics of monitoring, but in this case the operator must be an expert. Therefore, in the structure of the decision-making system proposed here the operator's actions are reduced to control of parameters $n$, $\alpha$, and $\beta$.

### 5.4.3 Elements of the system of the automated procedure for making statistical decisions

The theory of decision-making necessitates consideration of the above-formulated problem in the use of sequential analysis when choosing between competing

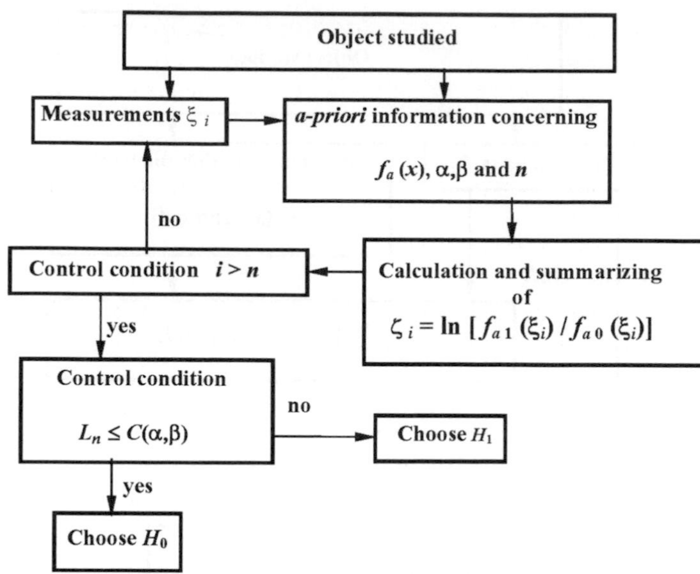


**Figure 5.4.** The Neumann–Pierson classical decision-making procedure to check the alternative hypotheses $H_0$ and $H_1$.

hypotheses in a more general form and considering all aspects that appear when assessing the risk of the decisions made. The problems that arise here cover the following themes:

- selection of a sufficiently effective criterion to evaluate the parameters;
- determination of the probabilistic characteristics of the process under study;
- a priori evaluation of potential losses in the accuracy of decisions made; and
- prediction of the dynamic stability of the results of the decision made.

A specific feature of the automated system for making statistical decisions foresees the presence of numerous functions that should, as far as possible, provide a solution to most intermediate and final problems in assessing the situation in the monitoring regime. Therefore, the system should include the following units:

- visualization of observation data as direct images, histograms, sums, and frequencies by time intervals;
- calculation of the statistical characteristics of observation results (average, disposition, moments of third and higher orders, excess, asymmetry coefficient, entropy, etc.);
- construction of empirical density and function of probability distribution;
- construction of the theoretical distribution of probabilities with its evaluation by one of the statistical criteria;
- calculation of the characteristics of the procedure of acceptance of hypotheses

using the classical method of Neumann–Pearson and the resulting assessment of the procedure by the current volume of sampling;
- calculation of the characteristics of the sequential analysis procedure and visualization of its state;
- realization of the functions of operator access to system units at any step of the system's functioning with possibilities of making decisions either on a change in procedure parameters or cessation of measurements; and
- visualization of the decision made.

This series of units facilitates formation of a measurement procedure model that depends on a priori available information about the parameters and character of the process under study. At the same time, the operator can check input information reliability and promptly change the monitoring strategy. On the whole, the totality of these units constitutes an automated system for making statistical decisions. The system is constructed following the standard scheme of man–machine dialog—that is, one using a hierarchical menu. The principal scheme of the system takes the form of a transformation $F$, whose structure and content are determined by the totality of units characterized above. The system's input receives measurement data from $n$ sources. Decisions as to the reliability of hypotheses $H_0$ or $H_1$ can be made either for each source or for their totality $m \leq n$. Different modes of operator–system interaction can be foreseen. The end of the decision-making procedure is followed by information about the length of the used data sequence for each working information channel.

With two or more channels, final acceptance of one of the hypotheses is accomplished by the weighted majority of decisions made for individual channels. When there are obstacles to decision-making the operator is warned about situations of uncertainty and can change the weights of channels and correct other parameters.

To carry out the procedure of acceptance of complicated hypotheses when the number of possible decisions is beyond binary situation, a recurrent procedure is constructed to solve the multi-criterion problem of decision-making by setting up simple binary situations.

## 5.5 PREDICTED RISKS OF GLOBAL CHANGE

The notion of risk is closely connected with the notion of uncertainty in prediction of the development of events in the NSS that are undesirable for humans. By definition (Burgman, 2005), risk is the chance of occurrence of an unfavorable event within some time interval: for the NSS such intervals are measured in centuries. Society primarily wants to know the prospects for improving living standards. However, the state of current environmental science does not guarantee solution of this problem, and therefore it is necessary to find constructive technologies of risk assessment. A reliable and constructive approach to solution of this problem could be brought about by expanding GMNSS functions. In particular, the number of uncertainties hindering reliable prediction of global ecodynamics is so large that without their purpose-oriented analysis and selection all their interactions cannot be formally taken into account in the global model. For instance, of the millions of chemical compounds and species of living organisms, only thousands have been adequately studied.

### 5.5.1 Biospheric evolution and natural disasters

One of the many factors involved in evolutionary process development in the NSS is sudden change in environmental characteristics that induces stress in living organisms or even death. In various periods of evolution, the scale and significance of individual factors are known to change. The role of catastrophic processes taking place in the 11,500-yr lifetime of our current civilization is being studied within two international ICSU programs: "Dark nature: Rapid natural change and human responses" (started in 2004) and "The role of Holocene environmental catastrophes in human history" (2003–2007). The main goals of these projects are to:

- examine how past societies and communities reacted in the face of harmful change;
- explore the implications of rapid natural change for current environmental and public policies; and
- focus on the inter-disciplinary investigation of Holocene geological catastrophes, which are of importance for civilizations and ecosystems.

A longer period in the evolution of the Earth system was considered by Condie (2005), who analyzed the interaction of various components of the planet for the last 4 million years, especially those that have affected the history of land ecosystems, oceans, and the atmosphere. The current period is characterized by an increasing trend of anthropogenically caused natural disasters, such as floods, forest and peat-bog fires, deforestation, desertification, and epidemics. Of course, it is not always possible to distinguish the cause of a natural disaster. But, one fact is apparent: the present ecodynamics is followed by an increase in the number of extreme situations in the environment. For instance, in 2000 and 2001, the number of human deaths from natural disasters constituted 17,400 and 25,000 despite the attempts of many countries to take preventive measures to protect their population. Maximum economic losses during the last years were recorded in 1995 ($180 billion) and in 2004 ($220 billion). The worst years for natural disasters were 1998 and 2004 resulting in 50,000 and 232,000 deaths, respectively. The contribution of different types of disasters to loss of human life and economic damage varies from year to year, but most victims die as a consequence of earthquakes, floods, snow storms, and tsunamis. For instance, of the 700 natural disasters recorded in 1998, winter storms and floods occurred in 240 and 170 cases, respectively, and the resulting economic losses constituted 85% of all losses. In Europe, in mid-November 1998 more than 215 people died from hypothermia.

The significance and scale of critical situations in the environment depends on the state of many NSS components. As a rule, natural disasters are evaluated in the context of damage to human life and economic activity. However, it is well known that natural disasters are an important factor of ecodynamics at both local and global scales. Lindenmayer *et al.* (2004) noted that natural disasters are of key importance in regulating many processes that determine ecosystem dynamics. Wright and Erickson (2003) discussed taking the impact of natural disasters into account when numerically modeling environmental change. Special attention has recently been paid to forest fires which, being an important component of ecodynamics, are mainly of anthropogenic origin and cause both material damage and loss of life.

Forest fires occur regularly in various regions and have increasingly become a factor in ecosystem dynamics, evidenced as fire-induced emissions of greenhouse gases and aerosols to the atmosphere. According to available estimates, about 30% of tropospheric ozone, carbon monoxide, and carbon dioxide contained in the atmosphere are from forest fires. Aerosol emissions to the atmosphere as a result of forest fires can substantially affect the microphysical and optical characteristics of the cloud cover, leading to climate change. For instance, satellite observations over Indonesia have demonstrated that the presence of smoke in the atmosphere due to prolonged fires has had the effect of precipitation suppression, which has further favored the development of fires. In this context, Ji and Stocker (2003) statistically processed the data of the TRMM satellite measuring rainfall in the tropics as well as of TOMS (total ozone mapping spectrometer) data for the period January 1998 to December 2001 in order to analyze the laws governing annual, intraseasonal, and interannual variability in the number of forest fires at the global scale. During this period, there was a clear annual course of fires in South-East Asia reaching a maximum in March. This was the case in Africa, North America, and South America in August. Analysis of these data further revealed the interannual variability in forest fires in Indonesia and Central America correlating with the El Niño/Southern Oscillation (ENSO) cycle in 1998–1999. In 1998, boreal forests burned over vast territories of Russia and North America. Fires covered some 4.8 million ha of boreal forest in Canada and the U.S.A. and 2.1 million ha in Russia.

There appears a clear correlation between variability in atmospheric aerosol content and variations in the frequency and intensity of forest fires. One exception is southwestern Australia where intensive fires recorded from TRMM data were not followed by smoke layer formation (from TOMS data). Excluding this Australian region, the coefficient of correlation between the number of fires and AI (from TOMS data) constitutes 0.55. The statistical analysis of data using the calculated empirical orthogonal function (EOF) revealed a contrast between the Northern and Southern Hemispheres as well as the existence of intercontinental transfer of aerosols originating from fires in Africa and America. Statistical analysis data point to the presence of 25-day to 60-day intraseasonal variations superimposed on annual change in the number of fires and aerosol content. A similarity was found between interseasonal variability in the number of fires and dynamics of the Madden–Julian Oscillation (Kondratyev and Grigoryev, 2004).

As McConnel (2004) justly noted, it still remains unknown why some forests have suffered fragmentation, degradation, and loss of biodiversity while other forests remain in a good state and even expand. There is no doubt that increasing population sizes, further growth in market economics, and intensive development of various economic infrastructures are important factors in such deforestation. Factors acting against these processes are measures taken for nature protection such as preservation of forests. Finally, the dynamics of forest cover is determined by a complicated and interactive totality of such factors as biogeophysical processes, increasing population density, market economics, various disturbing forcings (e.g., forest fires), and institutional microstructures.

Forest fires affect the way in which the global carbon cycle forms. In reality, the global scale of forest fires have recently become equivalent in area to Australia. The atmosphere receives almost 40% of global $CO_2$ emissions, and 90% of forest fires are anthropogenic in origin. This means that the balance of nature is no longer in equilibrium and the laws of natural evolution are strongly violated.

Among other natural phenomena affecting environmental dynamics are volcanic eruptions, dust storms, and thunderstorms. One of the consequences of a large-scale volcanic eruption is massive atmospheric pollution, which can lead to climate change both regionally and globally. For example, aerosols connected in many respects with emissions of sulfur dioxide during the 1991 Pinatubo eruption led to a global fall in temperature in the lower atmosphere in 1992 by about 0.5°C. In addition, it was noted that in the following two years the level of global ozone content decreased by 4% compared with the preceding 12-year period. Despite there being no drastic climate changes due to volcanic eruptions in the 20th century, historically such situations have taken place. For example, 73,500 years ago global atmospheric temperature fell by 3–5°C as a result of the Toba eruption on Sumatra. Glaciation increased globally and the Earth's population decreased (Oppenheimer, 1996). Keeping such historical data in mind, Earth's current 600 volcanoes are potentially a global threat to population, since they can seriously change climate and thereby affect further development of the NSS.

Dust storms occur in places where there is a lot of dust and strong winds and play a similar role to volvanoes in environmental change. Dust can propagate great distances affecting plants and soils negatively as well as strongly polluting the lower atmosphere for long periods. In regions with sparse vegetation cover and arid climate, dust storms remove the fertile soil layer, thereby damaging the plants. Dust storms occur periodically, for instance, on the plains of the central and western states of the U.S.A. during high-category hurricanes. Sometimes they last for several days, raise dust up to 1.5–1.8 km (some particles reach 5–6 km), and transfer it hundreds and sometimes even thousands of kilometers towards the Atlantic Ocean. Catastrophic dust storms are connected with removal of the soil–plant cover in steppes and partially-wooded steppes, as well as in regions of intensive agriculture. Despite short-duration impacts and marked damage, dust storms are a regular constituent of global ecodynamics which can manifest itself decades and even centuries later.

Thunderstorms play a special role in environmental change. Lightning discharges occur at practically all latitudes, affect photochemical reactions in the atmosphere, and are a factor of fire risk. From available estimates, 1,800 thunderstorms on average occur at any given moment on Earth, each discharging 200 lightning flashes per hour (or 3.3 flashes per minute). From satellite observations, the global frequency of lightning flashes averages 22–65 flashes per second. The Microlab-1 satellite observed 44 ± 5 flashes per second, which corresponds to 1.4 billion flashes per year. Microlab-1 data have made it possible to draw global maps of the frequency of lightning discharges in different seasons. Analysis of these maps shows that lightning discharges occur mainly over land; the ratio between land and ocean averages approximately 10 : 1. About 78% of lightning discharges are within the latitudinal band 30°N–70°S. The year-round regime of lightning discharges is highest in the River Congo Basin—where the average frequency of flashes reaches (in Rwanda) 80 flashes per $km^2$ per year—and in central Florida. Such a year-round regime of lightning flashes is characteristic of the northern sector of the Pacific Ocean where under the influence of cold air advection over the warm ocean surface the atmosphere becomes unstable. In the eastern sector of the tropical Pacific and in the Indian Ocean, where the atmosphere is warmer, lightning discharges are not that frequent. The maximum frequency of lightning occurrence in the Northern Hemisphere falls in summer, whereas in the tropics lightning discharges are characterized by a semi-annual cycle.

In northern regions, frosts are an important constituent of the mechanism controlling evolution; their impact on vegetation cover depends on plant hardiness. Jönsson *et al.* (2004)—taking the Norwegian spruce *Picea abies* as an example—studied the response of boreal forests to temperature changes. They showed that sudden temperature variations cause changes in wood compactness and that expected climate change alters the vegetation cover.

### 5.5.2 Problem of global warming and natural disasters

In recent years, when discussing the causes of potential climate change of key importance have been estimates of uncertainties in the values that serve as the basis for making conclusions about climate change and taking measures to prevent it. Particularly important is the problem of assessing the levels of GHG emissions to the atmosphere connected primarily to solution of the problem of the global carbon cycle. It is evident that without reliable verification of available estimates of emissions all discussions concerning the ecological benefit of various measures and expenditure to carry them out are abstract (Nilsson *et al.*, 2002). For instance, how can fines for not meeting recommendations on GHG emission reduction be substantiated if it is impossible to prove that emissions in 2012 will differ from those in 1990? So far, in the course of discussions on the Kyoto Protocol problem, quantitative estimates of uncertainties in the levels of GHG sinks have been ignored (especially those concerning the biosphere). Uncertainties in the estimates of total $CO_2$ fluxes are, however, very substantial (exceeding 100% in Russia). Calculations of errors in the estimates of total GHG fluxes give about ±5–25%, whereas the levels of GHG emission reduction foreseen by the Kyoto Protocol average about 5%. An average global situation is illustrated, for instance, by the fact that uncertainties in the estimates of GHG emissions as a result of energy production are approximately equal to uncertainties in the estimates of $CO_2$ assimilation by the biosphere and land.

Keeping this in mind, solution of the problem of estimate uncertainty (mainly concerning reliable information on carbon cycles) and verification are particularly important. Solution of the verification problem requires a common opinion of its mechanisms, which is also of great importance from the financial point of view. The results of simulation modeling indicate, for instance, that by raising from 50 to 95% the confidence level of GHG emission reduction by 5.2%, it will lead to an increase in expenditure on measures to reduce GHG emissions by factors of 3 to 4. The main conclusion is that science should serve as the compass for recommendations on measures in ecological policy (Nilsson *et al.*, 2002).

The serious uncertainty in the estimates of potential anthropogenically induced climate change and its determining factors has fueled heated discussion of these problems both in the scientific literature and the mass media. For instance, in an interview with a journalist from *Scientific American*, Professor R. Lindzen from the Massachusetts Institute of Technology (Lindzen *et al.*, 2001; Lindzen and Giannitsis, 2003) said that his anxiety about speculative opinions of climate change first emerged in 1988 when J. Hansen (Director of the Goddard Institute for Space Studies, New York) announced in public that global climate warming was the consequence of an increase in $CO_2$ concentration due to emissions to the atmosphere from fossil fuel burning (Hansen and Sato, 2001; Hansen *et al.*, 1997–2000). This announcement prompted R. Lindzen to explain that climate science was only at an early stage of development and that there was no consensus on the causes of climate change. In early 2001 he reported his findings to a meeting of the U.S. Cabinet on the problem of climate change. The fact that global mean surface air temperature (SAT) increased by $\sim$0.5°C over 100 years and that atmospheric $CO_2$ concentration increased by $\sim$30% does not reflect a cause-and-effect feedback between $CO_2$ concentration increase and temperature increase. Lindzen *et al.* (2001) believed that the most reliable estimate of climatic sensitivity (SAT increase as a ratio of doubled $CO_2$ concentration) was 0.4°C, which meant there was no basis for anxiety about catastrophic global climate change in the future.

According to IPCC mandate, the problem posed was to prepare an overview of "any climate change with time, both natural and anthropogenic". Pielke (2002) noted, however, that at least two climate-forming factors have turned out to be either unreliably considered or not considered at all: (1) impact on global climate of anthropogenic changes in land surface characteristics; (2) biological impacts of growing $CO_2$ concentration in the atmosphere (e.g., effect on fertilization) (Tianhong *et al.*, 2003). If both these factors are really substantial, the conclusion suggests itself that any agreement in the results of the global climate numerical modeling is accidental (i.e., concerning global mean annual mean SAT).

In this connection, Pielke (2001a, b; 2002) discussed information confirming the significance of both these climate-forming factors and expressed his opinion about how best to check the basis for this conclusion. To do this he used data on the impact of anthropogenic changes of land surface characteristics on local, regional, and global climates, which illustrated this impact was just as important as the impact of $CO_2$ doubling in the atmosphere (as well as any increase in other GHG concentrations). Also important is the fact that the interaction between atmosphere and surface is characterized by the presence of various non-linear feedbacks, and therefore it may be impossible to predict climate change for periods longer than a season.

As for the potential biological impacts of growth in $CO_2$ concentration, they are manifested through short-range (biophysical), middle-range (biogeochemical), and long-range (biogeographical) impacts of the landscape-forming processes of weather and climate. The biophysical impact includes, for instance, the effect of transpiration on the relationship of latent and sensible heat fluxes as surface heat balance components. Biogeochemical impacts include the effect of the growth of plants ("fertilization effect") on leaf area index, from which evaporation takes place, on surface albedo, and carbon supply. One manifestation of biogeographical impacts is the change over time in the species composition of plant communities. Numerical modeling results indicate that without considering biophysical/biogeochemical impacts any assessment of climate change cannot be reliable (Bounoua *et al.*, 2002).

Further development of climate models need to take into account the following aspects of climate formation: (1) direct and indirect impacts of landscape dynamics through biophysical, biogeochemical, and biogeographical processes; (2) consideration of anthropogenic changes in land use on various (local, regional, and global) spatio-temporal scales; (3) assessment of possibilities of climate forecasts for longer periods than a season, bearing in mind the functioning of numerous non-linear feedbacks which determine the interaction between atmosphere and surface. The openness of these and other problems has reduced the significance of the IPCC-2001 Report and the U.S. National Report to little more than an assessment of global climate sensitivity to changes of some climate-forming factors.

In the long process of preparation of the three IPCC Reports (IPCC, 1995, 2005) currently published, increasingly more complete analyses of climate-forming factors (e.g., atmospheric aerosols were considered in addition to GHGs) and various feedbacks were carried out. Nevertheless, the available very complicated numerical climate models cannot be considered adequate at considering all substantial climate-forming factors. A new step forward in the IPCC-2001 Report was consideration of the forcing $F$ of anthropogenic land use changes on climate, but this was restricted to consideration of only the impact of land use dynamics—starting from 1750—on surface albedo. The obtained estimates gave an average $F$ value equal to $-0.2\,\mathrm{W\,m^{-2}}$ with an uncertainty interval between 0 and $-0.4\,\mathrm{W\,m^{-2}}$. Thus, such estimates are highly uncertain, and this concerns even the sign of $F$. A more complete consideration of biophysical, biogeochemical, and biogeographical impacts of land use evolution on climate has led to the conclusion that in this case $F > 0$.

So, it is important to bear in mind that the new range of potential global warming by 2100 assumed in the IPCC-2001 Report (1.4–5.8°C) is based only on numerical modeling data (and therefore it will inevitably change in the future). Also, the problem is that this new range cannot be directly compared with earlier similar values. From the viewpoint of reliability of the estimates of future climate change, the use in the IPCC-2001 Report of the term *projections* instead of *predictions* is significant, since the latter implies that factors left out of account will not be substantial in the future. The unacceptability of the latter supposition is the reason none of the specialists in the field of numerical modeling will assert that climate can be predicted

100 years in advance. This conclusion confirms the results of numerical modeling carried out by Andronova and Schlesinger (2001).

The complexity of the problem of prognostic assessment of climate—especially the contribution of the anthropogenic component—is illustrated by the remaining inconsistency in analyzing the climatic impact of clouds. Tsushima and Manabe (2001) analyzed the impact of cloud feedback (CF) on annual change in global mean SAT using data from space observations of the Earth's radiation budget (ERB), bearing in mind the adequacy of CF consideration in numerical climate models by comparing calculated and observed change in global mean SAT. It follows from the observed data that the global annual change in SAT is in phase with the annual change of SAT in the Northern Hemisphere, and its amplitude constitutes 3.3°C (this in-phase character is determined by concentration in the Northern Hemisphere of continents contributing most to the amplitude in annual SAT change).

Analysis of the data from ERB component observations from February 1985 to February 1990 shows that global mean values of both short-wave and long-wave radiative forcings (SWRF and LWRF) do not depend to any significant degree on annual change in global mean SAT (the ERB data considered refer only to the band 60°N–60°S). Thus, cloud cover dynamics causes neither increase nor decrease in the annual change of SAT. The data on SWRF and the optical properties of clouds show that it is not just albedo but the number and height reached by the top of clouds that depend weakly on SAT; hence, CF does not affect annual change in global mean SAT to any significant degree.

It could be drawn from this conclusion that the impact of CF on annual change in global mean SAT or in global warming is negligible. However, this speculative opinion is dangerous in view of the complexity of the spatial field of SAT. Calculations using three climate models—taking into account the dynamics of the interaction of the microphysical properties of clouds—concluded there was a substantial increase in cloud top albedo with increasing SAT, which disagrees with observational data. This situation reflects the fact of the inexact estimation of the role of cloud feedbacks.

### 5.5.3 An adaptive geoinformation technology-based approach to the monitoring and prediction of natural disasters

Solution of the problem of reliably forecasting natural disasters requires the creation of efficient information technology for use in the geoinformation monitoring system (GIMS). This technology includes areas that are responsible for:

(1) measurement planning;
(2) development of algorithms for complex data-processing from different spheres of knowledge;
(3) creation of methods for decision-making on the basis of dynamic information analysis; and
(4) risk assessment from realization of these decisions.

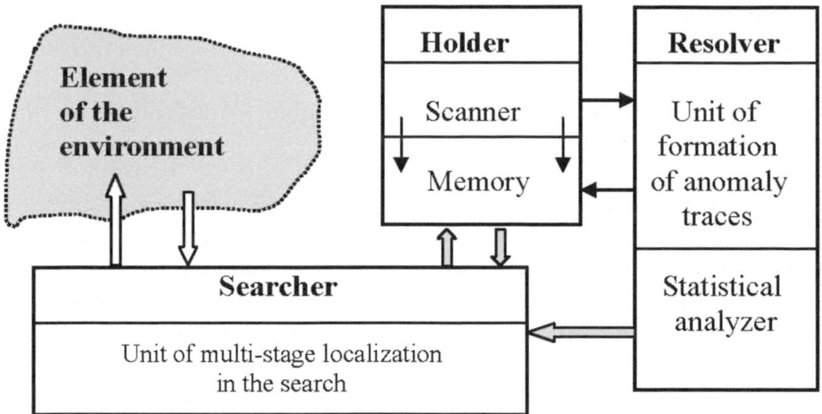


**Figure 5.5.** Monitoring system to detect anomalies in the environment (Kondratyev and Krapivin, 2004).

Risk assessment of the outcomes of natural disasters (when the disaster is a function of regional parameters) is a priority for future investigation. Let us now analyze the problems of natural disaster prediction and of GIMS synthesis for receiving, storing, and processing the necessary information for solution of these problems. An adaptive procedure is proposed for predicting when a natural disaster might occur. It is based on the mathematical model of the environment and evaluation of relative indicators. This procedure brings together the search for and identification of extreme ecological and human-made situations based on remote observations of environmental parameters. The general functional scheme of a standard monitoring system (SMS) is presented as a set of interacting sub-systems: holder, resolver, searcher, and fixer (Figure 5.5). The holder is a sub-system for cartographic data (image) acquisition and storing. The fixer records data from different remote and contact sensors (radar, radiometers, etc.) and memorizes "suspicious" elements in accordance with the algorithms included in the system. The resolver analyzes the input data from the fixer and holder and makes a decision about the signal (anomaly) or the noise (background) origin of these elements. The searcher, at the resolver's command, locates the anomaly and coordinates the functions of the SMS by specifying data on the observed anomaly.

Functional, structural, and algorithmic problems arising due to limited technical means, memory volume, and fast operating of onboard devices were considered by Kondratyev and Krapivin (2005c). These authors proposed a structure for a monitoring system based on realization of the following functions:

(1) periodic observation and examination of environmental (Earth surface) elements;
(2) memorizing suspicious elements;

(3) accumulation over time of data on fixed elements of the environment (Earth surface) for statistical analysis to resolve whether fixed suspicious elements are noise or signal in character; and
(4) location of environment anomalies and possible identification of where the catastrophe might strike.

As a result of the dynamic variation in anomalies, the monitoring system should perform a periodic survey.

There are many parameters describing the environmental conditions on Earth—for example, soil moisture and moisture-related parameters such as the depth of a shallow water table and contours of wetlands and marshy areas. Knowledge about these parameters and conditions is very important for agricultural needs, water management and land reclamation, for measuring and forecasting trends in regional to global hydrological regimes, and for obtaining reliable information about water conservation estimates.

In principle, the information required may be obtained by using on-site measurements and remote-sensing or by getting access to a priori knowledge-based data in GIS databases (Maguire *et al.*, 1991). However, there are problems arising here that consist in solving the following:

- What kinds of instruments are to be used for conducting so-called ground-truth and remote-sensing measurements?
- What is the cost of on-site and remote-sensing information?
- What kind of balance is to be struck between the information content of on-site and remote-sensing observations and their cost?
- What kinds of mathematical models can be used both for interpolation of data and their extrapolation in terms of time and space in attempts to reduce the frequency and thus the cost of observations and to increase the reliability of forecasting the environmental behavior of the items observed?

These and other problems can be solved by using a monitoring system based on combining the functions of environmental data acquisition, control of data archives, data analysis, and forecasting the characteristics of the most important processes in the environment. In other words, unifying all of these elements in this way has brought about a new information technology: GIMS technology. The term GIMS and the relationship between GIMS and GIS can be described by the formula: $GIMS = GIS + Model$. There are two ways of looking at the GIMS. In the first view the term "GIMS" is synonymous with "GIS". In the second view the definition of GIMS expands on the GIS. In keeping with the second view the main units of the GIMS are considered below.

The basic component of the GIMS is considered as a natural sub-system interacting through biospheric, climatic, and socio-economic connections with the global nature–society system (NSS). A model is created describing this interaction and the functioning of various levels of the space–time hierarchy of the whole combination of processes in the sub-system. The model encompasses the characteristic features of typical elements in natural and anthropogenic processes, and model development is based on the existing information base. The model structure is oriented toward an adaptive regime in its use.

Combining an environmental information acquisition system, a model of how a typical geoecosystem functions, a computer cartography system, and a means of artificial intelligence will result in creation of a geoinformation monitoring system of a typical natural element capable of solving the following tasks:

- evaluation of the global change effects on the environment of a typical element of the NSS;
- evaluation of the role of environmental change occurring in a typical element of the climatic and biospheric changes on Earth;
- evaluation of the environmental state of the atmosphere, hydrosphere, and soil–plant formations;
- formation and renewal of information structures on ecological, climatic, demographic, and economic parameters;
- operative cartography of the situation of the landscape;
- forecasting the ecological consequences of realizing certain anthropogenic scenarios;
- typifying land covers, natural phenomena, populated landscapes, surface contamination of landscapes, hydrological systems, and forests; and
- evaluation of population security.

Construction of a GIMS is connected with consideration of the components of the biosphere, climate, and social medium characterized by a given level of spatial hierarchy. Model evaluation of the interactions between these components permits any negative environmental change in the given area to be discovered.

The approach of the moment of the origin of a natural disaster can be characterized by the vector $\{x_i\}$ getting into some cluster of the multi-dimensional phase space $X_c$. In other words, proceeding from verbal reasoning to a quantitative estimation of this process, we introduce a generalized characteristic $I(t)$ of a natural disaster and identify it with the calibrated scale $\Xi$, for which we postulate the presence of relationships of the type $\Xi_1 < \Xi_2$, $\Xi_1 > \Xi_2$, or $\Xi_1 \equiv \Xi_2$. This means that there always exists a value of $I(t) = \rho$ which determines when a natural disaster of a given type can be expected: $\Xi \to \rho = f(\Xi)$ where $f$ is some conversion of the notion "natural disaster" into a number. As a result, the magnitude $\theta = |I(t) - \rho|$ determines the expected time interval before the catastrophe occurs.

Let us search for a satisfactory model to transform the portrait of a natural catastrophe into verbal notions and indicators and subject it to formalized description and transformation. With this aim in view, we select $m$ element sub-systems at the lowest level in the $N \cup H$ system, any interaction between which we can determine with the matrix function $A = \|a_{ij}\|$, where $a_{ij}$ is an indicator of the level of dependence of relationships between sub-systems $i$ and $j$. Then the $I(t)$ parameter can be estimated as the sum:

$$I(t) = \sum_{i=1}^{m} \sum_{j>i}^{m} a_{ij}$$

In general, we get $I = I(\varphi, \lambda, t)$. For a limited territory $\Omega$ with an area $\sigma$ the indicator $I$ is determined as an average value:

$$I_{\Omega}(t) = (1/\sigma) \int_{(\varphi,\lambda) \in \Omega} I(\varphi, \lambda, t)\, d\varphi\, d\lambda$$

Entering the indicator $I$ unveils a scheme for monitoring and forecasting natural catastrophes. A possible structure for such a monitoring system is presented in Figure 5.5; this structure has functions that search for, forecast, and escort natural catastrophes. Three levels are selected in the system—holder, resolver, and searcher—blocks of which have the following functions:

- periodic examination of Earth's surface elements;
- memorizing suspicious elements;
- forming of traces of the moving anomaly from suspicious elements;
- accumulation over time of data on fixed elements of the Earth's surface for statistical analysis to resolve whether fixed suspicious elements are noise or signal in character; and
- stage-by-stage localization of the procedure in the search for an anomaly.

The efficiency of such a monitoring system depends on the parameters involved in technical measuring and algorithms for observational data-processing. It is important here to use a model of the environment in parallel with the formation and statistical analysis of the series $\{I_{\Omega}(t)\}$ and adapted to the regime of monitoring according to Figure 5.5.

As can be seen from our criterion of an approaching natural catastrophe, the form and behavior of $I_{\Omega}(t)$ are special for each type of process in the environment. A complicated problem consists in determining these forms and their respective classification. For instance, such frequent dangerous natural events as landslides and mudflows have characteristic features, like a preliminary change in relief and landscape, which can be successfully recorded from satellites in the optical range. This together with data of aerial photography and surface measurements of relief slopes, exposition of slopes and the state of the hydro-system should make it possible to predict them several days beforehand. However, the restricted capabilities of the optical range under conditions of clouds or vegetation cover need to be broadened by introducing remote-sensing systems in the microwave range of the electromagnetic spectrum. Then, in addition to specific indicators of landslides and mudflows it is possible to add such informative parameters as soil moisture and biomass, since increase in soil moisture is conducive to the appearance of landslides, but increase in biomass is indicative of the increasing restraining role of vegetation cover against mountain soils and rocks moving. This is important when looking for snow–stone or simply snow avalanches. Building up a catalog of such indicators for all possible natural catastrophes and contributing their knowledge base can increase the efficiency of the monitoring system.

Knowledge of the set of informative indicators $\{x_i^j\}$ characterizing a natural catastrophe of *j*th type and a priori determination of the cluster $X^j$ in space of these indicators allows calculation of the velocity $v_j$ of approach of the point $\{x_i^j\}$ to the center of $X^j$ during the satellite observation and, thereby, evaluation of the time of arrival of the catastrophe. Other algorithms for forecasting natural catastrophes are possible. For instance, forecasting a forest fire is possible because of the dependencies of forest radiothermal radiation on the different wavelengths that occur as a result of moisture in combustible timber, which as a rule is situated between the layers. Knowledge of such dependencies gives a real possibility to assess the danger of forest fire by taking into account the water content in the plant cover and in the upper layer of the soil.

The studies of many authors have shown that there is a real possibility to assess fire danger in waterlogged and wet forests by considering the water content in the vegetation cover and upper layer of soil using microwave measurements within the range of 0.8–30 cm. Multi-channel observations facilitate forest classification by the fire stage categories based on the algorithms of cluster analysis. The efficiency of such methods depends on detailed descriptions of forest structure in the model, reflecting the state of the canopy and density of trees. The most dangerous and difficult to detect are low (litter) fires. In this case, the three-layer model of the soil–stems–canopy system is efficient when coordinated with the fire danger indicator $I(\lambda_1, \lambda_2) = [T_b(\lambda_1) - T_b(\lambda_2)]/[T_b(\lambda_1) + T_b(\lambda_2)]$ (Kondratyev *et al.*, 2004). For instance, between $\lambda_1 = 0.8$ cm and $\lambda_2 = 3.2$ cm indicator $I$ changes from approximately "$-0.25$" in zones where fire is absent to "0.54" in zones where there is fire. When $I \approx 0.23$ appears in a zone it is the first indicator of litter burning. The value $I$ also depends (albeit weakly) on the distribution of combustible layers in the forest such as lichens, mosses, grassy patches, and dead pine-needles or leaves.

Use of such a three-level model for decision-making about the approach of a natural catastrophe (in this case fire) depends on the consensus between the spatial and temporal scales of the monitoring system and the corresponding characteristics of the natural phenomena. Delayed-action natural catastrophes are the most complex to forecast, because such a delay may take decades. Ozone holes, global warming, desertification, biodiversion reduction, overpopulation of the Earth, etc. belong to this category of catastrophe. Reliably forecasting the appearance of such catastrophes or undesirable natural phenomena indicative of their approach at a regional scale can be accomplished by means of the GMNSS (Kondratyev *et al.*, 2004), which uses information from updated global databases as well as current satellite and land-based measurements.

GMNSS application to a set of global studies has shown that such technology permits not only prediction of delayed-action catastrophes, but also suggests scenarios of their prevention. One such scenario considered by Krapivin and Phillips (2001) involved reconstructing the water regime for the Aral–Caspian aquageosystem. What the GMNSS came up with is shown in Figure 5.6. It can be seen that allowing rain to fall on the lowlands on the east coast of the Caspian Sea without subsequent

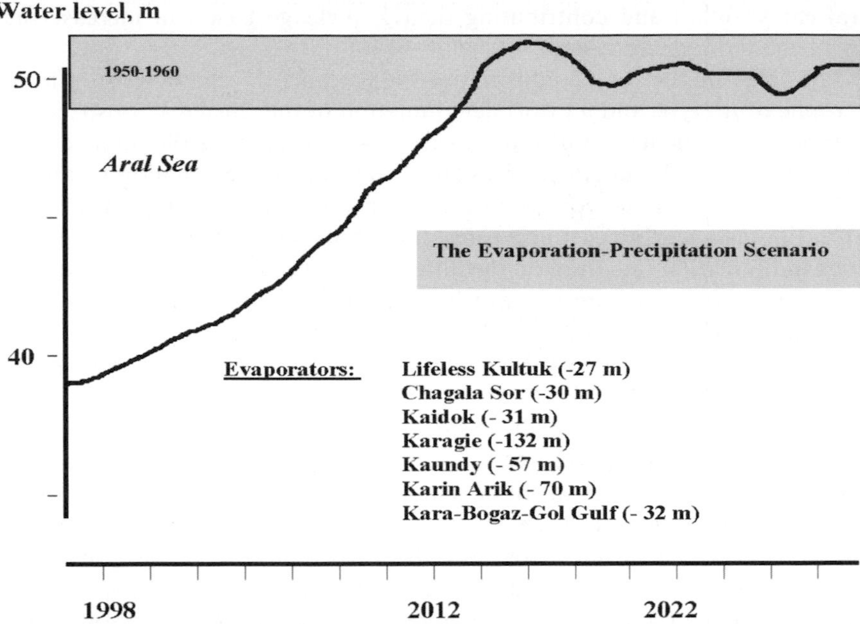


**Figure 5.6.** Possible dynamics of Aral Sea levels (in meters with respect to the World Ocean level) as a consequence of forced evaporators on the hydrological regime of the territory of the Aral–Caspian aquageosystem since 1998 (Brebbia, 2004).

anthropogenic influence can sharply change the hydrology of the territory between the Aral and Caspian Seas. Of course, this is merely a demonstration of the possibility of using the GMNSS to evaluate the consequences of the realization of such a scenario and its influence on the environment. This brings with it an ensemble of problems in how studies of such a scenario are organized, solution of which is possible within the framework of a complex research program by monitoring the zone of Aral and Caspian Sea influence.

Natural disasters can be classified in different categories, the value $I(t)$ of which is defined by the set $x = \{x_1, \dots, x_n\}$, where $x_i$ is the identifying characteristic of the $i$th sub-system or process of the environment. As has been shown, preliminary calculations of hurricane likelihood is indicated by the increase of $I(t)$ in a given area. Satellite observations of aquatories should facilitate the search for possible zones of the World Ocean where hurricane likelihood can be discovered by means of $I(t)$ variation. For example, Hurricane Gustav (Perrie *et al.*, 2004) had the following variables: changes in atmospheric pressure, increase in $CO_2$ exchange, change in vertical water velocity by 0.0001 m/s, and decrease in water temperature of 2–3°C. These variables—as components of $I(t)$—increased $I(t)$ by 120% with respect to the background level. Analogous use of $I(t)$ can help to predict earthquakes. In the case of earthquakes an indicator $I(t)$ is formed from characteristics of the vegetation canopy environment. Satellite techniques as well as devices located in the canopy

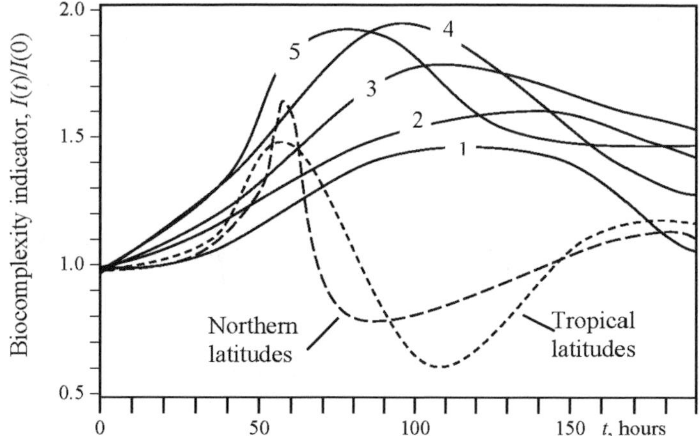


**Figure 5.7.** Biocomplexity indicator of the dynamics in hurricane zones (solid curves) and earthquake zones (dashed curves). Numbers indicate the hurricane category: wind speed (m/s) is changed in the interval 1 (33–42), 2 (43–49), 3 (50–58), 4 (59–69), and 5 (>70).

permit assessment of the temperature, concentration of $CO_2$, and other physical parameters within the canopy. Figure 5.7 demonstrates some calculations that characterize the dynamics of biocomplexity indicators. We can see that a hurricane increases the biocomplexity indicator depending on its category. The form of biocomplexity indicator in the case of earthquakes is more complex.

Reliably forecasting natural catastrophes requires the creation of an efficient information technology for introduction in environmental monitoring systems. This technology must include areas responsible for measurement planning, development of algorithms for complex data-processing from different spheres of knowledge, creation of methods for decision-making based on dynamic information analysis, and risk assessment of the accomplishment of such decisions. Risk assessment of fatalities, in the build-up to a natural catastrophe, as a function of regional parameters is a top priority for future investigations. It is known that such risk is heterogenously distributed in space. For example, average annual risk for different territories is evaluated by the following average factors (ppm): Globe 1, Bangladesh (floods) 50, China (floods and earthquakes) 25, Turkey/Iran/Turkmenistan (earthquakes) 20, Japan (earthquakes) 15, Caribbean region/Central America (storms, earthquakes, volcano eruptions) 10, Europe <1, and U.S.A./Canada <0.1. Consequently, account of this unevenness in the model of natural catastrophes will permit us to optimize the functions of the future system of monitoring.

There are many reviews dealing with hurricane seasons in different tropical zones of the World Ocean (Avila and Brown, 2006; Pasch, 2006; Pasch *et al.*, 2004). The main problem during the active tropical cyclone season in the Atlantic Basin concerns predicting the onset of a hurricane. Such prediction models have been developed by many experts (Elsner and Jagger, 2006; Lehmiller *et al.*, 1997). There are unusual approaches to the solution of this problem—see Rathford (2006), for example.

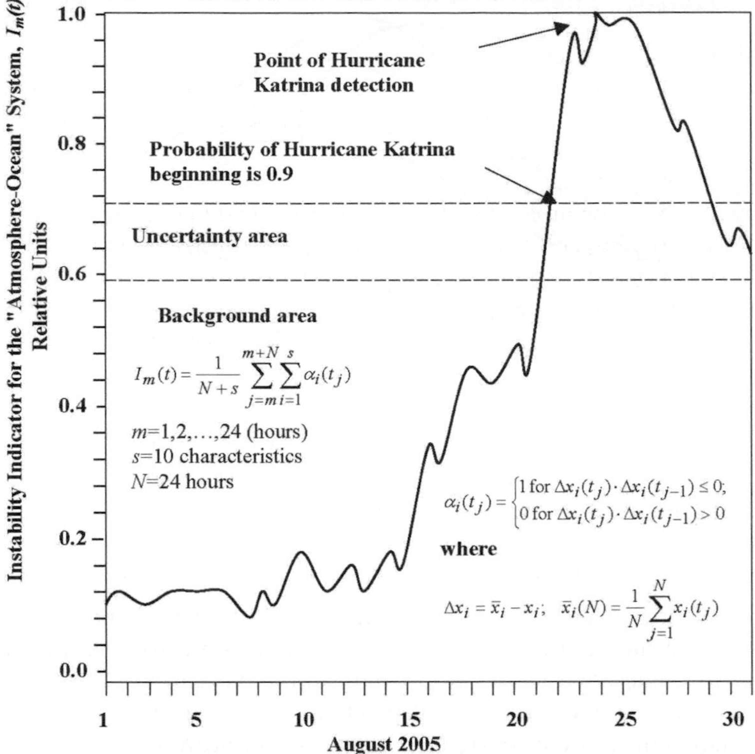


**Figure 5.8.** Hurricane indicator assessment.

Figure 5.8 gives an example of hurricane indicator dynamics calculated in this instance as the indicator of environment instability.

The approach proposed by Kondratyev and Krapivin (2005b) is principally aimed at creating such a technology based on assessment of the environment instability index. However, to progress toward achievement of this goal, a set of fundamental studies needs to be carried out and an ensemble of organizing and engineering problems needs to be solved. A priority here is the need for ranked systematization of natural catastrophes by highlighting their typical signs. This is a principle precondition for realization of the stages of the above-mentioned three-level decision-making procedure concerning the approach of a natural disaster. Maps compiled by Doomsday (2004) of dangerous natural zones in the U.S.A. with indication of the types and levels of dangers are good examples of the solution of a similar task. Another important task is modernization of the GMNSS: this should be done by adapting it to the problem of forecasting natural disasters and by parametrically coordinating it both with current environmental monitoring systems and those planned. Finally, full entailment of these ideas requires the concentration of knowledge-based, economic, and technical resources in an international united center of global geoinformation monitoring.

### 5.5.4 Potential future natural disasters

The dynamics of NSS development indicates that the diversity of potential natural disasters can only broaden due to the appearance of new types of disasters connected with broadening the spheres of new technology application (Table 5.8). The number of anthropogenically induced disasters will increase despite the growing efficiency of nature monitoring systems. The problem of how natural and anthropogenic constituents interact in NSS dynamics will be multi-dimensional in character in the space of characteristics of the impact of various types of catastrophes on society. The geological, climatic, ecological, and medical areas of this space will be quite different from how they are currently perceived (the early 21st century) by the planetary population. This will be connected with a change in the rhythm of catastrophes and transformation of the notions of real and subjective risk.

As Wright and Erickson (2003) noted, future natural disasters will mainly cause economic damage. Potential disasters in the 21st century include:

- instant greenhouse effect due to positive feedbacks of methane;
- rapid rise in sea level due to polar ice melting; and
- global change in World Ocean circulation.

These geophysical catastrophes can be both natural and anthropogenic in origin and have a number of different causes. For instance, accumulated supplies of methane held in sediments, especially in northern latitudes, can be destabilized in case of climate warming and cause a sharp increase in its concentration in the atmosphere. Clathrate compounds buried in the permafrost down to depths of 200 m will remain isolated unless the permafrost melts. Accumulated methane currently being released could enhance the greenhouse effect by 25%. However, as in the case of $CO_2$, there may not be a problem here since released methane entering the global biogeochemical cycle, together with other gases, can be transformed and partially assimilated by the ocean. There is a similar uncertainty in assessment of the consequences of methane release from deposits in the form of methane hydrates and other gases.

The consequences of Antarctic ice melting, whose volume is estimated at 3.8 million km$^3$, can be reduced to assessment of the scale of the rise in the World Ocean level and enumeration of flooded land territories. With $CO_2$ doubling in the atmosphere this could happen. The most pessimistic scenarios show that a rise in the World Ocean level by 4–7 m due to Antarctic and Greenland ice melting has a 5% probability of occurring in the next 200 years. Even existing climate models—with their many imperfections—demonstrate the multitude of unconsidered factors that could crop up with realization of such scenarios including new and enhanced known feedbacks that could stabilize the climate.

Thermohaline circulation in the World Ocean is a key mechanism of the global climate system. Its horizontal constituents are responsible for almost half the heat exchange between equatorial and polar zones, and its vertical components govern the fluxes of heat, nutrients, and gases (including $CO_2$) between the atmosphere and deep layers of the ocean. Realization of the global scenario of warming will apparently lead

**Table 5.8.** Possible future global catastrophes (Holtz, 2005; Zavarzin, 2003).

| Natural and human-made catastrophes | Some comments |
|---|---|
| Earthquake | Possible super-earthquake could cause trillions of dollars in damage and could trigger a worldwide depression |
| Pandemic | A naturally-arising pathogen could kill millions of people. Evolutionary pressures tend to make pathogens less virulent over time, and newly-arising pathogens rarely kill their host species even at initial outbreak. Genetically-engineered pathogens have unpredicted consequences for people |
| Alien aggression | Human history has witnessed many occasions when the arrival of one race of *Homo sapiens* from another territory heralded the end of an indigenous *Homo sapiens* race |
| Interplanetary impact | The impact on Earth of an asteroid or comet a few miles in diameter would have devastating effects—blast waves, tidal waves, fire, and smoke. The amount of smoke and debris held in the atmosphere would decrease the amount of sunlight for one or two agricultural seasons, which could lead to the starvation of billions of people |
| Supernova | A supernova would have to be within a few tens of light years of Earth for its radiation to endanger life. There are no known stars that close to Earth that are likely to become supernovas in the next few million years |
| Ice age | The next ice age would turn many problems that exist now on their head |
| Magnetic field reversal | This process is believed to happen gradually and probably would not affect humanity catastrophically |
| Nuclear catastrophe | There are three risks: radioactive pollution (fallout) from accidents or testing, limited nuclear bombing (i.e., Hiroshima, Nagasaki), and a full-blown nuclear war. Each of these events could set back human progress by different degrees |
| Cultural decline | Vice, crime, and corruption currently on the increase in many regions of the world could lead to social collapse and cultural stagnation |
| Bio-terrorism | Uncontrolled use of different chemical and biological processes could result in extinction for many species (humans included) |
| Robot aggression | A combination of robotics and artificial intelligence could lead to the creation of a new dominant species that would not tolerate human control |
| Nanoplague | The use of nanotechnology could change energy flows on Earth and lead to unpredicted consequences |
| Volcanoes | Possible super-volcano eruption could cause a global catastrophe. The sky would darken, black rain would fall, and the Earth would be plunged into the equivalent of a nuclear winter. The last super-volcano to erupt was Toba, 74,000 years ago in Sumatra. Ten thousand times bigger than Mt St Helens, it represented a global catastrophe—dramatically affecting life on Earth |
| Tsunamis | Super-tsunamis come along from time to time. They can kill thousands of people and destroy economic structures over enormous areas. |

to a change in the vertical profiles of the temperature and salinity of water, and hence will change the structure of ecological systems. Wright and Erickson (2003) noted that in discussions of potential changes in the structure of the oceanic fields numerous questions arise, the answers to which are problematic at the present level of knowledge. The role played by the inertia of the Northern Hemisphere ice cover dynamics in formation of mechanisms of the World Ocean circulation change is not clear. Also, there remains great uncertainty in assessment of the respective variability of atmospheric circulation. All this confirms the conclusion of Kondratyev (1999a) that a study of global dynamics that takes geophysical and ecological constituents into account is only possible with an interdisciplinary approach that considers NSS multi-dimensionality. Of course, using different scenarios of anthropogenic development should remain as one approach to parametrization of some NSS functions, but the spectrum of these scenarios should be confined to consideration of economic influences on the environment.

Consideration of potential natural disasters using available forecasts for social development is important for understanding the strategy of the interaction between social development and nature. Humankind as an element of the biosphere will inevitably become an object of biophysical catastrophes with dire consequences and, maybe, with irreversible results. Therefore, when studying global ecodynamics and working out scenarios of development, it is necessary to take NSS multi-dimensionality into account and remember there is a boundary beyond which human existence becomes problematic.

In the future the character of natural disasters may change due to the impact of new technologies on processes in the environment. In fact, catastrophes for future generations may happen as a result of transformation of mechanisms in the biological regulation in ecosystems. Therefore, the theory of catastrophes and risk is faced with principally new problems of predicting the way in which civilization will develop. In discussions of global dynamics, present sociologists and philosophers—in contrast to specialists in exact sciences who employ notions of "information society" or "technosphere"—use the concept of "risk society". The safety of people against natural disasters will likely decrease because of a broadening spectrum of dangers and their possible scale, despite the development of sciences and industry. Therefore, in the future, development of the theory of risk will mainly be determined by political decisions. This will mainly be connected with forecasting new risks, which is possible with better use of the nature–science NSS component, in order to reduce the level of danger.

### 5.5.5 Assessments of the accomplishment of some scenarios

After the tragedy in Asia on 26 December 2004 as a result of the magnitude 9 earthquake near the northwestern coast of Sumatra, the problem of extreme natural phenomena prediction has become urgent. It is clear that present geophysical science can only comment on the causes of earthquakes, putting forward various hypotheses that explain them as shifts in the Earth's crust. The most complicated problem facing present-day science is arguably earthquake prediction. Despite the existence of specialist centers that can identify minute oscillations in the Earth's crust, the progress of the scientific community in studying the laws dictating planetary development is still negligible. Nevertheless, some progress in predicting other types of natural disasters has been achieved due to development of the theory of climate and global ecodynamics. However, assessments and predictions are only possible with certain scenarios of climate and strategies of humankind development. Therefore, it is important that these scenarios are based on NSS historical considerations (Jolliffe and Stephenson, 2003; Lawrence, 2003; Yue *et al.*, 2005; Vaitheeswaran, 2005; Turner *et al.*, 2001; Thomashow, 2003; Singh, 2005; Schellnhuber and Stock, 2000).

One approach to the prediction of earthquakes and volcanic eruptions involves using the statistics from natural disasters as input information for the GMNSS. The prediction of random successions with the help of the method of evolutionary modeling enables one to determine with some probability the time of occurrence of the next event. As a result this approach gives the following values. In the next 20 years one should expect on average two to three earthquakes every year of magnitude 7.0 and four to five earthquakes of magnitude 5.0–7.0. As for volcanic eruptions, the GMNSS predicts five events by 2020.

Recent years have witnessed the development of the InSAR technique which increases the efficiency of the systems of environmental monitoring at early stages of different natural catastrophes such as earthquakes, landslides, wildfires, floods, etc. (Lanari *et al.*, 1996; Sarti *et al.*, 2006; Verbesselt *et al.*, 2006; Zhan *et al.*, 2006). The origin of many natural disasters can be found by changes in Earth's surface topography. The InSAR technique is based on generation of an interferogram by using SAR images of the same area with two different look angles. One of the main applications of such a polarimetric interferometric method is urban building analysis (Guillaso *et al.*, 2005). Stebler *et al.* (2005) showed that—despite the difficulties in processing and accurately calibrating airborne SAR measurements affected by high topographic gradients—there is a definite need for higher resolution and multiparameter imaging for glaciological applications. Dierking and Busche (2006) showed that the use of L-band radar increases the precision of the InSAR technique when sea ice cover in the polar regions is studied. Sea ice influences a number of important processes such as Earth's radiation balance, the exchange of heat and momentum between the ocean and the atmosphere, and deep ocean water formation. Sea ice plays a significant role in numerous environmental processes at high latitudes. Long-term changes in the sea ice cover state could therefore be an indicator of broader climate change such as global warming. This is the reason imaging radars are an important component of the net of spaceborne and airborne sensors used to monitor the sea ice cover in the Arctic and Antarctic.

Prediction of other types of natural disasters using the GMNSS is possible because the model considers all direct connections and feedbacks in the biosphere–climate system. To make such predictions it is necessary to prescribe scenarios of the potential development of the interaction between society and the environment. The diversity of such scenarios complicates the problem, though using evolutionary technology here makes it possible to reveal the most probable trends in this inter-

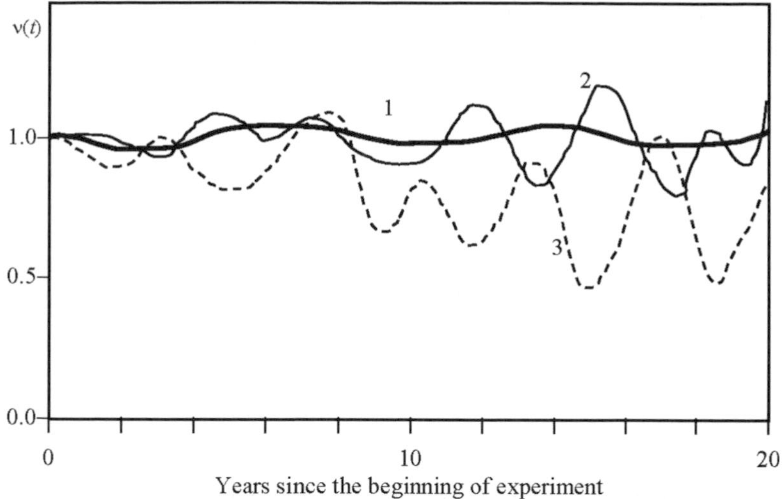


**Figure 5.9.** Assessment of survivability of the Peruvian upwelling ecosystem with different scenarios of global ecodynamics. Notations: 1, A1T Message; 2, A1 AIM; 3, A2 ASF (Edmonds *et al.*, 2004).

action. Let us consider some outcomes of SRES scenarios (Edmonds *et al.*, 2004). The most pessimistic scenarios are A1G MiniCAM and A2ASF which lead to an increase in the partial pressure of $CO_2$ in the atmosphere by 2020 up to 390–410 ppm and by 2100 up to 520–550 ppm. As a result, the pH of the upper layer of the oceans, especially its coastal basins, decreases, which leads to changes in trophic relationships between ecosystem elements. A characteristic example is the ecosystem of the Peruvian upwelling whose trophic pyramid under standard conditions is characterized by its spatial binary nature. The curves in Figure 5.9 demonstrate the state of survivability of this ecosystem evaluated by the criterion:

$$\nu(t) = \sum_{i=1}^{m} B_i(t) \Big/ \sum_{i=1}^{m} B_i(t_0),$$

where $m$ is the number of trophic levels, $t_0 = 1999$, and $B_i(t)$ is the total biomass of the $i$th trophic level over the water basin. It is assumed that minimum concentrations of nutrients not assimilated at other levels constitute 10% of their initial values.

Figure 5.9 demonstrates the response of the system to an increase in upper-layer temperature. Calculations show that an increase in temperature of 0.4°C is harmless, and that a greater increase would result in the system changing its phase state. In the latter case, the effect of spatial binary relations in the trophic pyramid disappears and the system enters the phase of unstable functioning. In general, the GMNSS makes it possible to study the behavior of land and marine ecosystems in the various scenarios indicated in Table 5.9. In the case of scenarios A1T Message and B1 Mage the climatic situation in the Peruvian upwelling basin does not change substantially and the ecosystem changes trophic structure only in the coastal zone between

**Table 5.9.** General characteristics of scenarios in the SRES series by the rate of development of technologies of extraction, re-equipment, and distribution of energy resources (Arnell, 2004; Fenhann, 2000; Nakicenovic and Swart, 2000; Nicholls, 2004).

| Class of scenario | Coal | Oil | Gas | Non-fossil fuels |
|---|---|---|---|---|
| A1B | Average | High | High | High |
| A2 | Average | Low | Low | Low |
| B1 | Average | Average | Average | Medium–high |
| B2 | Low | Below average | Medium–high | Average |
| A1G | Low | Very high | Very high | Average |
| A1C | High | Low | Low | Low |
| A1T | Low | High | High | Very high |

El Niño periods (Nitu *et al.*, 2004). In the case of scenarios B2 Message and A1 AIM, periods occur when there is a prolonged increase in upper-layer temperature, which causes some imbalance between energy fluxes in the ecosystem, but on the whole its stability is preserved. In the third case, when scenarios A1G MiniCAM or A2 ASF are included, the ecosystem starts moving into another state characterized by long-term reduction in total biomass. Additional experiments show that considerable water temperature oscillations principally change the state of the ecosystem. Phase trajectories of the ecosystem form quasi-periodic structures of the type of standing waves with a shift of the center of masses toward a decrease in $v(t)$. The system can withstand an increase in temperature of more than 5°C for no longer than 190 days. Oscillations in the concentration of dissolved oxygen—which decreases with increasing temperature—should not be beyond $0.2\,\text{ml}\,\text{l}^{-1}$ for longer than 100 days, and the rate of vertical advection should not be below $0.5 \times 10^{-4}\,\text{cm}\,\text{s}^{-1}$. On the whole, assessment of the vitality of the Peruvian upwelling ecosystem shows that with long-term slow changes in environmental conditions the community re-arranges the structure and intensity of energy fluxes between trophic levels. One of the factors of the ecosystem's high stability is the vertical shift of the biomasses of ecosystem components, which makes it possible to preserve the phase pattern of the community for a long time even with substantial changes in environmental parameters as, for instance, in the case of the A2 ASF scenario.

Studies accomplished by many authors (Kondratyev *et al.*, 2004, 2003d; Krapivin and Kondratyev, 2002) show that the study of global ecodynamics requires development of a mathematical tool that can fulfill the interdisciplinary needs of biology, geophysics, economy, sociology, climatology, and biocenology. The GMNSS only partially meets these requirements. One feature of the GMNSS is the possibility to study the processes of interaction between natural and anthropogenic factors that take into account the broad spectrum of direct connections and feedbacks between NSS components. The principal non-linearity in the parametric presentation of these connections complicates the analysis of the laws of global ecodynamics and poses additional problems for evaluation of the numerous parameters that depend on time and space coordinates. Therefore, the reliability of any values and predictions depends on the accuracy of assumptions and scenarios.

Another prediction that can be made using the GMNSS is assessment of the variability in global water balance components. Taking the IPCC IS92a scenario that foresees the growth of population size by 2100 up to 11 billion people as our basis, we can predict that increased rain rates will be observed in northwestern Europe by 2020, which will cause a decrease in atmospheric moisture flow from the European continent to America of about 400 $km^3$ $day^{-1}$. In other regions, the water cycle will vary within ±7% with a gradual increase in amplitude by 2100. As a result, by the end of the century the rain rate will increase near the Pacific coastline of the U.S.A., northeastern India, southwestern China, and the zone of heavy rains in Europe will extend northward. Hence, floods in these regions will be more frequent. At the same time, the rain rate will decrease along the eastern coastline of North America, in the countries of Middle Asia and the Near East, and the regime of contrasting alternation of wet and dry seasons will change in southeastern Asia. For the European continent, a negative fact will be a marked decrease in rainfall in Greece, Italy, and the Caucasus. In Central Europe the regime of precipitation will change by no more than 3%.

The GMNSS can also assess the potential risks of any greenhouse effect. These assessments are exemplified in Figures 5.10–5.12. Comparison of the results of predicted temperature changes obtained with the Hadley Centre model and the

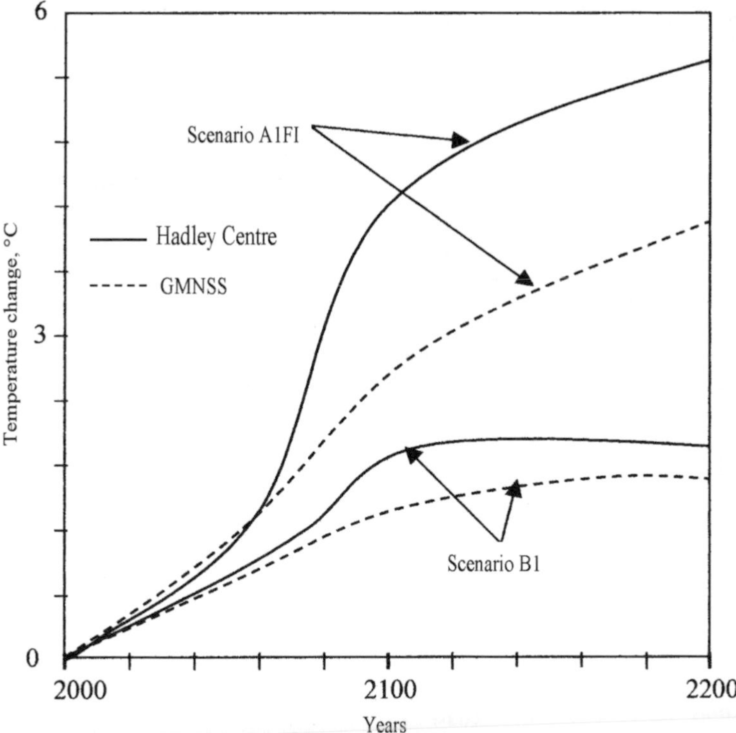


**Figure 5.10.** Forecasts of global mean temperature change using the Hadley Centre climate model and the GMNSS with two scenarios of energy use.

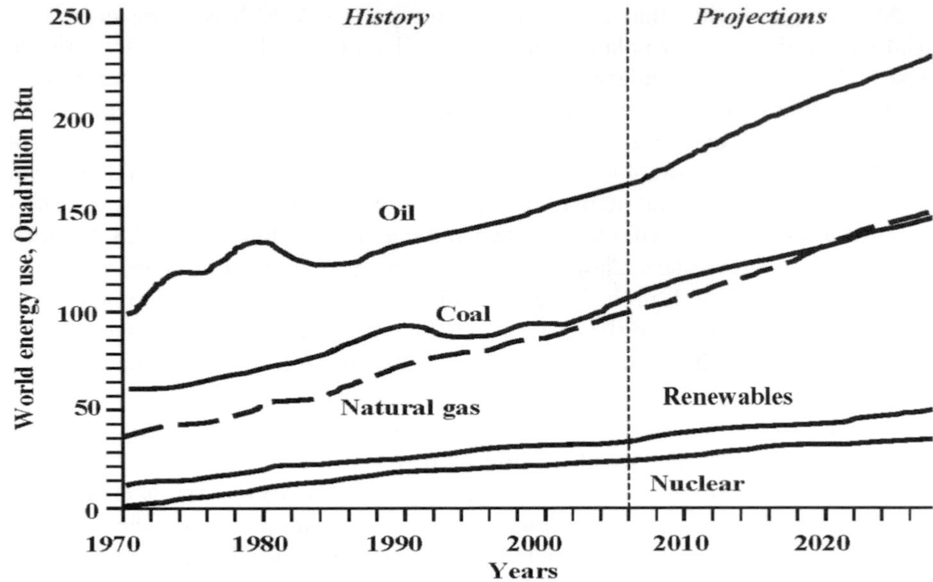


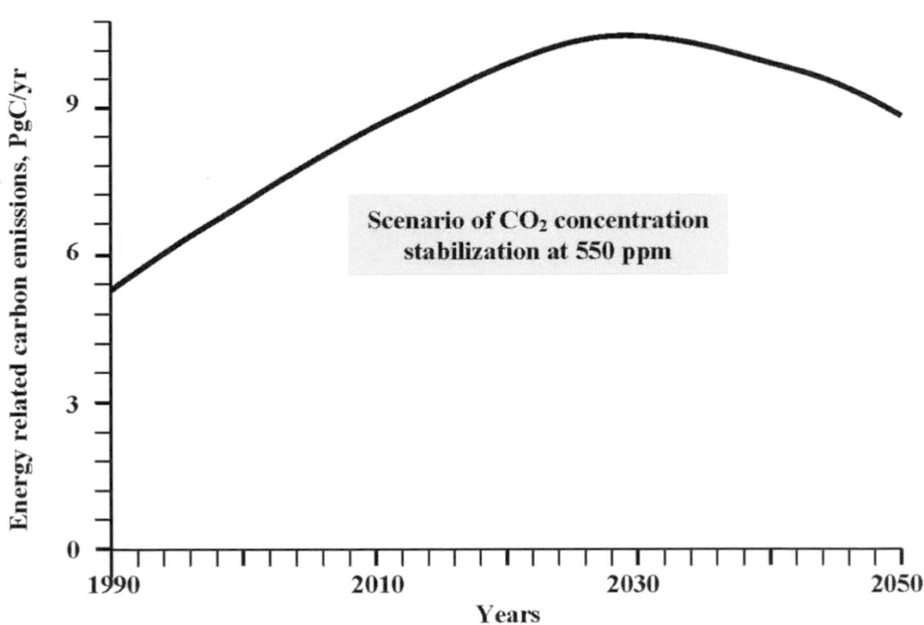


**Figure 5.11.** Possible future perspectives related to climate change.

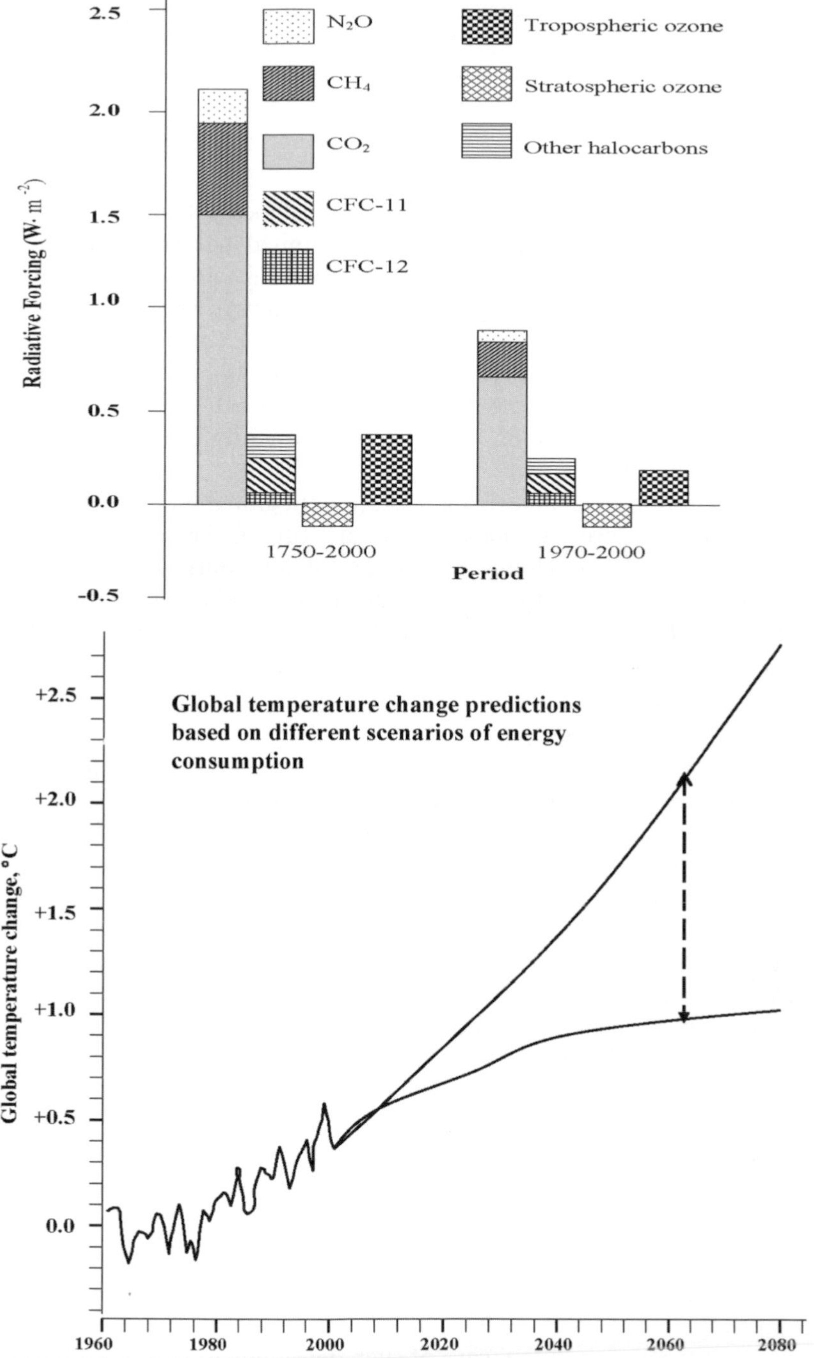


**Figure 5.12.** Forecasts of atmospheric temperature change in the future based on different scenarios of energy consumption.

GMNSS demonstrates the efficiency of GIMS technology and points to the need of further GMNSS modernization by extending its units, especially those connected with parametrizing the interactive mechanisms involved in climate regulation.

### 5.5.6 Concluding remarks

Lomborg (2001, 2004) is of course right to reject the apocalyptic predictions of global ecodynamics based on an exaggerated fear of limited natural resources and the environmental state. Lomborg's opinions and assessments are confirmed by the data in Table 1.21 compiled by Holdren (2003), which characterize both real and potential global energy resources. Energy units are expressed here (in case of non-renewable energy sources) in TW-year, which is equivalent to 31 exaJ (1 TW = 1 TW-yr $yr^{-1}$ = 31.5 exaJ $yr^{-1}$). It should be added that global energy consumption in 2000 constituted about 15 TW or 15 TW-yr $yr^{-1}$ which is expected to increase up to 60 TW-yr $yr^{-1}$ by 2100.

Despite the optimistic data in Table 1.21, present global ecodynamics shows that existing consumption levels in society have no future. Therefore, at the World Summit on Sustainable Development held in Johannesburg in 2002, the need to carry out 10-year programs to bring about stable production and consumption was emphasized; the following recommendations were made (Starke, 2004):

- developed countries should take the leading role to bring about stability between production and consumption;
- these goals should be achieved on the basis of common responsibility;
- stability between production and consumption should play the key role;
- the young must take part in solution of the problem of sustainable development;
- the "polluter pays" principle should be practiced;
- control over the complete cycle in a product's evolution from production, consumption, right through to disposal in order to raise production efficiency;
- support should be given to political parties favoring the output of ecologically acceptable products and rendering ecologically adequate services;
- to develop more ecological and effective methods of energy provision and eliminate energy subsidies;
- to support the free-will initiatives of industry aimed at raising its social and ecological responsibility; and
- to study and introduce means of ecologically pure production, especially in developing countries as well as in small- and medium-sized businesses.

Though these recommendations are rather declarative, they still clearly point to the necessity to change the paradigm of socio-economic development (primarily in developed countries) from a consumption society to priorities of public and spiritual values. A concrete analysis of the ways of such development requires a participation of specialists in the field of social sciences. Some related opinions were expressed by Corcoran (2005) in the Introduction to *The Earth Charter in Action*. Therefore, the question as to whether humans can change climate still needs further studies (Borisov, 2005).

Finally, one can draw the following conclusions:

- existing climate models cannot be used to make decisions and assess the risk of future anthropogenic scenarios becoming reality;
- the level of uncertainty in climate forecasts can be reduced by giving broader consideration in global models to interactive bonds in the NSS and to the mechanisms of biotic regulation of the environment, in addition to improving global monitoring systems; and
- the use of hydrocarbon energy sources in the 21st century will not lead to catastrophic climate change if the Earth's land covers are preserved and the World Ocean is protected from pollution.

# References


Agueman J., Bullard R.D., and Evans B. (eds.) (2003). *Just Sustainabilities: Development in an Unequal World.* The MIT Press, Cambridge, MA, 352 pp.

Aires F., Prigent C., and Rossow W.B. (2005). Sensitivity of satellite microwave and infrared observations to soil moisture at a global scale: 2. Global statistical relationships. *J. Geophys. Res.*, **110**(D11103), doi:10.1029/2004JD005094.

Alessio S., Longhetto A., and Richiardone R. (2004). Evolutionary spectral analysis of European climatic series. *Il Nuovo Cimento C.*, **27**(1), 73–98.

Allen M.R., Stott P.A., Mitchell J.F.B., Schnur R., and Delworth T.L. (1999). Uncertainty in forecasts of anthropogenic climate change. RAL Techn. Rept., No. RAL-TR-084, 10 pp.

Altinbilek D., Karin S., and Taylor R. (2005). Hydropower's role in delivering sustainability. *Energy and Environment*, **16**(5), 815–624.

Alverson K.D., Bradley R.S., and Pedersen T.F. (2003). *Paleoclimate, Global Change and Future.* Springer-Verlag, Heidelberg, Germany, 221 pp.

Amartya Sen and Gates H.L. (eds.) (2006). *Identity and Violence: The Illusion of Destiny.* W.W. Norton, New York, 224 pp.

Anderson P., Petrino R., Halpern P., and Tintinalli J. (2006). The globalization of emergency medicine and its importance for public health. *Bulletin of the WHO*, **84**(10), 835–839.

Anderson W. and Anderson W.T. (2001). *All Connected Now: Life in the First Global Civilization.* Westview Press, New York, 320 pp.

Andoh R.Y.G. and Iwugo K.O. (2002). Sustainable urban drainage systems: A UK perspective. In: E.W. Strecker and W.C. Huber (eds.), *Urban Drainage 2002: Proceedings of the 9th International Conference on Urban Drainage, 8–13 September 2002, Portland, OR*, pp. 350–355.

Andreev I.L. (2003). Russia: View of the future. *Herald of RAS*, **73**(4), 320–339 [in Russian].

Andronova N.G. and Schlesinger M.E. (2001). Objective estimation of the probability density function for climate sensitivity. *J. Geophys. Res.*, **106**(D19), 22605–22611.

Annan K. (2003). Science for all nations. *Science*, **303**(5660), 4–8.

Aota M., Shirasawa K., Krapivin V.F., and Mkrtchyan F.A. (1993). A project of the Okhotsk Sea GIMS. *Proc. of the 8th Int. Sympos. on Okhotsk Sea & Sea Ice and ISY/Polar Ice*

*Extent Workshop, 1–5 February 1993, Mombetsu (Japan)*. Okhotsk Sea & Cold Ocean Research Association, Mombetsu. 1993, pp. 498–500.

Armand N.A., Krapivin V.F., and Shutko A.M. (1997). GIMS-technology as a new approach to dataware for environmental studies. *Problems of the Environment and Natural Resources*, **3**, 31–50 [in Russian].

Arnell N.W. (2004). Climate change and global water resources: SRES emissions and socio-economic scenarios. *Global Environmental Change*, **14**(1), 31–52.

Arsky Yu.M., Zakharov Yu.F., Kalutskov V.A., and Sokolov V.E. (1992). *Ecoinformatics*. Gidrometeoizdat, St. Petersburg, 520 pp. [in Russian].

Arsky Yu.M., Krapivin V.F., and Potapov I.I. (2000). Dataware for ecological studies into problems of environmental diagnostics. *Scientific and Technical Information*, **7**, 7–11 [in Russian].

Artiola J.F., Pepper I.L., and Brusseau M.L. (2004). *Environmental Monitoring and Characterization*. Elsevier, San Diego, CA, 300 pp.

Ashby W.R. (1956). *An Introduction to Cybernetics*. Chapman & Hall, London, 334 pp.

Avila L.A. and Brown D.P. (2006). *Tropical Cyclone Report: Storm Alberto*. NHC, Report No. AL012006, Washington, 17 pp.

Bak P. (1996). *How Nature Works: The Science of Self-organized Criticality*. Copernicus Books, New York, 320 pp.

Balmford A., Bennun L., ten Brink B., Cooper D., Coté J.M., Crane P., Dobson A., Dudley N., Dutton I., Green R.E. *et al.* (2005). The Convention on Biological Diversity's 2010 target. *Science*, **307**, 212–213.

Bagheri A. and Hjorth P. (2005). Monitoring for sustainable development: A systemic framework. *The IJSD*, **8**(4), 280–301.

Basharinov A.E. and Fleishman B.S. (1962). *Methods of Statistical Successive Analysis*. Soviet Radio, Moscow, 352 pp. [in Russian].

Basharinov A.E., Zotova E.N., Naumov M.I., and Chukhlantsev A.A. (1979). Radiation characteristics of plant cover in the microwave range. *Radiotechnics*, **34**(5), 16–20 [in Russian].

Bates D.G. (2005). *Human Adaptive Strategies: Culture, Ecology, and Politics* (278 pp.). Allyn & Bacon, Boston.

Bates T., Scholes M., Doherty S., and Young B. (eds.) (2006). *IGAC Science Plan and Implementation Strategy*. IGBP Report 56, IGBP Secretariat, Stockholm, 44 pp.

Bauer M. and Mar E. (2005). Transport and energy demand in the developing world: The urgent alternatives. *Energy and Environment*, **16**(5), 825–843.

Bazilevich N.I. and Rodin L.E. (1967). The map-schemes of productivity and of the biological cycle of the most significant types of land vegetation. *Bull. of the All-Union Geographical Soc.*, **99**(3), 190–194 [in Russian].

Beder S. (2005). *Setting the Global Agenda: Corporate Coalitions, Coercion, and Control*. Earthscan, London, 224 pp.

Bell S. and Morse S. (2000). *Sustainable Indicators: Measuring the Immeasurable*. Earthscan, London, 175 pp.

Bell S. and Morse S. (2003). *Measuring Sustainability: Learning from Doing*. Earthscan, London, 189 pp.

Belov G.A. (1999). *Political Science*. CheRo, Moscow, 304 pp. [in Russian].

Bengtsson L. (1999). *Climate Modelling and Prediction: Achievements and Challenges*. WCRP/WMO, **954**, 59–73.

Beven K. and Blazkova S. (1998). Estimating changes in flood frequency under climate change by continuous simulation (with uncertainty). In: P. Bronstert and R.P. Samuels (eds.),

*Proceedings of the RIBAMOD Final Workshop, 26–27 February 1998, Wallingford, U.K.*, pp. 269–279.

Beyer A. and Mitthies M. (2001). Long-range transport potential of semivolatile organic chemicals in coupled air–water systems. *Environ. Sci. & Pollut. Res.*, **8**(3), 173–179.

Bhaduri A. and Nayyar D. (1996). *Intelligent Person's Guide to Liberalization*. Penguin Books India, New Delhi, 200 pp.

Blancarte P.R. (2006). *Religion, Culture, and Sustainable Development* (297 pp.). El Colegio de México, México.

Bobylev L.P., Kondratyev K.Ya., and Johannessen O. (eds.) (2003). *Arctic Environment Variability in the Context of Global Change*. Springer-Praxis, Chichester, U.K., 471 pp.

Bodri L. and Čermák V. (1999). Climate change of last millennium inferred from borehole temperatures: Regional patterns of climate changes in the Czech Republic. Part III. *Glob. and Planet. Change*, **21**(4), 225–235.

Boehmer-Christiansen S. (1997). Who is driving climate change policy? *IEA Stud. Educ.*, **10**(1), 53–72.

Boehmer-Christiansen S. and Kellow A. (2002). *International Environmental Policy*. Edward Elgar, Abingdon, U.K., 232 pp.

Böhringer C., Finus M., and Vogt C. (2002). *Controlling Global Warming*. Edward Elgar, Abingdon, U.K., 320 pp.

Boiko I.P. (ed.) (2005). *Proceedings of 7th International Conference of the Russian Society of Ecological Economy: Globalization, New Economy and the Environment. Problems of Society and Business on the Way to Sustainable Development, St. Petersburg, 23–25 June 2005*. St. Petersburg State University, St. Petersburg, 395 pp. [in Russian].

Bolin B. (1999). Global environmental change and the need for international research programmes. *WCRP/WMO*, **954**, 11–14.

Bonan G.B. (2002). *Ecological Climatology: Concepts and Applications*. Cambridge University Press, Cambridge, U.K., 678 pp.

Borisov P.M. (2005). *Can Humans Change Climate?* Science, Moscow, 270 pp. [in Russian].

Borodin L.F., Krapivin V.F., Golfeld G.B., and Nazarian N.A. (1997). A search and identification of extreme ecological and technogenic situations. *Problems of the Environment and Natural Resources*, **10**, 2–20 [in Russian].

Bortnik V.N., Lopatina S.A. and Krapivin V.F. (1994). Simulation system for the study of hydrophysical fields in the Aral Sea. *Meteorology and Hydrology (Moscow)*, **9**, 102–108 [in Russian].

Botkin D. and Keller E. (2002). *Environmental Science: Earth as a Living Planet*, 4th Edition. Wiley, London, 672 pp.

Bounoua L., Defries R., Collatz G.J., Sellers P., and Khan H. (2002). Effects of land cover conversion on surface climate. *Clim. Change*, **52**(1–2), 29–64.

Bousserez N., Attié J.L., Peuch V.H., Nédélec P., Edwards D., Emmons L., Pfister G., and Ziskin D. (2005). First results from assimilation of MOPITT CO into MOCAGE CTM during ITOP experiment. *Geophysical Research Abstracts*, **7**, 9875–9879.

Bowen D.Q. (2000). Tracing climate evolution. *Earth Heritage*, Millennium Issue, 8–9.

Boysen M. (ed.) (2000). *Biennial Report 1998/1999*. Potsdam Institute for Climate Impact Research, Potsdam, 130 pp.

BP (2005). Putting energy in the spotlight. *BP Statistical Review of World Energy June 2005*. BP, London, 44 pp.

Brasseur G.P., Prinn R.G., and Pszenny A.A.P. (eds.) (2002). *Atmospheric Chemistry in a Changing World*. Springer-Verlag, Heidelberg, Germany, 330 pp.

Brebbia C.A. (2004). *Risk Analysis*, IV. WIT Press, Southampton, U.K., 400 pp.

Brebbia C.A. and Sakellaris I. (eds.) (2003). *Energy and the Environment*. WIT Press, Southampton, U.K., 384 pp.

Brewer T.L., Brenton P.A., and Boyd G. (eds.) (2002). *Globalizing Europe. Deepening Integration, Alliance Capitalism and Structural Statecraft*. Edward Elgar, London, 368 pp.

Brown L.R. (2005). Outgrowing the Earth. *The Food Security Challenge in an Age of Falling Water Tables and Rising Temperatures*. Earthscan, London, 224 pp.

Bruntland G. (ed.) (1987). *Our Common Future: The World Commission on Environment and Development*. Oxford University Press, Oxford, U.K., 318 pp.

Bui T.L. (1998). Data processing in the geoinformation monitoring systems in Vietnam. Ph.D. thesis, Institute of Radioengineering and Electronics, Moscow, 354 pp [in Russian].

Bui T.L. and Krapivin V.F. (1998). Technology of simulation modeling in the systems of geoinformation monitoring of megapolises. *Problems of the Environment and Natural Resources*, **2**, 79–85 [in Russian].

Bukatova I.L. and Makrusev V.V. (2004). *Theory of the Integral-Evolutionary Intellectualization of Social Systems*. MIGCM, Moscow, 126 pp. [in Russian].

Bukatova I.L. and Rogozhnikov E.A. (2002). The integral-evolutionary conception of the information system directed at thematic processing of the Earth's remote sounding data. *Problems of the Environment and Natural Resources*, **4**, 31–43 [in Russian].

Bull M., Matthews J., Moroney C., and Smyth M. (2005). *MISR Data Products Specifications*. JPL D-13963, California, 380 pp.

Bunyard P. (1999). Eradicating the Amazon rainforest will wreak havoc on climate. *Ecologist*, **29**(2), 81–84.

Burgman M. (2005). *Risks and Decisions for Conservation and Environmental Management*. Cambridge University Press., Cambridge, U.K., 488 pp.

Callinicos A. (2001). *Against the Third Way*. Polity Press, London, 160 pp.

Camillo P.J., O'Neill P.E., and Gurney R.J. (1986). Estimating soil hydraulic parameters using passive microwave data. *IEEE Trans. on Geosci. and Remote Sensing*, **GE-24(6)**, 930–936.

CDCP (2006). *Environmental Public Health Indicators*. CDCP, Atlanta, GA, 40 pp.

Chambers R. (2005). *Ideas for Development*. Earthscan. London, 2005, 320 pp.

Chang C.-P., Zhang Y., and Li T. (2000). Interannual and interdecadal variations of the East Asian summer monsoon and tropical Pacific SSTs. Part 1: Roles of the subtropical ridge. *J. Climate*, **13**, 4310–4325.

Chapin F.S. III, Matson P., and Mooney K.A. (2004). *Principles of Terrestrial Ecosystem Ecology*. Springer for Science, Amsterdam, The Netherlands, 436 pp.

Chapman W.L. and Walsh J.E. (1993). Recent variations of sea ice and air temperature in high latitudes. *Bulletin of the American Meteorological Society*, **74**, 33–47.

Charitonova V.I. (2004). Spiritual factor in the present life of races of the north and Siberia (39 pp.). *Proceedings of the Institute for Ethnology and Anthropology, Moscow*, **167** [in Russian].

Charitonova V.I. (2006). *Feniks from the Ashes: Siberian Shamanism on the Boundary of Millenniums* (372 pp.). Science, Moscow [in Russian].

Charman D.J. (2002). *Peatlands and Environmental Change*. Wiley, New York, 312 pp.

Chen J.M., Liu J., Leblanc S.G., Lacaze R., and Roujean J.-L. (2003). Multi-angular optical remote sensing for assessing vegetation structure and carbon absorption. *Remote Sensing of Environment*, **84**(4), 516–525.

Chen W., Chen J., and Cihlar J. (2000). An integrated terrestrial ecosystem carbon-budget model based on changes in disturbance, climate, and atmospheric chemistry. *Ecological Modelling*, **135**(1), 55–79.

Chernavsky D.S. (ed.) (2004). *Recognition, Autodiagnostics, Thinking: Synergetics and Human Science*. Radiotekhnika, Moscow, 272 pp. [in Russian].

Chernavsky D.S., Rodshtadt I.V., and Karp V.P. (2005). Artificial neuro-networks and conception of the person auto-diagnostic system. *Artificial Intellect News*, **2**, 63–76 [in Russian].

Cheru F. and Bradford C. (eds.) (2005). *The Millenium Development Goals: Raising the Resources to Tackle World Poverty*. Zed Books, London, 256 pp.

Chin M., Ginoux P., Kinne S., Holben B.N., Duncan B.N., Martin R.V., Logan J.A., Higurashi A., and Nakajima T. (2002). Tropospheric aerosol optical thickness from the GOCART model and comparisons with satellite and sunphotometer measurements. *J. Atmos. Sci.*, **59**, 461–483.

Chistobayev A.I. (2001). Sustainable development: Global context and local specificity. *Proceedings of Russian Geographic Society*, **133**(4), 22–27 [in Russian].

Chistobayev A.I. and Solodovnikov A.Yu. (2005). Assessment of the impact of future projects on the environment: State and problems. *Proceedings of Russian Geographic Society*, **137**(5), 1–11 [in Russian].

Choi S.-D. and Chang Y.-S. (2006). Carbon monoxide monitoring in Northeast Asia using MOPITT: Effects of biomass burning and regional pollution in April 2000. *Atmospheric Environment*, **40**(4), 686–697.

Chukhlantsev A.A. (2006). *Microwave Radiometry of Vegetation Canopies*. Springer-Verlag, Berlin, 2006, 287 pp.

Chukhlantsev A.A. and Shutko A.M. (1988). Microwave radiometry of the Earth's surface: effect of vegetation. *Research of the Earth from Space*, **2**, 67-72 [in Russian].

Chukhlantsev A.A., Shutko A.M., and Golovachev S.P. (2003). *Attenuation of electromagnetic waves by vegetation canopies in the 100–1000 MHz frequency band*. ISTC/IRE Technical Report, #2059-1, Moscow, 59 pp.

Chukhlantsev A.A., Golovachev S.P., Krapivin V.F., and Shutko A.M. (2004). A remote sensing-based modeling system to study the Aral/Caspian water regime. *Proc. of the 25th ACRS, 22–26 November 2004, Chiang Mai, Thailand*, Vol. 1, pp. 506–511.

Churin G.Yu. (2005). Legal mechanisms for transition of society to sustainable development on global and regional levels. In: V.N. Troyan and I.A. Dementiev (eds.), *Sustainable Development and Ecological Management*, St. Petersburf State University, St. Petersburg, pp. 369–372 [n Russian].

Clark P.U. and Mix A.C. (2002). Ice sheets and sea level of the last glacial maximum. *Quaternary Science Reviews*, **21**, 1–7.

Clarke R. and King J. (2004). *The Atlas of Water: Mapping the Global Crisis in Graphic Facts and Figures*. Earthscan, London, 2004, 128 pp.

Condie K.C. (2005). *Earth as an Evolving Planetary System*. Elsevier Academic, Burlington, MA, 461 pp.

Corcoran P.P. (ed.) (2005). *The Earth Charter in Action: Toward a Sustainable World* (192 pp.). KIT Publ., Amsterdam, The Netherlands.

Cordova A.M., Longo K., Freitas S., Gatti L.V., Artaxo P., Procópio A., Silva Dias M.A.F., and Freitas E.D. (2004). Nitrogen oxides measurements in an Amazon site and enhancements associated with a cold front. *Atmos. Chem. Phys. Discuss.*, **4**, 2301–2331.

Corning P. (2003). *Nature's Magic. Synergy in Evolution and the Fate of Humankind*. Cambridge University Press, Cambridge, 464 pp.

Cozzi L. (2003). *World Energy Outlook Insights: Global Energy Investment Outlook*. IEA, Paris, France, 20 pp.

Craik W. (ed.) (2005). *Weather, Climate, Water, and Sustainable Development*. WMO, Geneva, 24 pp.

Cropp R.A., Gabric A.J., McTainsh G.H., Braddock R.D., and Tindale N. (2005). Coupling between ocean biota and atmospheric aerosols: Dust, dimethylsulphide, or artifact? *Global Biogeochemical Cycles*, **19**(GB4002), doi: 10.1029/2004GB002436.

Crossland C.J., Kremer H.H., Lindeboom H.J., Crossland J.I.M., and Le Tissier M.D. (eds.) (2005). *Coastal Fluxes in the Anthropocene: The Land–Ocean Interactions in the Coastal Zone Project of the International Geosphere–Biosphere Programme*. Springer-Verlag, Heidelberg, Germany, 121 pp.

Crowley T.J. (2000). Causes of climate change over the past 1000 years. *Science*, **289**(5477), 270–277.

Cuyvers L. (2001). *Globalisation and Southeast Asian Evidence*. Edward Elgar, London, 368 pp.

Dai A., Washington W.M., Meehl G.A., Bettge T.W., and Stand W.G. (2004). The ACPI climate change simulations. *Climatic Change*, **62**, 29–43.

Dalby S. (2002). *Environment Security*. University of Minnesota Press, Minneapolis, 312 pp.

Daly H.E. (2002). *Sustainable Development: Definitions, Principles, Policies*. World Bank, New York (*http://www.earthrights.net/docs/daly.html*), 12 pp.

Danilov-Danilyan V.I. and Losev K.S. (2000). *Ecological Challenge and Sustainable Development*. Progress-Tradition, Moscow, 416 pp. [in Russian].

Danilov-Danilyan V.I., Losev K.S., and Reif I.E. (2005). *Lead-up to Major Challenge to Civilization: View from Russia*. INFRA-M, Moscow, 224 pp. [in Russian].

Davidenkov I.V. and Kesler Ya.A. (2005). *Civilization Resources*. Eksmo, Moscow, 544 pp. [in Russian].

Davies P.S. (2003). *Norms of Nature: Naturalism and the Nature of Functions*. MIT Press, Cambridge, MA., 250 pp.

Dearden P. and Mitchel B. (2005). *Environmental Change and Challenge: A Canadian Perspective*, Second Edition. Oxford University Press, London, 512 pp.

Decarte R. (1994). *Works in Two Volumes*. Thought, Moscow, 569 pp. [in Russian].

Del Frate F., Ferrazzoli P., and Schiavon G. (2003). Retrieving soil moisture and agricultural variables by microwave radiometry using neural networks. *Remote Sensing of Environment*, **84**(2), 174–183.

Delworth T.L. and Knutson T.R. (2000). Simulation of early 20th century global warming. *Science*, **287**(5461), 2246–2250.

Demirchian K.S. (2001). Once again Russia is misted up. What is to be done? Human aspects of an anti-crisis strategy. *Industrial Bulletin*, **8–9**, 1–4; **10–11**, 3–4 [in Russian].

Demirchian K.S. and Kondratyev K.Ya. (1998a). Development of energetics and environment. *Annals of RAS, Energetics*, **6**, 3–46 [in Russian].

Demirchian K.S. and Kondratyev K.Ya. (1998b). Energetics development and the environment. *Annals of RAS, Energetics*, **6**, 3–27 [in Russian].

Demirchian K.S. and Kondratyev K.Ya. (1999). Scientific substantiation of forecasts of the climatic impact of energetics. *Annals of RAS, Energetics*, **6**, 3–27 [in Russian].

Demirchian K.S., Demirchian K.K., Danilevich Ya.B., and Kondratyev K.Ya. (2002). Global warming, energetics, and geopolitics. *Annals of RAS, Energetics*, **3**, 221–235 [in Russian].

Devkota S.R. (2005). Is strong sustainability operational? An example from Nepal. *Sustainable Development*, **13**(5), 297–310.

DeWitt D.P. and Nutter G.D. (1989). *Theory and Practice of Radiation Thermometry*. Wiley, New York, 1989, 1138 pp.

Diamond J. (2005). *Collapse: How Societies Choose to Fail or Succeed*. Viking Press, New York, 2005, 591 pp.

Dierking W. and Busche T. (2006). Sea ice monitoring by L-band SAR: An assessment based on literature and comparisons of JERS-1 and ERS-1 imaging. *IEEE Trans. on Geosci. and Remote Sensing*, **44**(2), 957–970.

Diffey, B.L. (1991). Solar ultraviolet radiation effects on biological systems. *Review in Physics in Medicine and Biology*, **36**(3), 299–328.

Dodds F. and Pippard T. (eds.) (2005). *Human and Environmental Security: An Agenda for Change*. Earthscan, London, 270 pp.

Dolšak N. and Ostrom, E. (eds.) (2003). *The Commons in the New Millennium: Challenges and Adaptation*. MIT Press, Cambridge, MA., 392 pp.

Dong J., Kaufmann R.K., Myneni R.B., Tucker C.J., Kauppi P.E., Liski J., Buermann W., Alexeyev V., and Hughes M.K. (2003). Remote sensing estimates of boreal and temperate forests woody biomass: Carbon pools, sources, and sinks. *Remote Sensing of Environment*, **84**(3), 393–410.

Doomsday M.A. (2004). *The Science of Catastrophic Events*. Praeger, Westport, CT, 194 pp.

Drabek Z. (ed.) (2001). *Globalisation under Threat: The Stability of Trade Policy and Multilateral Agreements*. Edward Elgar, London, 256 pp.

Dresner S. (2002). *The Principles of Sustainability*. Earthscan, London, 200 pp.

Ebel A., Eckhardt S., Feldmann H., Forster C., Jäger H., Hermann J., James P., Kanter H.-J., Kerschgens M., Kreipl S. *et al.* (2004). Atmospheric long-range transport and its impact on the trace-gas concentrations in the free troposphere over Central Europe (ATMOFAST). *AFO 2000 Newsletter*, **9**, 7–10.

Edmonds J., Joos F., Nakicenovic N., Richels R.G., and Sarmiento J.L. (2004). Scenarios, targets, gaps, and costs. In: C.B. Field and M.R. Raupach (eds.), *Global Carbon Cycle: Integrating Humans, Climate, and the Natural World*. Island Press, Washington, pp. 77–102.

Ehleringer J.R., Cerling T.E., and Dearing M.D. (2005). *A History of Atmospheric $CO_2$ and Its Effects on Plants, Animals, and Ecosystems*. Springer-Verlag, Heidelberg, Germany, 534 pp.

Ehlers E. and Kraft T. (eds.) (2005). *Earth System Science in the Anthropocene: Emerging Issues and Problems*. Springer-Verlag, Heidelberg, Germany, 300 pp.

Ehrlich P.R. and Ehrlich A.H. (2004). *One with Nineveh: Politics, Consumption, and the Human Future*. Island Press, Washington, D.C., 459 pp.

Ehrlich P.R. and Kennedy D. (2005). Millennium assessment of human behavior. *Science*, **309**, 562–563.

EIA (2005) *International Energy Outlook 2005*. DOE/EIA-0484, Washington, D.C., 194 pp.

Elinson M.I. and Sukhanov A.A. (1984). Problems of interconnections in modern microelectronics. *Microelectronics*, **13**(3), 179–195 [in Russian].

Elliot D. (2003). *Energy, Society, and Environment*. Routledge, London, 400 pp.

Elsner J.B. and Jagger T.H. (2006). Prediction models for annual U.S. hurricane counts. *J. Climate*, **19**, 2935–2952.

Elster J. (1986). *An Introduction to Karl Marx*. Cambridge University Press, Cambridge, 200 pp.

Engels F. (1880). Socialisme utopique et socialisme scientifique. *MECW*, **24**, 281-306.

Engels F. (1940). *Dialectics of Nature*. International, New York, 383 pp.

Engman E.T. and Chauhan N. (1995). Status of microwave soil moisture measurements with remote sensing. *Remote Sensing of Environment*, **51**(1), 189–198.

Enz R. (ed.) (2006). *Natural Catastrophes and Man-made Disasters 2005: High Earthquake Casualties, New Dimension in Windstorm Losses* (40 pp.). Swiss Reinsurance, Zurich, Switzerland.

EPA (2001). *Non-$CO_2$ Greenhouse Gas Emissions from Developed Countries: 1990–2010*. EPA-430-R-01-007. U.S. Environmental Protection Agency, Washington, D.C., 132 pp.

EPA (2005). *Proceedings of EPA Science Forum 2005: Collaborative Science for Environmental Solutions, May 16–18 2005?* U.S. Environmental Protection Agency, Washington, D.C., 162 pp.

EPA (2006). *Inventory of U.S. Greenhouse Gas Emissions and Sinks: 1990–2004*. Report No. EPA 430-R-06-002. U.S. Environmental Protection Agency, Washington, D.C., 459 pp.

Evans J.W. (2005). *Overview: Environment Matters. Annual Review*. World Bank, Washington, D.C., pp. 3–7.

Ewing M. (2005). *Access to Information on the Environment*. TIPT, Ireland, 52 pp.

Fall L. (2005). Energy access: Illusion or reality for the poor? *Energy and Environment*, **16**(5), 743–761.

Faludi A. (ed.) (2002). *European Spatial Planning* (235 pp.). Lincoln Institute of Land Policy, Cambridge, MA.

Fasham M.J.R. (ed.) (2002). *Ocean Biogeochemistry: The Role of the Ocean Carbon Cycle in Global Change*. Springer-Verlag, Heidelberg, Germany, 320 pp.

Fedotov A.P. (2002). *Globalistics: Principles behind Present World Science*, 2nd edition. Aspect Press, Moscow, 224 pp. [in Russian].

Fenhann J. (2000). Industrial non-energy, non-$CO_2$ greenhouse gas emissions. *Technological Forecasting and Social Change*, **63**(2–3), 313–334.

Ferrazzoli P. and Guerriero L. (1996). Passive microwave remote sensing of forests: A model investigation. *IEEE Trans. on Geosci. and Remote Sensing*, **34**(2), 433–443.

Field C.B. and Raupach M.R. (2004). *The Global Carbon Cycle: Integrating Humans, Climate, and the Natural World*. Island Press, London, 326 pp.

Filatov N., Pozdnyakov D., Johannessen O.M., Petterson L.H., and Bobylev L.P. (2005). *White Sea: Its Marine Environment and Ecosystem Dynamics Influenced by Global Change*. Springer-Verlag, Heidelberg, Germany, 504 pp.

Fleishman B.S. (1982). *The Principles of Systemology*. Radio & Communication, Moscow, 250 pp. [in Russian].

Fogel L., Owens A., and Walsh M. (1966). *Artificial Intelligence through Simulated Evolution*. Wiley, New York, 250 pp.

Fogg P. and Sangster J. (2003). *Chemicals in the Atmosphere: Solubility, Sources and Reactivity*. Wiley, New York, 468 pp.

Folland C., Frich P., Basnett T., Rayner N., Parker D., and Horton B. (2000). Uncertainties in climate datasets: A challenge for WMO. *WMO Bull.*, **49**(1), 59–68.

Follett R.F. and Kimble J.M. (2000). *The Potential of U.S. Grazing Lands to Sequester Carbon and Mitigate the Greenhouse Effect*. CRC Press, London, 472 pp.

Formenti P., Elbert W., Maenhaut W., Haywood J., and Andreae M. O. (2003). Chemical composition of mineral dust aerosol during the Saharan Dust Experiment (SHADE) airborne campaign in the Cape Verde region, September 2000: The Saharan Dust Experiment (SHADE). *J. Geophys. Res.*, **108**(D18), SAH3.1–SAH3.15.

Forrester J.W. (1971). *World Dynamics*. Wright-Allen Press, Cambridge, 189 pp.

Foster R.G. and Kreitzman L. (2004). *Rhythms of Life: The Biological Clocks that Control the Daily Lives of Every Living Thing*. Yale University Press, London, 288 pp.

Fourier C. (1876). *Theory of Social Organization*. C.P. Somerby, New York, 167 pp.

Frances H. (ed.) (2004). *Global Environmental Issues*. Wiley, Washington, D.C., 312 pp.

Frankham R. (1995). Inbreeding and extinction: A threshold effect. *Conservation Biology*, 9(4), 792–799.

Freeman A.M. III. (2005). *The Measurement of Environment and Resource Values: Theory and Methods*, Second Edition. RFF Press, Washington, D.C., 496 pp.

Freeman M. (2005). *China Takes the Lead in Nuclear Energy*. 21st Century Science & Technology Books, Washington, D.C., pp. 65–67.

Freeman P.K. (2000). Infrastructure, natural disasters, and poverty. *Proc. of the Euro-Conference on Global Change and Catastrophe Risk Management: Flood Risks in Europe, Luxemburg, Austria, 6–9 June 2000*. International Institute for Applied Systems Analysis, Luxemburg, Austria, pp. 234–238.

Freitas S.R., Longo K.M., Silva Dias M.A.F., Silva Dias P.L., Chatfield R., Prins E., Artaxo P., Grell G.A., and Recuero F.S. (2005). Monitoring the transport of biomass burning emissions in South America. *Environmental Fluid Mechanics*, **5**(1–2), 135–167.

Friedman T.L. (2005). *The World Is Flat: A Brief History of the Twenty-first Century*. Farrar, Straus & Giroux, New York, 496 pp.

Friend A.D. and Kiang N.Y. (2005). Land surface model development for the GISS GCM: Effects of improved canopy physiology on simulated climate. *J. Climate*, **18**(15), 2883–2902.

Furth J.H. (1965). Professor James on the theory of monetary policy. *J. Economics*, **25**(1–2), 199–203.

Galitsky V.V. (1985). The horizontal structure and dynamics of an even-aged vegetation community: Numerical modelling. In: Yu.M. Svirezhev (ed.), *Mathematical Modelling of Biogeocenotic Processes*. Science, Moscow, 59–69 [in Russian].

Ganti T. (2003). *The Principles of Life*. Wiley, New York, 220 pp.

Garaguso G. and Marchisio S. (1993). *Rio 1992: Vertice per la Terra*. Franco Angeli, Milano, 752 pp.

Garcia-Barrón L. and Pita M.F. (2004). Stochastic analysis of time series of temperatures in the south-west of the Iberian Peninsula. *Atmosfera*, **17**(4), 225–244.

Gavin B. (2001). *The European Union and Globalisation. Towards Global Democratic Governance*. Edward Elgar, London, 264 pp.

Gibson R.B., Hassan S., Holtz S., Tansey J., and Whitelaw G. (2005). *Sustainability Assessment: Criteria, Processes and Applications*. Earthscan, London, 254 pp.

Gillaspie A. (2001). *The Illusion of Progress: Unsustainable Development in International Law and Policy*. Earthscan, London, 244 pp.

Gitelson I.I., Bartsev S.I., Okhonin V.A., Sukhovilsky V.G., and Khlebopros R.S. (1997). What is the best strategy for development? *Herald of RAS*, **67**(5), 405-412 [in Russian].

Giupponi C., Jakeman A., Kassenberg D., and Hare M. (eds.) (2006). *Sustainable Management in Water Resources: An Integrated Approach*. Edward Elgar, Cheltenhan, U.K., 368 pp.

Glazovsky N.F. (2005). State, problems, and prospects for education development in the sphere of sustainable development abroad. In: V.N. Troyan and I.A. Dementyev (eds.), *Sustainable Development and Ecological Management, Issue 1: Proceedings of International Conference, St. Petersburg, 17–18 November 2005*. St. Petersburg State University, St. Petersburg, pp. 3–28 [in Russian].

Gloersen P., Parkinson C.L., Cavalieri D.J., Comiso J.C., and Zwally H.J. (1999). Spatial distribution of trends and seasonality in the hemispheric ice covers: 1978–1996. *J. Geophys. Res.*, **104**(C9), 20827–20835.

Goldemberg J. and Johansson T.B. (eds.) (2004). *World Energy Assessment*. UNDP, New York, 88 pp.

Goldman L. and Coussens C.M. (eds.) (2004). *Environmental Health Indicators: Bridging the Chasm of Public Health and the Environment*. National Academy Press, Washington, D.C., 136 pp.

Golfeld G.B. and Krapivin V.F. (1997). Remarks on technology of assessment and forecast of the national ecological security of Russia. *Problems of the Environment and Natural Resources*, **2**, 28–40 [in Russian].

Golitsyn G.S. (1995). Using the level of the Caspian Sea as a means of diagnosis and forecast of regional climate change. *Physics of the Atmos. and Ocean*, **31**(3), 385–391 [in Russian].

Golubeva G.N. (ed.) (2002). *Global Ecological Perspective, 3: Past, Present, and Prospect for the Future*. Interdialekt, Moscow, 2002, 504 pp. [in Russian].

Gorshkov V.G. (1990). *Biospheric Energetics and Environmental Stability: Progress in Science and Technology* (Theoretical and General Problems of Geography series). ARISTI, Moscow, 237 pp. [in Russian].

Gorshkov V.G. (1995). *Physical and Biological Basis of Life Stability: Man, Biota, Environment*. Springer-Verlag, Berlin, 350 pp.

Gorshkov V.G. and Makarieva A.M. (1999). Impact of virgin and farmed biota on global environment. *Studying the Earth from Space*, **5**, 3–11 [in Russian].

Gorshkov V.G., Kondratyev K.Ya., and Losev K.S. (1998). Global ecodynamics and sustainable development: Natural–science aspects and "human dimension". *Ecology*, **3**, 163–170 [in Russian].

Gorshkov V.G., Gorshkov V.V., and Makarieva A.M. (2000). *Biotic Regulation of the Environment: Key Issues of Global Change*. Springer-Praxis, Chichester, U.K., 367 pp.

Gorshkov V.G., Esenin B.K., Karibayeva K.N., Kurochkina L.Ya., Losev K.S., Makarieva A.M., and Shukurov E.D. (2004). Scientific base for strategic directions of the nature protection policy. *Ecology and Education*, **1**, 29–31 [in Russian].

Gottman, J. (1964). *Megalopolis: The Urbanized Northeastern Seaboard of the United States* (810 pp.). MIT Press, Cambridge, MA.

Gottman, J. (1987). *Megalopolis Revisited: 25 Years Later* (71 pp.). University of Maryland Institute for Urban Studies, College Park, MD.

Goudie A.S. (ed.) (2001). *Encyclopedia of Global Change*. Oxford University Press, London, 1440 pp.

Grassl H. (2000). Status and improvements of coupled general circulation models. *Science*, **288**, 1991–1997.

Grigoryev Al.A. and Kondratyev K.Ya. (2001). *Natural and Anthropogenic Ecological Disasters*. St. Petersburg Sci. Centre RAS, St. Petersburg, 688 pp. [in Russian].

Grimstead J. (ed.) (2003). *Affirmations and Denials Concerning Certain Essential Doctrines of the Church*. The International Church Council Project, CA, 93 pp.

Grossi P. and Kunreuther H. (eds.) (2005). *Catastrophe Modeling: A New Approach to Managing Risk*. Springer-Verlag, New York, Vol. 25, 252 pp.

Guhathakurta S. (ed.) (2003). *Integrated Land Use and Environmental Models: A Survey of Current Applications and Research*. Springer for Science, Amsterdam, The Netherlands, 271 pp.

Guieu C., Bonnet S., and Wagener T. (2005). Biomass burning as a source of dissolved iron to the open ocean? *Geophys. Res. Lett.*, **32**(L19608), doi: 10.1029/2005GL022962.

Guillaso S., Ferro-Famil L., Reigber A., and Pottier E. (2005). Building characterization using L-band polarimetric interferometric SAR data. *IEEE Geosci. and Remote Sensing Letters*, **2**(3), 347–351.

Guista M.L. and Kambhampati U.S. (eds.) (2006). *Critical Pespectives on Globalization*. Edward Elgar, London, 656 pp.

Gupta, J. (2001). *Our Simmering Planet: What to Do about Global Warming*. Zed Books, London, 178 pp.

Gurjar B.R. and Leliveld J. (2005). New directions: Megacities and global change. *Atmospheric Environment*, **39**(2), 391–393.

Guyatt N. (2000). *Another American Century? The United States and the World after 2000*. Zed Books, London.

Hansen J.E. and Sato M. (2001). Trends of measured climate forcing agents. *Proc. Natl. Acad. Sci. U.S.A.*, **98**(26), 14778–14783.

Hansen, J., Sato, M., and Ruedy, R. (1997). Radiative forcing and climate response. *J. Geophys. Res.*, **102**, 6831–6864.

Hansen J., Sato M., Lacis A., Ruedy R., Tegen L., and Matthews E. (1998). Climate forcing in the Industrial Era. *Proc. Natl. Acad. Sci. U.S.A.*, **95**(22), 12753–12758.

Hansen J., Ruedy R., Glascoe J., and Sato M. (1999). GISS analysis of surface temperature change. *J. Geophys. Res.*, **104**(D24), 30997–31022.

Hansen J.E., Sato M., Ruedy R., Lacis A., and Oinas V. (2000). Global warming in the twenty-first century: An alternative scenario. *Proc. U.S. Natl. Acad. Sci.*, **97**(18), 9875–9880.

Hawken P., Lovins A.B., and Lovins L.H. (2005). *Natural Capitalism: The Next Industrial Revolution*. Earthscan. London, 416 pp.

Hay R. (2005). Becoming ecosynchronous, Part I: The root causes of our unsustainability way of life. *Sustainable Development*, **13**(5), 311–325.

Held D., Goldblatt D., Makriu E., and Perraton D. (2004). *Global Transformations: Politics, Economy, and Culture*. Logos, Moscow, 576 pp. [in Russian].

Helming K. and Wiggering H. (eds.) (2003). *Sustainable Development of Multifunctional Landscapes*. Springer for Science, Amsterdam, The Netherlands, 286 pp.

Heynemann S.P. (2005). Organizations and social cohesions. *Peabody J. Education*, **80**(4), 1–7.

Heywood, V.H. (2004). The biodiversity crisis and global change. *Proc. of "100 Jahre in Berlin-Dahlem", International Scientific Symposium, Botanic Gardens: Awareness for Biodiversity, Programme and Abstracts*. Botanic Garden and Botanical Museum Berlin-Dahlem, Berlin, pp. 18–20.

Hillel D. (ed.) (2003). *Encyclopedia of Soils in the Environment* (four-volume set). Elsevier, San Diego, 2200 pp.

Holdren J.P. (2003). Environmental change and human condition. *Bull. Amer. Acad. Arts. Sci., New York*, **57**(1), 25–31.

Holtz B. (2005). *Human Knowledge: Foundations and Limits* (170 pp.). *http://humanknowledge.net/HumanKnowledge.pdf*

Hossay P. (2005). *Unsustainable: A Primer for Global Environmental and Social Justice*. Zed Books, London, 56 pp.

Houghton J.T. (2000). Global climate and human activities. *Encyclopedia Of Life Support System*, **1**, 1–13.

Houghton J.T. (2004). *Global Warming: The Complete Briefing*. Cambridge University Press, Cambridge, U.K., 382 pp.

Howe S. (ed.) (2004). *GEO Year Book 2003*. United Nations Environment Programme, Nairobi, Kenya, 76 pp.

Huang S., Pollack H.N., and Shen P.-Y. (2000). Temperature trends over the past five centuries reconstructed from borehole temperatures. *Nature*, **403**, 756–758.

Hudson R. (2005). Toward sustainable economic practices, flows and spaces: Or is the necessary impossible and the impossible necessary? *Sustainable Development*, **13**(4), 239–252.

Hulme M., Barrow E.M., Arnell N.W., Harrison P.A., Johns T.C., and Downing T.E. (1999). Relative impacts of human-induced climate change and climate variability. *Nature*, **397**, 689–691.

Hulst, R. van. (1979). The dynamics of vegetation: Succession in model communities. *Vegetatio*, **39**(2), 85–96.

IEA (2002). *Energy Policies of IEA Countries*. OECD/IEA Books, Paris, France, 137 pp.

IEA (2005a). *Energy Policies of IEA Countries*. IEA Books, Paris, France, 588 pp.

IEA (2005b). *Key World Energy Statistics*. IEA Books, Paris, France, 82 pp.

IEA (2005c). *World Energy Outlook: Middle East and North Africa Insights*. IEA Books, Paris, France, 600 pp.

ILO (2004). *A Fair Globalization: Creating Opportunities for All*. World Commission on the Social Dimension of Globalization, Geneva, 190 pp.

Ingegnoli V. (2002). *Landscape Ecology: A Widening Foundation*. Springer for Science, Amsterdam, The Netherlands, 357 pp.

Inomata Y., Iwasaka Y., Osada K., Hayashi M., Mori I., Kido M., Hara K., and Sakai T. (2006). Vertical distributions of particles and sulfur gases (volatile sulfur compounds and $SO_2$) over East Asia: Comparison with two aircraft-borne measurements under the Asian continental outflow in spring and winter. *Atmospheric Environment*, **40**(3), 430–444.

IPCC (1995). *IPCC Second Assessment Climate Change 1995* (a report). Intergovernmental Panel on Climate Change, Geneva, Switzerland, 64 pp.

IPCC (2005). *Special Report on Carbon Dioxide Capture and Storage* (final draft). Intergovernmental Panel on Climate Change, London, 571 pp.

Islam S.M.N. and Clarke M.F. (2005). The welfare economics of measuring sustainability: A new approach based on social choice theory and systems analysis. *Sustainable Development*, **13**(5), 282–296.

Ivanov-Rostovtsev A.G., Kolotilo L.G., Tarasiuk Yu.F., and Sherstiankin P.P. (2001). *Self-organization and Self-regulation of Natural Systems: Model, Method, and Fundamentals of the D-Self Theory*. Russian Geographic Society, St. Petersburg, 216 pp. [in Russian].

Izrael Yu.A. (ed.) (2001). *State and Complex Monitoring of the Environment and Climate Change Limits*. Science, Moscow, 243 pp. [in Russian].

Jenkins G., Betts R., Collins M., Griggs D., Lowe J., and Wood R. (2005). *Stabilizing Climate to Avoid Dangerous Climate Change: A Summary of Relevant Research at the Hadley Centre*. Met. Office, London, 19 pp.

Jennings P. (2006). East of Ego: The intersection of narcissistic personality and Buddhist practice. *J. Religion and Health*, **45**(4), 480–485.

Jessen C.A. (2006). *The Ups and Downs of the Holocene: Exploring Relationships between Global $CO_2$ and Climate Variability in the North Atlantic Region* (42 pp.). Lund University, Lund, Sweden.

Ji Y. and Stocker E. (2003). Seasonal, intraseasonal, and interannual variability of global land fires and their effects on atmospheric aerosol distribution. *J. Geophys. Res. D.*, **107**(23), ACH10/1–ACY10/11.

Johnson J. (2005). *Environment, Health, and Growth: Building on Renewed Commitments. Environment Matters. Annual Review*. World Bank, Washington, D.C., pp. 1–10.

Jolliffe I.T. and Stephenson D.B. (2003). *Forecast 2002 Verification*. Wiley, London, 254 pp.

Jones C., Baker M., Carter J., Jay S., Short M., and Wood C. (eds.) (2005). *Strategic Environment Assessment and Land Use Planning: An International Evaluation*. Earthscan, London, 300 pp.

Jones R. (1990). Review of current CASE technologies. In: S. Holloway (ed.), *The Distributed Development Environment*. Chapman & Hall, London, pp. 1–14.

Jönsson A.M., Linderson M.-L., Stjernquist I., Schlyter P., and Bärring L. (2004). Climate change and the effect of temperature backlashes causing frost damage in *Picea abies*. *Global and Planetary Change*, **44**(1–4), 195–207.

Kalb D., Pansters W., and Sibers H. (2004). *Globalisation and Development: Themes and Concepts in Current Research*. Kluwer Academic, Dordrecht, the Netherlands, 203 pp.

Karam M.A., Fung A.K., Lang R.H., and Chauhan N.S. (1992). A microwave scattering model for layered vegetation. *IEEE Trans. on Geosci. and Remote Sensing*, **30**(4), 767–784.

Karekezi S. and Sihag A.R. (eds.) (2004). *GNESD: Energy Assess Theme Results*. United Nations Environment Programme, Roskilde, Denmark, 59 pp.

Karl T.R., Hassol S.J., Miller C.D., and Murray W.L. (eds.) (2006). *Temperature Trends in the Lower Atmosphere* (180 pp.). U.S. Climate Change Science Program, NASA, Washington, D.C.

Kasimov N.S., Mazurov Yu.A., and Tikunov V.S. (2004). Sustainable development conception: Perception in Russia. *Herald of RAS*, **74**(1), 28–36 [in Russian].

Keigwin, L.D. and Boyle, E.A. (2000). Detecting Holocene changes in thermohaline circulation. *Proceedings of the National Academy of Sciences U.S.A., Washington*, **97**, 1343–1346.

Kelley J.J., Krapivin V.F., and Vilkova L.P. (1992a). Conception of a model for estimation of the role of Arctic ecosystems in the global carbon budget. *Proceedings of the International Symposium "Problems of Ecoinformatics", Zvenigorod, 14–18 December 1992*. IREE, Moscow, pp. 19–20 [in Russian].

Kelley J.J., Rochon G.L., Novoselova O.A., and Krapivin V.F. (1992b). The future of global geo-eco-information monitoring. *Proceedings of the International Symposium "Problems of Ecoinformatics", Zvenigorod, 14–18 December 1992*. IREE, Moscow, pp. 3–7 [in Russian].

Kennedy D. (2003). Sustainability and the commons (Editorial). *Science*, **302**, 1861–1863.

Kerr R.A. (2005). How hot will the greenhouse world be? *Science*, **309**, 100.

Khor M. (2001). *Rethinking Globalization: Critical Issues and Policy Choices*. Zed Books, London, 160 pp.

Kidder R.M. (1997). *An Agenda for the 21st Century*. MIT Press, Cambridge, MA, 240 pp.

Kidron M., Segal R., and Smith D. (eds.) (2004). *A Unique Survey of Current Events and Global Trends*. Earthscan, London, 28 pp.

Kiehl J., Hack J., Gent P., Large W., and Blackman M. (2003). *Community Climate System Model Strategic Business Plan*. NCAR, Washington, D.C., 28 pp.

King D. (2002). *The Science of Climate Change: Adapt, Mitigate or Ignore?* (The Ninth Zuckerman Lecture). Foundation for Science and Technology, London, 16 pp. (*www.ost.gov.uk*).

Kirchner J.W. (2003). The Gaia hypothesis: Conjectures and refutations. *Clim. Change*, **58**(1–2), 21–45.

Kirdiashev K.P., Chukhlantsev A.A., and Shutko A.M. (1979). Microwave radiation of the Earth's surface in the presence of vegetation cover. *Radio Eng. Electron., Phys. Engl. Transl.*, **24**, 256–264.

Knutson T.R., Delwoorth T.U., Dixon. K.W., and Stouffer R.J. (1999). Model assessment of regional temperature trends (1949–1997). *J. Geophys. Res.*, **104**(D24), 30981–30996.

Kobayashi K. (2005). Toward a sustainable Japan (Corporations at Work Article Series No. 28: A Key Player in Information and Culture—the Toppan Printing Co.). *Japan for Sustainability Newsletter*, **036** (August 31), 1–11.

Kogan I.M. (2004). *Biofield: Substrate of the Noosphere* (176 pp.). Graal, Moscow [in Russian].

Kogan I.M. (2006). *Biofield: Super-weak Interactions—Faith—Life* (20 pp.). Book and Business, Moscow [in Russian].

Kogan I.M. and Kruglova L.V. (2005). *Biofield's Factor of Globalization* (217 pp.). Sinergia, Moscow [in Russian].

Kolokolchikova N.B., Kartseva E.V., and Potapov I.I. (1997). Ecological information systems in Russian Federation. *Problems of the Environment and Natural Resources*, **7**, 2–61 [in Russian].

Kolomyts E.G. (2003). *A Regional Model of Global Environmental Changes*. Science, Moscow, 371 pp. [in Russian].

Kondratyev K.Ya. (1996). Global change and demographic dynamics. *Proceedings of Russian Geographic Society*, **128**(3), 1–12 [in Russian].

Kondratyev K.Ya. (1998). *Multidimensional Global Change*. Wiley-Praxis, Chichester, U.K., 771 pp.

Kondratyev K.Ya. (1999a). *Ecodynamics and Geopolitics. Vol. 1: Global Problems*. St. Petersburg State University, St. Petersburg, 1036 pp. [in Russian].

Kondratyev K.Ya. (1999b). Pivotal point: The end of the growth paradigm. *Proceedings of Russian Geographic Society*, **131**(2), 1–14 [in Russian].

Kondratyev K.Ya. (2000). Global change of nature and society at the turn of two millennia. *Proceedings of Russian Geographic Society*, **5**, 3–19 [in Russian].

Kondratyev K.Ya. (2001). Key issues of global change at the end of the second millenium. In: M.K. Tolba. (ed.), *Our Fragile World: Challenges and Opportunities for Sustainable Development*. EOLSS, Oxford, U.K., pp. 147–165.

Kondratyev K.Ya. (2002). Problems of globalization. *Proceedings of Russian Geographic Society*, **4**, 1–8 [in Russian].

Kondratyev K.Ya. (2004a). Global climate change: Observational data and numerical modelling results. *Research of the Earth from Space*, **2**, 61–96 [in Russian].

Kondratyev K.Ya. (2004b). Key aspects of global climate change. *Energy and Environment*, **15**(3), 469–503.

Kondratyev K.Ya. (2004c). Priorities of global climatology. *Proceedings of Russian Geographic Society*, **136**(2), 1–25 [in Russian].

Kondratyev K.Ya. (2005a). Key problems of global climate change. *Proceedings of Russian Geographic Society*, **137**(6), 78–85 [in Russian].

Kondratyev K.Ya. (2005b). Sustainable development and the consumption society. In: V.N. Troyan and I.A. Dementyev (eds.), *Sustainable Development and Ecological Management, Issue 1: Proc. Int. Scientific–Practical Conf., 17–18 November 2005, St. Petersburg*. St. Petersburg State University, St. Petersburg, pp. 29–62 [in Russian].

Kondratyev K.Ya. and Grigoryev Al.A. (2004). Forest fires as a global ecodynamics component. *Optics of the Atmos. and Ocean*, **136**(4), 279–292 [in Russian].

Kondratyev K.Ya. and Krapivin V.F. (2003). Global change: Real and possible in future. *Research of the Earth from Space*, **4**, 1–10 [in Russian].

Kondratyev K.Ya. and Krapivin V.F. (2004). Global carbon cycle: State, problems, and perspectives. *Research of the Earth from Space*, **3**, 1–10 [in Russian].

Kondratyev K.Ya. and Krapivin V.F. (2005a). Civilization development and its ecological limitations: Numerical modelling and monitoring. *Research of the Earth from Space*, **4**, 3–23 [in Russian].

Kondratyev K.Ya. and Krapivin V.F. (2005b). Key problems of global ecodynamics. *Research of the Earth from Space*, **4**, 3–33 [in Russian].

Kondratyev K. Ya. and Krapivin V. F. (2005c). Monitoring and prediction of natural disasters. *Il Nuovo Cimento*, **25C**(6), 657–672.

Kondratyev K.Ya. and Krapivin V.F. (2005d). Nature/society system and climate as its interactive element: Ecological crises and catastrophes. *Energy: Economics, Technics, Ecology*, **11**, 6–12 [in Russian].

Kondratyev K.Ya. and Krapivin V.F. (2005e). Numerical modelling of nature/society system dynamics. *Energy: Economics, Technics, Ecology*, **12**, 17–22 [in Russian].

Kondratyev K.Ya. and Krapivin V.F. (2005f). The present consumption society: Ecological limitations. *Proceedings of Russian Geographic Society*, **136**(2), 13–38 [in Russian].

Kondratyev K.Ya. and Krapivin V.F. (2006a). Earth's radiative balance as an indicator of global ecological stability. *Research of the Earth from Space*, **1**, 3–9 [in Russian].

Kondratyev K.Ya. and Krapivin V.F. (2006b). Global ecosystem dynamics: Natural changes and anthropogenic impacts. *Proceedings of Russian Geographic Society*, **136**(4), 1–14 [in Russian].

Kondratyev K.Ya. and Krapivin V.F. (2006c). *Natural Disasters as Interactive Component of Global Ecodynamics*. St. Petersburg University Press, St. Petersburg, 626 pp. [in Russian].

Kondratyev K.Ya. and Krapivin V.F. (2006d). Natural ecodynamics per observational data. *Energy: Economics, Technics, Ecology*, **1**, 19–23 [in Russian].

Kondratyev K.Ya. and Krapivin V.F. (2006e). On the Earth's heat balance. *Energy: Economics, Technics, Ecology*, **3**, 7–13 [in Russian].

Kondratyev K.Ya. and Krapivin V.F. (2006f). Present state and prospects for global energy development in the context of global ecodynamics. *Proceedings of Russian Geographic Society*, **138**(3), 14–30 [in Russian].

Kondratyev K.Ya. and Krapivin V.F. (2006g). Present state and prospects for world energy development. *Energy: Economics, Technics, and Ecology*,**2**, 17–23 [in Russian].

Kondratyev K.Ya. and Losev K.S. (2002). The present stage of civilization development and its prospects. *Astrakhan Bulletin of Ecological Education*, **1**(3), 5–24 [in Russian].

Kondratyev K.Ya. and Romaniuk L.P. (1996). Sustainable development: Conceptual aspects. *Proceedings of Russian Geographic Society*, **6**, 1–12 [in Russian].

Kondratyev K.Ya. and Varotsos C.A. (2001a). Global tropospheric ozone dynamics: Part I, Tropospheric ozone precursors; Part II, Numerical modelling of tropospheric ozone variability. *Environ. Sci. Pollut. Res.*, **8**(1), 57–62.

Kondratyev K.Ya. and Varotsos C.A. (2001b). Global tropospheric ozone dynamics: Part II, Numerical modelling of tropospheric ozone variability; Part I, Tropospheric ozone precursors [*ESPR* **8**(1), 57–62 (2001)]. *Environ. Sci. Pollut. Res.*, **8**(2), 113–119.

Kondratyev K.Ya., Ortner J., and Preining O. (1992). Priorities of global ecology now and in the next century. *Space Policy*, **8**(1), 39–48.

Kondratyev K.Ya., Moreno-Pena F., and Galindo I. (1994). *Global Change: Environment and Society*. Universidad de Coloma, Mexico, 147 pp.

Kondratyev K.Ya., Moreno-Pena F., and Galindo I. (1997). *Sustainable Development and Population Dynamics*. Universidad de Coloma, Mexico, 128 pp.

Kondratyev K.Ya., Krapivin V.F., and Phillips G.W. (2002). *Global Environmental Change: Modelling and Monitoring*. Springer-Verlag, Heidelberg, Germany, 316 pp.

Kondratyev K.Ya., Krapivin V.F., and Savinykh V.P. (2003a). *Prospects for Civilization Development: Multi-dimensional Analysis*. Logos, Moscow, 574 pp. [in Russian].

Kondratyev K.Ya., Krapivin V.F., and Varotsos C.A. (2003b). *Global Carbon Cycle and Climate Change*. Springer-Praxis, Chichester, U.K., 344 pp.

Kondratyev K.Ya., Losev K.S., Ananicheva M.D., and Chesnokova I.V. (2003c). *Natural–Science Fundamentals of Life Stability*. ARISTI, Moscow, 240 pp. [in Russian].

Kondratyev K.Ya., Losev K.S., Ananicheva M.D., and Chesnokova I.V. (2003d). *Stability of Life on Earth*. Springer-Praxis, Chichester, U.K., 152 pp.

Kondratyev K.Ya., Krapivin V.F., Savinykh V.P., and Varotsos C.A. (2004). *Global Ecodynamics: A Multidimensional Analysis*. Springer-Praxis, Chichester U.K., 658 pp.

Kondratyev K.Ya., Ivlev L.S., Krapivin V.F., and Varotsos C.A. (2006a). *Atmospheric Aerosol Properties: Formation, Processes and Impacts*. Springer-Praxis, Chichester, U.K., 572 pp.

Kondratyev K.Ya., Krapivin V.F., Lacasa H., and Savinikh V.P. (2006b). *Globalization and Sustainable Development—Ecological Aspects—Introdution*. Science, St. Petersburg, 241 pp. [in Russian].

Kondratyev K.Ya., Krapivin V.F., and Varotsos C.A. (2006c). *Natural Disasters as Components of Ecodynamics*. Springer-Praxis, Chichester, U.K., 625 pp.

Koniayev K.V., Zolotokrylin A.N., Vinogradova V.V., and Yitkova T.B. (2003). Determination from satellite data of the vegetation cover response to anomalies of climatic indicators. *Research of the Earth from Space*, **2**, 18–26 [in Russian].

Koroneos C.J. and Xydis G. (2005). Sustainable development of the Prefecture of Kephalonia. *IJSD*, 243–257.

Kotliakov V.M. (ed.) (2000). *Anomaly of Crises*. Science, Moscow, 239 pp. [in Russian].

Kramer H.J. (1995). *Observation of the Earth and Its Environment: Survey of Missions and Sensors*. Springer-Verlag, New York, 832 pp.

Krapivin V.F. and Kondratyev K.Ya. (2002). *Global Environmental Changes: Ecoinformatics*. St. Petersburg University, St. Petersburg, 724 pp [in Russian].

Krapivin V.F. and Marenkin D.S. (1999). Co-evolution of man and nature. *Problems of the Environment and Natural Resources*, **1**, 2–10 [in Russian].

Krapivin V.F. and Phillips G.W. (2001a). A remote sensing based expert system to study the Aral–Caspian aquageosystem water regime. *Remote Sensing of Environment*, **75**, 201–215.

Krapivin V.F. and Phillips G.W. (2001b). Application of a global model to the study of Arctic basin pollution: Radionuclides, heavy metals and oil carbohydrates. *Environmental Modelling and Software*, **16**, 1–17.

Krapivin V.F., Svirezhev Yu.M., and Tarko A.M. (1982). *Mathematical Modeling of Global Biospheric Processes*. Science, Moscow, 272 pp. [in Russian].

Krapivin V.F., Bui T.L., Rochon G.L., and Hicks D.R. (1996). A global simulation model as a method for estimation of the role of regional area in global change. *Proc. of the Second Ho Chi Minh City Conf. on Mechanics, 24–25 September 1996*. Instiitute of Applied Mechanics, Ho Chi Minh City, Vietnam, pp. 68–69.

Krapivin V.F., Nazarian N.A., and Potapov I.I. (1997). An adaptive system for decision making in ecological monitoring. *Problems of the Environment and Natural Resources*, **8**, 16–29 [in Russian].

Krapivin V.F., Graur A., Nitu C., and Oancea D. (1998a). A new approach for geo-eco-information monitoring. *Proceedings of the 4th International Conference on Development and Application Systems, 21–22 May 1998, Suceava (Romania)*. Suceava University, Suceava, pp. 3–10.

Krapivin V.F., Michalev M., Mkrtchyan F.A., Nitu C., and Shutko A.M. (1998b). Intelligent information technology and its application to agricultural systems monitoring. *Proc. of the Third Int. Sympos. on Ecoinformatics Problems, 8–9 December 1998*. Institute of Radioengineering and Electronics, Moscow, pp. 6–12 [in Russian].

Krapivin V.F., Shutko A.M., Chukhlantsev A.A., Golovachev S.P., and Phillips G.W. (2006). GIMS-based method for vegetation microwave monitoring. *Environmental Modelling and Software*, **21**, 330–345.

Kravtsov Yu.V. (2002). Study of the ecologo-geochemical state of underground and surface waters in Urengoy oil and gas extraction zone. PhD thesis, Tyumen State University, Tyumen, Russia, 200 pp.

Kreisler H. (2006). Action on global warming. *Conversation with Sir David King*, July 10, 2006 (*http://globetrotter.berkeley.edu/people5/King/king-con0.html*).

Kresl P.K. and Gallais S. (2002). *France Encounters Globalization*. Edward Elgar, London, 256 pp.

Kunzi K. (2003). *Environmental Physics, I: The Atmosphere*. Institute of Environmental Physics, University of Bremen, Bremen, Germany, 40 pp.

Kuznetsov S.L., Kuznetsov P.G., and Bolshakov B.E. (2000). *The Nature–Society–Human System: Sustainable Development*. Noosphere, Moscow, 392 pp. [in Russian].

Kuznetsov V.N. (1989). *German Classic Philosophy of the Second Part of the 18th Century and the First Part of the 19th centuries*. High School, Moscow, 480 pp. [in Russian].

Lanari R., Migliaccio M., Schwäbisch M., and Coltelli M. (1996). Generation of digital elevation models by using SIR-C/X-SAR multifrequency two-pass interferometry: The Etna case study. *IEEE Trans. on Geosci. and Remote Sensing*, **34**(5), 1097–1114.

Lang R.E. and Ghavale D. (2005). *Beyond Megalopolis: Exploring America's New "Megapolitan" Geography* (36 pp.). Megapolitan Institute Census Report Series, Census Report 05:01 (July 2005), Alexandria, VA.

Lautenbacher C.C. (ed.) (2005). *New Priorities for the 21st Century: NOAA's Strategic Plan Undated for FY 2006–FY 2011*. National Oceanic and Atmospheric Administration, Washington, D.C., 24 pp.

Lawrence D.P. (2003). *Environmental Impact Assessment: Practical Solutions to Recurrent Problems*. Wiley, New York, 562 pp.

Lean G. (1995). *Down to Earth: Convention to Combat Desertification*. UNCCD, Bonn, Germany, 37 pp.

Legendre P. and Legendre L. (1998). *Numerical Ecology*. Elsevier, Amsterdam, 853 pp.

Lehmiller G.S., Kimberlain T.B., and Elsner J.B. (1997). Seasonal prediction models for North Atlantic Basin hurricane location. *Monthly Weather Review*, **125**, 1780–1791.

Leslie, P.H. (1945). On the use of matrices in certain population mathematics. *Biometrika*, **33**(3), 183–212.

Letz S.E. (2000). Social, political, legal aspects of extreme situations. In: I.M. Makarov (ed.), *Risk Management: Risk, Sustainable Development, Synergetics*. Science, Moscow, pp. 331–365 [in Russian].

Levesque J. and King D.J. (2003). Spatial analysis of radiometric fractions from high-resolution multispectral imagery for modelling individual tree crown and forest canopy structure and health. *Remote Sensing of Environment*, **84**(4), 589–602.

Levinson D.H. and Waple A.M. (eds.) (2004). State of the climate in 2003. *Bull. Amer. Meteorol. Soc.*, **85**(6), 1–72.

Levitus S., Antonov J.I., Boyer T.P., and Stephens C. (2000). Warming of the World Ocean. *Science*, **287**, 2225–2229.

Lieth H. (1985). A dynamic model of the global carbon flux through the biosphere and its relations to climatic and soil parameters. *Int. J. Biometeorol.*, **29**(2), 17–31.

Lin C.-Y., Liu S.C., Chou C.-K., Huang S.-J., Liu C.-M., Kuo C.-H., and Young C.-Y. (2005). Long-range transport of aerosols and their impact on the air quality of Taiwan. *Atmospheric Environment*, **39**(33), 6066–6076.

Lin D.-L., Sakoda A., Shibasaki R., Goto N., and Suzuki M. (2000). Modelling a global biogeochemical nitrogen cycle in terrestrial ecosystems. *Ecological Modelling*, **135**(1), 89–110.

Lindenmayer D.B., Foster D.R., Franklin J.F., Hunter M.L., Noss R.F., Schmiegelow F.A., and Perry D. (2004). Salvage harvesting policies after natural disturbance. *Science*, **303**(5662), 1303.

Lindsey B. (2001). *Against the Dead Hand: The Uncertain Struggle for Global Capitalism*. Wiley, New York, 368 pp.

Lindzen R.S. and Giannitsis C. (2003). Reconciling observations of global temperature change. *Geophys. Res. Lett.*, **29**(10), X1–X3.

Lindzen R.S., Chou M.-D., and Hou A.Y. (2001). Does the Earth have an adaptive infrared iris? *Bull. Amer. Meteorol. Soc.*, **82**, 417–432.

Logofet D.O. (2002). Matrix population models: Construction, analysis and interpretation. *Ecological Modelling*, **148**(3), 307–310.

Lomborg B. (2001). *The Sceptical Environmentalist: Measuring the Real State of the World.* Cambridge University Press, Cambridge, U.K., 496 pp.

Lomborg B. (ed.) (2004). *Global Crisis, Global Solutions.* Cambridge University Press, Cambridge, U.K., 670 pp.

Losev K.S. (2001). *Ecological Problems and Prospects for Sustainable Development in Russia in the 21st Century.* Kosinform, Moscow, 400 pp. [in Russian].

Losev K.S. (2003). Natural–science basis of stable life. *Herald of RAS*, **73**(2), 110–116 [in Russian].

Loulou R., Waaub J.-P., and Zaccour G. (eds.) (2005). *Energy and Environment.* Springer-Verlag, Berlin, 282 pp.

Lovelock J.E. (2003). GAIA and emergence: A response to Kirchner and Volk. *Clim. Change*, **57**(1–2), 1–7.

Lubchenko J., Olson A.M., Brubaker L.B., Carpenter S.R., Holland M.M., Hubbell S.P., Levin S.A., MacMahon J.A., Matson P.A., Melillo J.M. *et al.* (1991). The sustainable biosphere initiative: An ecological research agenda. *Ecology*, **72**, 371–412.

Luke T.W. (2005). Neither sustainable nor development: Reconsidering sustainability in development. *Sustainable Development*, **13**(4), 228–238.

Luken R.A. and Hesp P. (eds.) (2004). *Towards Sustainable Development in Industry? Reports from Seven Developing and Transition Economies.* Edward Elgar, Cheltenham, U.K., 288 pp.

Madeley J. (2000). *Hungry for Trade: How the Poor Pay for Free Trade.* Zed Books, London, 178 pp.

Maguire D.J., Goodchild M.F., and Rhind D.W. (eds.) (1991). *Geographical Information Systems*, Vols. 1 and 2. Longman Scientific & Technical, New York, 649 and 447 pp.

Mahlman J.D. (1998). Science and nonscience concerning human-caused climate warming. *Ann. Rev. Energy and Environ., Palo Alto (CA)*, **23**, 83–105.

Mäkiaho J.-P. (2005). *Development of Shoreline and Topography in the Olkiluoto Area, Western Finland, 2000BP–8000AP* (Working Report 2005-79). University of Helsinki, Helsinki, 54 pp.

Malinetsky G.G., Medvedev I.G., Mayevsky V.I., Osipov V.I., Zalikhanov M.Ch., Frolov K.V., Makhutov N.A., L'vov D.S., Levashov V.K., Rimashevskata N.M. *et al.* (2003). Crises in present Russia and scientific monitoring. *Herald of RAS*, **73**(7), 579–597 [in Russian].

Malthus T.R. (1798, 1976). *An Essay on Principle of Population.* W.W. Norton, New York, 61 pp.

Mander J. and Goldsmith E. (eds.) (2006). *The Case against the Global Economy.* IFG, San Francisco, CA, 560 pp.

Marchuk G.I. and Kondratyev K.Ya. (1992). *Priorities of Global Ecology.* Science, Moscow, 264 pp. [in Russian].

Marfenin N.N. and Stepanov S.A. (eds.) (2005). *Russia in the Environment: 2004 (Analytical Year Book).* Modus-K-Eterna, Moscow, 318 pp. [in Russian].

Martens P. and Rotmans J. (eds.) (2002). *Transitions in a Globalizing World.* Swets & Zeitlinger, Amsterdam, The Netherlands, 135 pp.

Mason C. (2003). *The 2030 Spike: Countdown to Global Catastrophe*. Earthscan, London, 256 pp.

Matson P., Lohse K.A., and Yall S.J. (2002). The globalization of nitrogen deposition: consequences for terrestrial ecosystems. *Ambio*, **31**(2), 113–119.

Matthies M. and Scheringer M. (eds.) (2001). Long-range transport in the environment. *Environ. Sci. & Pollut. Res.*, **8**(3), 149–150.

Mayers J.C. (2004). London's wettest summer and wettest year—1903. *Weather*, **59**(10), 274–278.

McCarthy J.J., Canziani O.F., Leary N.A., Dokken D.J., and White K.S. (2001). *Climate Change 2001: Impacts, Adaptation, and Vulnerability* (contribution of Working Group II to the Third Assessment Report of the IPCC). Intergovernmental Panel on Climate Change, Washington, D.C., 1032 pp.

McConnel W.J. (2004). Forest cover change: Tales of the unexpected. *Global Change Newsletter*, **57**, 8–11.

McElroy M.B. (2002). *The Atmospheric Environment: Effects of Human Activity*. Princeton University Press, Princeton, NJ, 360 pp.

McHarry J., Strachan J., Callway R., and Ayre G. (2004). *The Plain Language Guide to the World Summit on Sustainable Development*. Earthscan, London, 168 pp.

McMichael A. J., Butler C. D., and Folke C.(2003). New visions for addressing sustainability. *Science*, **302**, 1919–1920.

MEA (2005a). *Ecosystems and Human Well-Being: Opportunities and Challenges for Business and Industry*. World Resource Institute, Washington, D.C., 36 pp.

MEA (2005b). *Scientific Facts on Ecosystem Change*. GreenFacts, Brussels, Belgium, 55 pp.

Meadows D.H., Meadows D.L., Rangers J., and Behrens W.W. (1972). *The Limits of Growth*. New York University Books, New York, 208 pp.

Meadows D.H., Randers J., and Meadows D.L. (2004). *The Limits to Growth: The 30-year Update*. Earthscan, London, 368 pp.

Mearns L.O., Bogardi I., Giorgi F., Matyasovsky I., and Palecki M. (1999). Comparison of climate change scenarios generated from regional climate model experiments and statistical downscaling. *J. Geophys. Res.*, **104**(D6), 6603–6621.

Meler P. and Munasinghe M. (2005). *Sustainable Energy in Developing Countries: Policy Analysis and Case Studies*. Edward Elgar, Cheltenham, U.K., 304 pp.

Melillo J.M., Field C.B., and Moldan B. (2003). *Interactions of the Major Biogeochemical Cycles: Global Change and Human Impacts*. Island Press, Washington, D.C., 357 pp.

Merian E., Anke M., Inhat M., and Stoeppler M. (2004). *Elements and Their Compounds in the Environment: Occurence, Analysis and Biological Relevance* (three-volume set). Wiley, New York, 1600 pp.

Michie J. (ed.) (2003). *The Handbook of Globalization*. Edward Elgar, Cheltenham, U.K., 448 pp.

Mill J.S. (1909). *Principles of Political Economy with Some of Their Applications to Social Philosophy*. Longmans, London, 734 pp.

Millstone E. and Lang T. (2003). *The Atlas of Food*. Earthscan, London, 128 pp.

Mintzer I.M. (1987). *A Matter of Degrees: The Potential for Controlling the Greenhouse Effect*. World Resources Institute, New York, Res. Rep. No. 15, 70 pp.

Mirkin B.M. (1986). *What Are Plant Communities?* Nauka, Moscow, 160 pp. [in Russian].

Mkrtchian F.A. (1982). *Optimal Recognition of Signals and Monitoring Problems*. Science, Moscow, 185 pp. [in Russian].

Mkrtchian F.A. (ed.) (1992). *Proceedings of the International Symposium "Problems of Ecoinformatics", Zvenigorod, 14–18 December 1992*. IREE, Moscow, 223 pp.

MMSD (2001). *Report of the First Workshop on Mining and Biodiversity, London, 11–12 June 2001, No. 201*. IIED, London, 42 pp.

Moiseyev N.N. (1995). *Present Rationalism*. IGBP, Moscow, 376 pp. [in Russian].

Moiseyev N.N. (1999). *Humankind: To Be or Not To Be ...?* Science, Moscow, 288 pp. [in Russian].

Moiseyev N.N. (2003). *Selected works, Vol. 2: Interdisciplinary Studies of Global Problems. Publicism and Public Problems*. Science, Moscow, 263 pp. [in Russian)].

Moiseyev N.N., Aleksandrov V.V., and Tarko A.M. (1985). *Man and the Biosphere*. Science, Moscow, 272 pp. [in Russian].

Monin A.S. and Shishkov Yu.A. (1990). *Global Ecological Problems*. Knowledge, Moscow, 48 pp. [in Russian].

Mont O. and Dalhammar C. (2005). Sustainable consumption: At the cross-road of environmental and consumer policies. *IJSD*, 258–279.

Morén L. and Påsse T. (2001). *Climate and Shoreline in Sweden during Weichsel and the Next 150000 Years* (SKB Technical Report, TR-01-19). Swedish Nuclear Fuel and Waste Management Company, Stockholm, 67 pp.

Morgenstern R.D. and Portuey P.R. (eds.) (2005). *New Approaches on Energy and the Environment: Policy Advice for the President*. RFF Press, Washington, D.C., 239 pp.

Moron V., Vautard R., and Ghil M. (1998). Trends, interdecadal and interannual oscillations in global sea-surface temperatures. *Climate Dynamics*, **7–8**, 545–469.

Morris J. (1997). Introduction: Climate change—prevention or adaptation? *IEA Stud. Educ.*, **10**, 13–37.

Morris J. (ed.) (2002). *Sustainable Development: Promoting Progress or Perpetuating Poverty?* Profile Books, London, 370 pp.

Munasinghe M. (2002). *Towards Sustainomics: Making Development More Sustainable*. Edward Elgar, London, 400 pp.

Munasinghe M. and Swart R. (2005). *Primer on Climate Change and Sustainable Development: Facts, Policy, Analysis, and Applications*. Cambridge University Press, Cambridge, U.K., 445 pp.

Munasinghe M., Sunkel O., and de Miguel C. (eds.) (2001). *The Sustainability of Long-term Growth*. Edward Elgar, London, 488 pp.

Nakamura H. and Yamagata T. (1998a). Oceans and climate shifts (Letters to the Editor). *Science*, **281**(8), 1144–1145.

Nakamura H. and Yamagata T. (1998b). Structure and time evolution of decadal/interdecadal climatic events observed in the North Pacific. *Proc. of the International Symposium "Triangle '98" in Kyoto, Japan, 29 September–2 October, 1998*. JAMSTEC, Tokyo, pp. 304–313.

Nakicenovic, N. and Swart, R. (eds.) (2000). *Special Report on Emissions Scenarios*. Cambridge University Press, Cambridge, U.K., 570 pp.

Nazaretian A.P. (2004). Anthropogenic crises: An hypothesis on techno–human balance. *Herald of RAS*, **74**(4), 319–330 [in Russian].

Neumayer E. (2003). *Weak versus Strong Sustainability: Exploring the Limits of Two Opposing Paradigms* (Second Edition). Edward Elgar, Cheltenham, U.K., 288 pp.

Newbold K.B. (2002). *Six Billion Plus: Population Issues in the Twenty-First Century*. Rowman & Littlefield, Lanham, MD, 213 pp.

Nicholls R.J. (2004). Coastal flooding and wetland loss in the 21st century: Changes under the SRES climate and socio-economic scenarios. *Global Environmental Change*, **14**(1), 69–86.

Nikanorov A.M. (2005). *Scientific Basis for Water Quality Monitoring*. Gidrometeoizdat, St. Petersburg, 576 pp. [in Russian].

Nikanorov A.M. and Khoruzhaya T.A. (2000). *Global Ecology*. PISDR, Moscow, 286 pp. [in Russian].

Nilsson S., Jonas M., and Obersteiner M. (2002). COP-6: A healing shock? (Editorial). *Clim. Change*, **52**(1–2), 25–28.

Nitu C., Krapivin V.F., and Bruno A. (2000a). *Intelligent Techniques in Ecology*. Printech, Bucharest, 150 pp.

Nitu C., Krapivin V.F., and Bruno A. (2000b). *System Modelling in Ecology*. Printech, Bucharest, 260 pp.

Nitu C., Krapivin V.F., and Pruteanu E. (2004). *Ecoinformatics: Intelligent Systems in Ecology*. Magic Print, Onesti, Bucharest, 411 pp.

NOAA (2003). *New Priorities for the 21st Century: NOAA's Strategic Plan for FY 2003–FY 2008 and Beyond*. National Oceanic and Atmospheric Administration, Washington, D.C., 16 pp.

NRC (1999). *Our Common Journey: A Transition toward Sustainability*. National Academy Press, Washington, D.C., 384 pp.

NSF (2003). *Complex Environmental Systems: Synthesis for Earth, Life, and Society in the 21st Century—A 10-year Outlook*. National Science Foundation, ERIC, Washington, D.C., 82 pp.

NSI (2005). *Civil Society and the Global Agenda: From Evaluation to Action*. WFUNA, New York, 8 pp.

Obaid T.A. (ed.) (2003). *State of World Population 2002: People, Poverty, and Possibilities*. UNFPA (United Nations Population Fund), New York, 84 pp.

Ocampo J.A. (2005). The path to sustainability: Improving access to energy. *Energy and Environment*, **16**(5), 737–742.

Odell P.R. (2004). *Why Carbon Fuels Will Dominate the 21st Century's Global Energy Economy*. Multi-Science, Brentwood, U.K., 192 pp.

Oganesian V.V. (2004). Changes in Moscow's climate from 1879 to 2002 expressed as extremums of temperature and precipitation. *Meteorology and Hydrology*, **9**, 31–37 [in Russian].

Oldfield F. (2005). *Environmental Change: Key Issues and Alternative Approaches*. Cambridge University Press, Cambridge, U.K., 392 pp.

Olenyev V.V. and Fedotov A.P. (2003). Globalistics on the verge of the 21st century. *Problems of Philosophy*, **4**, 18–30 [in Russian].

Oppenheimer C. (1996). Volcanism. *Geography*, **81**(1), 65–81.

Orsato R.J. and Cregg S.R. (2005). Radical reformism: Towards critical ecological modernization. *Sustainable Development*, **13**(4), 253–267.

Overpeck J.T., Sturm M., Francis J.A., Perovich D.K., Serreze M.C., Benner R., Carmack E.C., Chapin III E.S., Gerlach S.C., Hamilton L.C. *et al.* (2005). Arctic system on trajectory to new, seasonally ice-free state. *EOS*, **86**(34), 309–312.

Paley M.D. (1993). The last man: Apocalypse without millennium. In: A. Fisch, A.K. Mellor, and E.H. Schor (eds.), *The Other Mary Shelley: Beyond Frankenstein*. Oxford University Press, New York, pp. 107–123.

Palmer T.N., Alessandri A., Anderson U., Cantelaube P., Davey M., Délécluse P., Déqué M., Diez E., Doblas-Reyes F. J., Feddersen H. *et al.* (2004). Development of a European Multimodel Ensemble System for Seasonal-to-Interannual Prediction (DEMETER). *Bull. Amer. Meteorol. Soc.*, **85**(6), 853–872.

Palutikof J.P., Goodess C.M., Watkins S.J., and Burgess P.E. (1999). Long-term climate change. *Progr. Environ. Sci.*, **1**(1), 89–96.

Panarin A.S. (2004). Globalization as a challenge to life. *Herald of RAS*, **74**(7), 619–626 [in Russian].

Parkinson C.L. (2002). *Aqua: AIRS, AMSU, HSB, AMSR-E, CERES, MODIS*. NASA Goddard Space Flight Center, Greenbelt, MD, 44 pp.

Parkinson C.L. (2003). Aqua: An Earth-observing satellite mission to examine water and other climate variables. *IEEE Trans. on Geosci. and Remote Sensing*, **41**(2), 173–183.

Parson E.A. and Fisher-Vanden K. (1997). Integrated assessment models of global climate change. *Annual Review of Energy and the Environment*, **22**, 589–628.

Parson E.A. and Fisher-Vanden K. (1999). Joint implementation of greenhouse gas abatement under the Kyoto Protocol's "Clean Development Mechanism"': Its scope and limits. *Policy Sciences*, **32**(3), 207–224.

Pasch R.J. (2006). *Tropical Cyclone Report: Tropical Storm Beryl* (Report No. AL022006). NHC, Washington, D.C., 10 pp.

Pasch R.J., Lawrence M.B., Avila L.A., Beven J.L., Franklin J.L., and Stewart S.R. (2004). Atlantic hurricane season of 2002. *Monthly Weather Review*, **132**, 1829–1859.

Pelling M. (2003). *Natural Disaster and Development in a Globalizing World*. Routledge, London, 272 pp.

Perrie W., Zhang W., Ren X., Long Z., and Hare J. (2004). The role of midlatitudes storms on air–sea exchange of $CO_2$. *Geophys. Res. Letters*, **31**(L09306), doi: 10.1029/2003GL019212.

Petzold D.E. and Goward S.N. (1988). Reflectance spectra of subarctic lichens. *Remote Sensing of Environment*, **24**, 481–492.

Phelan J.P. (2004). *Topics: Annual Review of North American Natural Catastrophes 2003*. American Re, Princeton, NJ, 48 pp.

Pielke R.A., Jr. (2001a). Carbon sequestration: The need for an integrated climate system approach. *Bull. Amer. Meteorol. Soc.*, **82**(11), 20–21.

Pielke R.A., Jr. (2001b). Earth system modeling: An integrated assessment tool for environmental studies. In: T. Matsuno and H. Kida (eds.), *Present and Future of Modeling Global Environmental Change: Toward Integrated Modeling*. TERRAPUB, Tokyo, pp. 311–337.

Pielke R.A., Jr. (2002). Overlooked issues in the U. S. national climate and IPCC assessments. *Clim. Change*, **52**(1–2), 1–11.

Pielke R.A., Jr., Schellnhuber H.J., and Sahagian D. (2003). Non-linearities in the Earth system. *Global Change Newsletter*, **55**, 11–15.

Pittock A.B. (2005). *Climate Change: Turning up the Heat*. Earthscan, London, 272 pp.

Pollack H.N. (2003). *Uncertain Science ... Uncertain World*. Cambridge University Press, Cambridge, U.K., 256 pp.

Pollack R. (2006). A place for religion in science? Winter 2005/2006 *CSSR Newsletter*, 4–5.

Porritt J. (2005). *Capitalism: As If the World Matters*. Earthscan, London, 304 pp.

Potapov I.I., Krapivin V.F., and Soldatov V.Yu. (2006). An assessment of the risk in the regime of geoinformation monitoring. *Ecological Systems and Devices*, **8**, 11–18 [in Russian].

PRECIS (2002). *The Hadley Centre Regional Climate Modelling System*. Met Office Hadley Centre, London, 20 pp.

Prescott-Allen R. (2001). *Wellbeing of Nations: A Country-by-country Index of Quality of Life and the Environment* (350 pp.). Island Press, Washington, D.C.

Princen T., Maniates M., and Conca K. (2002). *Confronting Consumption*. MIT Press, Cambridge, MA, 415 pp.

Prins E.M., Feltz J.M., Menzel W.P., and Ward D.E. (1998). An overview of GOES-8 diurnal fire and smoke results for SCAR-B and the 1995 fire season in South America. *J. Geophys. Res.*, **103**(D24), 31821–31836.

Pulido E., Allen A., and Reaves N. (eds.) (2005). *2005 Vital Signs*. United Way of Central Oklahoma, OK, 42 pp.

Purvis M. and Grainger A. (eds.) (2004). *Exploring Sustainable Development: Geographical Perspectives*. Earthscan, London, 400 pp.

Quaddus M.A. and Siddique M.A.B. (2004). *Handbook on Sustainable Development Planning: Studies in Modelling and Decision Support*. Edward Elgar, Cheltenham, U.K., 360 pp.

Ramsar COP8 (2002). *The 8th Conference of the Contracting Parties to the Ramsar Convention, Valencia, Spain, 18–26 November 2002*. UNCCD, Gland, Switzerland (*http://www.ramsar.org/cop8/cop8_regionalmeetings_schedule.htm#list)*

Ramsar COP9 (2005). *Wetlands and Water: Supporting Life, Sustaining Livelihoods—Report of 9th Meeting of the Conference of the Contracting Parties to the Convention on Wetlands, Kampala, Uganda, 8–15 November 2005*. UNCCD, Bonn, Germany, 50 pp.

Rathford M. (2006). *The Nostradamus Code: World War III 2007–2012*. Morris, Kearney, NE, 94 pp.

Raven P.(2005). Biodiversity and our common future. *Bull. Amer. Acad. Arts Sci.*, **58**(3), 20–24.

Redclifte M.R. (2005). Sustainable development (1987–2005): An oxymoron comes of age. *Sustainable Development*, **13**(4), 212–227.

Redclifte M.R. and Woodgate G. (eds.) (2005). *New Developments in Environmental Sociology*. Edward Elgar, Cheltenham, U.K., 672 pp.

Rees M. (2003). *Our Final Century: The 50/50 Threat to Humanity's Survival*. William Heinemann, London, 238 pp.

Reid W.V. (ed.) (2005). *Ecosystems and Human Well-Being: Synthesis*. Island Press, Washington, D.C., 155 pp.

Riccio A., Barone G., Chianese E., and Giunta G. (2006). A hierarchical Bayesian approach to the spatio-temporal modeling of air quality data. *Atmospheric Environment*, **40**(3), 554–566.

Richardson L.L. and Le Drew E.F. (eds.) (2006). *Remote Sensing of Aquatic Coastal Ecosystem Processes: Science and Management Applications*. Springer-Verlag, Heidelberg, Germany, 325 pp.

Rimashevskaya N.M. (ed.) (2002). *Population and Globalization*. Science, Moscow, 324 pp. [in Russian].

Rivero O. (2001). *The Myth of Development: The Non-viable Economies of the 21st Century*. Zed Books, London, 192 pp.

Robertson D. and Kellow A. (eds.) (2001). *Globalization and the Environment: Risk Assessment and the WTO*. Edward Elgar, London, 288 pp.

Roels H. (2005). Customer needs, political expectations and energy economics: Sustainability as a challenge for European energy companies (keynote speech). *Energy and Environment*, **16**(5), 729–735.

Rosenzweig M. (2003). *Win–Win Ecology: How the Earth's Species Can Survive in the Midst of Human Enterprise*. Oxford University Press, New York, 221 pp.

Rotmans J. and Rothman D.S. (eds.) (2003). *Scaling in Integrated Assessment: Integrated Assessment Studies 2*. Swets & Zeitlinger, Amsterdam, The Netherlands, 355 pp.

Rousseau J.-J. (1969). *Treatises*. Cannon Press, Moscow, 544 pp. [in Russian].

Ruck M. (2002). *Natural Catastrophes 2002: Annual Review*. Munich Re Topics, Dresden, Germany, 50 pp.

Russel G.L., Miller J.R., Rind D., Ruedy R.A., Schmidt G.A., and Sheth S. (2000). Comparison of model and observed regional temperature changes during the past 40 years. *J. Geophys. Res. (Atmospheres)*, **105**(G11), 14891–14898.

Russo G., Eva C., Palau C., Caneva A., and Saechini A. (2000). The recent abrupt increase in the precipitation rate, as seen in an ultra-centennial series of precipiation. *Il Nuovo Cimento*, **23C**(1), 39–51.

Sachs J.D. (ed.) (2005). *Investing in Development: A Practical Plan to Achieve the Millennium Development Goals*. Earthscan, London, 329 pp.

Saiko E.V. (ed.) (2003). *Civilization: Ascent and Breaking. The Structure-forming Factors and Subjects of Civilization Process*. Science, Moscow, 453 pp. [in Russian].

Santer B.D., Wigley T.M.L, Boyle J.S., Gaffen D.J., Hnilo J.J., Nychka D., Parker D.E., and Taylor K.E. (2000). Statistical significance of trends and trend differences in layer average atmospheric temperature time series. *J. Geophys. Res.*, **105**(D6), 7337–7356.

Santos M.A. (2005). Environmental stability and sustainable development. *Sustainable Development*, **13**(5), 326–336.

Sarti F., Briole P., and Oirri M. (2006). Coseismic fault rupture detection and slip measurement by ASAR precise correlation using coherence maximization: Application to a north–south blind fault in the vicinity of Bam (Iran). *IEEE Geosci. and Remote Sensing Letters*, **3**(2), 187–191.

Sasaki Y., Asanuma I., Muneyama K., Naito G., and Suzuki T. (1987). A simplified microwave model and its application to the determination of some oceanic environmental parametric values, using multi-channel microwave radiometric observation. *IEEE Trans. on Geosci. and Remote Sensing*, **GE-25**(3), 384–392.

Sasaki Y., Asanuma I., Muneyama K., Naito G., and Suzuki T. (1988). Microwave emission and reflection from the wind-roughned sea surface at 6.7 and 18.6 GHz. *IEEE Trans. on Geoscience and Remote Sensing*, **26**(6), 860–868.

Saul J.S. (2005). *Development after Globalization: Theory and Practice for the Embattled South in a New Imperial Age*. Zed Books, London, 144 pp.

Scheer H. (2005). *A Solar Manifesto*. Earthscan, London, 272 pp.

Schellnhuber H.-J. and Stock M. (2000). Introduction, overview and perspectives. In: M. Boysen (ed.), *Biennial Report 1998 and 1999*. PIC Press, Potsdam, Germany, pp. 1–8.

Schlesinger M.E., Ramankutty N., and Andronova N. (2000). Temperature oscillations in the North Atlantic. *Science*, **289**(5479), 547.

Schmugge T. (1990). Measurements of surface soil moisture and temperature. In: R.J. Hobbs and H.A. Mooney (eds.), *Remote Sensing of Biosphere Functioning*. Springer-Verlag, New York, pp. 31–62.

Schröter D., Cramer W., Leemans R., Prentice C.I., Araujo M.B., Arnell N.W., Bondean A., Bugmann H., Carter T.R., Gracia C.A. *et al.* (2005). Ecosystem service supply and vulnerability to global change in Europe. *Science*, **310**(5752), 1333–1337.

Schultz J. (2005). *The Ecozones of the World: The Ecological Divisions of the Geosphere*, 2nd rev. and updated edn. Springer-Verlag, Heidelberg, Germany, XII, 252 pp.

Schulze E.-D. (2000). The carbon and nitrogen cycle of forest ecosystems. *Ecol. Studies*, **142**, 3–13.

Sedwick P.N., Church T.M., Bowie A.R., Marsay C.M., Ussher S.J., Achilles K.M., Lethaby P.J., Johnson R.J., Sarin M.M., and McGillicuddy D.J. (2005). Iron in the Sargasso Sea (Bermuda Atlantic time-series study region) during summer: Eolian imprint, spatiotemporal variability, and ecological implications. *Global Biogeochemical Cycles*, **19**(GB4006), doi: 10.1029/2004GB002445.

Sellers P.J., Los S.O., Tucker C.J., Justice C.O., Dazlich D.A., Collatz G.J., and Randall D.A. (1996). A revised land surface parametrization (SiB2) for atmospheric GCMs, Part II: The generation of global fields of terrestrial biophysical parameters from satellite data. *J. Climate*, **9**(4), 708–737.

Shelley M. (1826). *The Last Man* (three-volume set). Henry Colburn, London, 330 pp.

Shelley M. (1992). *Frankenstein, or the Modern Prometheus*. Oxford University Press, New York, 261 pp.

Shelley M. (1994). *The Last Man*. Oxford University Press, New York, 280 pp.

Shuffelton F. (1990). From Jefferson to Thoreau: The possibilities of discourse. *Arizona Quarterly*, **46**, 1–16.

Shuffelton F. (1995). Presenting Jefferson. *Early American Literature*, **30**, 275–285.

Shutko A.M. (1986). *Microwave Radiometry of Water Surface and Soils*. Science Press, Moscow, 188 pp. [in Russian].

Shutt H. (2001). *A New Democracy: Alternatives to a Bankrupt World Order*. Zed Books, London, 2001, 176 pp.

Si F.C. and Hai L.D. (1997). Control of environmental pollution from industrial and domestic waste in the key region of economic development. *Proc. of the Workshop on Environmental Technology and Management in Ho Chi Minh City, 28–29 May 1997*. Institute of Applied Mechanics, Ho Chi Minh City, Vietnam, pp. 32–42 [in Vietnamise].

Sibruk J. (2006). *Nobrow: Marketing of Culture*. Ad Marginem, Moscow, 305 pp. [in Russian].

Simpson R. and Ganstang M. (eds.) (2003). *Hurricane: Coping with Disaster*. American Geophysical Union, Washington, D.C., 360 pp.

Sims D.A. and Gamon J.A. (2003). Estimation of vegetation water content and photosynthetic tissue area from spectral reflectance: A comparison of indices based on liquid water and chrolophyll absorption features. *Remote Sensing of Environment*, **84**(4), 526–537.

Singh H.B. and Jacob D.J. (2000). Future directions: Satellite observations of tropospheric chemistry. *Atmospheric Environment*, **34**, 4399–4401.

Singh K. (2005). *Questioning Globalization*. Zed Books, London, 160 pp.

Sinreich R., Volkamer R., Filsinger F., Frieß U., Kern C., Platt U., Sebastián O., and Wagner T.(2006). MAX-DOAS detection of glyoxal during ICARTT 2004. *Atmos. Chem. Phys. Discuss.*, **6**, 9459–9481.

Smil V. (2002). *The Earth's Biosphere: Evolution, Dynamics, and Change*. MIT Press, Cambridge, MA, 346 pp.

Smith K.(2001). *Environmental Hazards: Assessing Risk and Reducing Disaster*. Routledge, London, 432 pp.

Solozhentsev E.D. (2004). *The Scenario of Logical–Probabilistic Risk Control in Business and Engineering*. Business Press, St. Petersburg, 416 pp. [in Russian].

Soon W. and Baliunas S. (2003). Proxy climatic and environmental changes of the past 1000 years. *Climate Research*, **23**, 89–110.

Soon W., Baliunas S., Idso C., Idso S., and Legates D.R. (2003). Reconstructing climatic and environmental changes of the past 1000 years: A reappraisal. *Energy and Environment*, **14**, 233–296.

Southerton D., Chapells H., and van Vliet B. (eds.) (2004). *Sustainable Consumption: The Implications of Changing Infrastructures of Provision*. Edward Elgar, London, U.K., 192 pp.

Spangenberg J. H. (2004). Reconciling sustainability and growth: Criteria, indicators, policies. *Sustainable Development*, **12**(2), 74–86.

Spellerberg I.F. (2005). *Monitoring Ecological Change*, Second Edition. Cambridge University Press, Cambridge, U.K., 409 pp.

Speth J.G. (2004). *Red Sky at Morning: America and the Crisis of the Global Environment*. Yale University Press, New Haven, CT, 320 pp.

Springett D. (2005). Critical perspectives on sustainable development. *Sustainable Development*, **13**(4), 209–211.

Stanton W. (2003). *The Rapid Growth of Human Populations, 1750–2000: Histories, Consequences, Issues—Nation by Nation*. Multi-Science, Brentwood, U.K., 230 pp.

Starke L. (ed.) (2004). *State of the World—2004: Progress towards a Sustainable Society*. Earthscan, London, 246 pp.

Stebler O., Schwerzmann A., Lüthi M., Meier E., and Nüesch D. (2005). Pol-InSAR observations from an Alpine glacier in the cold infiltration zone at L- and P-band. *IEEE Geosci. and Remote Sensing Letters*, **2**(3), 357–361.

Steffen W. and Tyson P. (eds.) (2001). *Global Change and the Earth System: A Planet under Pressure*. IGBP Science 4, Stockholm, 32 pp.

Steffen W., Jäger J., Carson D.J., and Bradshaw C. (eds.) (2002). *Challenges of a Changing Earth*. Springer-Verlag, Heidelberg, Germany, 216 pp.

Stempell D. (1985). *Weltbevolkerung*. Springer-Verlag, Berlin, 205 pp.

Stokstad E. (2005). Will Malthus continue to be wrong? *Science*, **309**, 102.

Stone R.S. (1998). Monitoring ferosol optical depth at Barrow, Alaska and South Pole: Historical overview, recent results, and future goals. *J. Geophys. Res.*, **103**, 16565–16579.

Storm S. and Naastepad C.W.M. (2001). *Globalization and Economic Development*. Edward Elgar, London, 432 pp.

Straub C.P. (ed.) (1989). *Practical Handbook of Environmental Control* (537 pp.). CRC Press., Boca Raton, FL.

Streets D.G., Jiang K., Hu X., Sinton J.E., Zhang Z.-Q., Xu D., Jacobson M.Z., and Hansen H.E. (2001). Recent reductions in China's greenhouse gas emissions. *Science*, **294**(5548), 1835–1837.

Strong A.E., Kearns E.J., and Gjovig K.K. (2000). Sea surface temperature signals from satellites update. *Geophys. Res. Lett.*, **27**(11), 1667–1670.

Subbotin A. (ed.) (1977). *F. Beckon's Works in Two Volumes* (Philosophical Heritage series, Vol. 2). Thought, Leningrad, 576 pp. [in Russian].

Sungurov A. (2005). The conception of the noosphere and formation of civil society. *Green World*, **23/24**, 469–470 [in Russian].

Svirezheva-Hopkins A. and Schellnhuber H.-J. (2005). *Urbanized Territories as a Specific Component of the Global Carbon Cycle* (PIC Report, No. 94). PIC, Potsdam, 184 pp.

Tabb W.K. (2001). *The Amoral Elephant: Globalization and the Struggle for Social Justice in the Twenty-First Century*. Monthly Review Press, New York, 288 pp.

Tett S.F.B., Stott P.A., Allen M.R., Ingram W.J., and Mitchell J.F.B. (1999). Causes of twentieth-century temperature change near the Earth's surface. *Nature*, **399**, 569–572.

Thomashow M. (2003). *Bringing the Biosphere Home: Learning to Percieve Global Environmental Change*. MIT Press, Cambridge, MA, 256 pp.

Tianhong L., Yanxin S., and An X. (2003). Integration of large scale fertilizing models with GIS using minimum unit. *Environmental Modelling*, **18**(3), 221–229.

Tiezzi E., Brebbia C.A., and Uso J.-L. (2003). *Ecosystems and Sustainable Development IV*. WIT Press, Southampton, U.K., Vol. 1 (722 pp.), Vol. 2 (584 pp.).

Timoshevskii A., Yeremin V., and Kalkuta S. (2003). New method for ecological monitoring based on the method of self-organising mathematical models. *Ecological Modelling*, **162**(2–3), 1–13.

Tolba M.K. (ed.) (2001). *Our Fragile World: Challenges and Opportunities for Sustainable Development*. EOLSS, Oxford, U.K., 2001, Vol. I (1051 pp.), Vol. II (2263 pp.).

Troyan V. N. and Dementyev I. A. (eds.) (2005). *Sustainable Development and Ecological Management, Issue 1: Proc. Int. Sci.–Practical Conf., St. Petersburg, 17–18 November 2005*. St. Petersburg State University, St. Petersburg, 484 pp. [in Russian].

Tsushima Y. and Manabe S. (2001). Influence of cloud feedback on annual variation global mean surface temperature. *J. Geophys. Res.*, **106**(D19), 22646–22655.

Turner M.G., Gardner R.H., and O'Neill R.V. (2001). *Landscape Ecology in Theory and Practice: Pattern and Process*. Springer for Science, Amsterdam, The Netherlands, 404 pp.

UN (2000). *Report of the Secretary-General: "We the Peoples: The Role of the United Nations in the 21st Century", Millennium Summit, 6–8 September 2000*. United Nations, New York, 82 pp.

UNEP (2002). *Global Environmental Outlook, 3: Past, Present and Future Perspectives*. United Nations Environment Programme, Nairobi, Kenya, 20 pp.

UNEP (2004). *UNEP Science Initiative. Environmental Change and Human Needs: Assessing Inter-linkages. An Input to the Fourth Global Environment Outlook (GEO-4)* (concept paper). United Nations Environment Programme, Nairobi, Kenya, 36 pp.

UNEP (2005). *Convention on Migratory Species: 8th Meeting of the Conference of the Parties to CMS (COP 8), Nairobi, 20–25 November 2005, Resolution 8.24*. UNEP/CMS, Nairobi, Kenya, 2 pp.

UNEP (2006). *Convention on Biological Diversity: Proc. of the Conference on the Parties to the Convention on Biological Diversity, Curitiba, Brazil, 20–31 March 2006*. UNEP/CBD/COP, New York, 374 pp.

Vaitheeswaran V.V. (2005). *Power to the People: How the Coming Energy Revolution Will Transform Industry, Change Our Lives, and Maybe Even Save the Planet*. Earthscan, London, 368 pp.

Vakulenko N.V., Kotliakov V.N., Monin A.S., and Sonechkin D.M. (2004). Proof of increase in greenhouse gas concentration by temperature variations from Vostok data. *Proceedings of RAS*, **396**(5), 686–689 [in Russian].

Vald A. (1960). *Successive Analysis*. Phys.-Math. Literature, Moscow, 328 pp. [in Russian].

Van den Bulcke D. and Verbeke A. (2001). *Globalization and the Small Open Economy*. Edward Elgar, London, 256 pp.

Varotsos C. (2002). The southern hemisphere ozone hole split in 2002. *Environmental Science and Pollution Research*, **9**(6), 375–376.

Varotsos C. (2005). News on the Antarctic Ozone Hole. *Environmental Science and Pollution Research*, **12**(6), 322–323.

Varotsos C., Assimakopoulos, M-N., and Efstathiou M. (2006). Technical note: Long-term memory effect in the atmospheric $CO_2$ concentration at Mauna Loa. *Atmos. Chem. Phys. Discuss.*, **6**, 11957–11970.

Vasilevich V.I. (1983). *Essays on Theoretical Phytocenology*. Science, Leningrad, 226 pp. [in Russian].

Verbesselt J., Jönsson P., Lhermitte S., van Aardt J., and Coppin P. (2006). Evaluating satellite and climate data-derived indices as fire risk indicators in savanna ecosystems. *IEEE Trans. on Geosci. and Remote Sensing*, **40**(6), 1622–1632.

Vernadsky V.I. (1944). Several words about the noosphere. *Present Biology Successes*, **18**(2), 49–93 [in Russian].

Vernadsky V.I. (2002). *Biosphere and Noosphere*. Airis Press, Moscow, 2002, 572 pp. [in Russian].

Victor B.G., Raustiala K., and Skolnikoff E.B. (eds.) (1998). *The Implementation and Effectiveness of International Environmental Commitments: Theory and Practice*. MIT Press, Cambridge, MA, 737 pp.

Viscusi W.K. and Gayer T. (2004). *Classics in Risk Management* (two-volume set). Edward Elgar, Northampton,. MA, 1203 pp.

Vladimirov V.A., Vorobyev Yu.L., Salov S.S., Faleye M.I., Arkhipova N.I., Kapustin M.A., Kashchenko S.A., Kosiachenko S.A., Kuznetsov I.V., Kul'ba V.V. *et al.* (2000). *Risk Management and Sustainable Development*. Science, Moscow, 431 pp. [in Russian].

Volk T. (2003). *Gaia's Body: Toward a Physiology of Earth*. MIT Press, Cambridge, MA., 296 pp.

Voltaire F.-M.A. (1996). *Philosophical Works*. Science, Moscow, 560 pp. [in Russian].

Wagner A. (2005). *Robustness and Evolvability in Living Systems*. Princeton University Press, Princeton, NJ, 383 pp.

Wallace J.M. (1998). Observed climatic variability: Spatial structure. In: D.L.T. Anderson and J. Willebrand (eds.), *Decadal Climate Variability: Dynamics and Predictability*. Springer-Verlag, Berlin, pp. 31–81.

Wallström M., Bolin B., Crutzen P., and Steffen W. (2004). The Earth's life-support system in peril. *Global Change Newsletter*, **57**, 22–23.

Ward B. and Dubos R. (1972). *Only One Earth: The Care and Maintenance of a Small Planet*. Penguin, London, 304 pp.

Watson R.T. (ed.) (2002). *Climate Change 2001: Synthesis Report* (Third Assessment Report). Intergovernmental Panel on Climate Change, Washington, D.C., 408 pp.

Watson, R.T., Noble, I.R., Bolin, B., Ravindranath, N.H., Verardo, D.J, and Dokken, D.J. (eds.) (2000). *Land Use, Land-use Change, and forrestry*. Cambridge University Press, Cambridge, U.K., 377 pp.

WB (2003). *Global Economic Prospects 2003: Investing to Unlock Global Opportunities*. World Bank, Washington, D.C., 244 pp.

WB (2006). *World Development Report*. World Bank, Washington, D.C., 34 pp.

WCC (2005). *Alternative Globalization Addressing Peoples and Earth*. World Council of Churches, Geneva, Switzerland, 68 pp.

WCSDG (2004). *A Fair Globalization: Creating Possibilities for All* (report of the World Commission on the Social Aspects of Globalization). International Trade Organization, Geneva, Switzerland, 190 pp.

Weber G.R. (1992). *Global Warming: The Rest of the Story*. Dr. Boettiger Verlag, Wiesbaden, Germany, 188 pp.

Weinstein J.R. (2001). *On Adam Smith*. Wadsworth, Belmont, CA, 243 pp.

Wen J.G., Wei S.-J., and Zou H. (eds.) (2002). *The Globalization of the Chinese Economy*. Edward Elgar, London, 360 pp.

WEO (2002). *World Energy Outlook* (533 pp.). OECD/IEA, Paris, France.

WEO (2004). *World Energy Outlook* (550 pp.). OECD/IEA, Paris, France.

WEO (2005). *Middle East and North Africa Insights* (600 pp.). IEA, London.

WEO (2006). *Energy Market Reform; Energy Policy; Energy Projections* (600 pp.). IEA, London.

Werner D. and Newton W.E. (eds.) (2005). *Nitrogen Fixation in Agriculture: Forestry, Ecology, and the Environment*. Springer-Verlag, Heidelberg, Germany, 348 pp.

Wicks R. and Keay M. (2005). Can coal contribute to sustainable development? *Energy and Environment*, **16**(5), 767–779.

Widmann M., Jones J.M., and von Storch H. (2004). Reconstruction of large-scale atmospheric circulation and data assimilation in paleoclimatology. *Pages News*, **12**(2), 12–13.

Wigley T.M.L. (1999). *The Science of Climate Change: Global and U.S. Perspectives*. Pew Center on Global Climate Change, Arlington, VA, 48 pp.

Wijen F., Zoeteman K., and Pieters J. (eds.) (2004). *A Handbook of Globalization and Environmental Policy*. Edward Elgar, London, 550 pp.

Wijen F., Zoeteman K., and Pieters J. (eds.) (2005). *A Handbook of Globalization and Environmental Policy: National Government Interventions in a Global Arena*. Edward Elgar, London, 768 pp.

Wisner B., Toulmin C., and Chitiga R. (2005). *Toward a New Map of Africa*. Earthscan, London, 336 pp.

WMO (2005). *Saving Paradise: Ensuring Sustainable Development* (WMO Bull. No. 973). World Meteorological Organization, Geneva, 28 pp.

Woodcock A. (1999a). Global warming: A natural event? *Weather*, **54**(5), 162–163.

Woodcock A. (1999b). Global warming: The debate heats up. *Weather*, **55**(4), 143–144.

Workman J. (2005). *Pragmatic Solutions: An Assessment of Progress 2005*. Atar Roto Presse, Geneva, Switzerland, 80 pp.

Wright E.L. and Erickson J.D. (2003). Incorporating catastrophes into integrated assessment: Science, impacts, and adaptation. *Climatic Change*, **57**, 265–286.

Wright J. (2003). *Environmental Chemistry*. Routledge, London, 432 pp.

WWI (2003). *Vital Signs: 2003–2004. The Trends that Are Shaping Our Future*. Earthscan, London, 192 pp.

WWI (2004). *State of the World 2005: Global Security*. Earthscan, London, 288 pp.

WWI (2005). *Vital Signs: 2005–2006. The Trends that Are Shaping Our Future*. Earthscan, London, 128 pp.

WWI (2006). *State of the World 2006: China, India, and the Future of the World*. Earthscan, London, 242 pp.

Yakovlev O.I. (2001). *Space Radio Science*. Taylor & Francis, London, 320 pp.

Yamagata T., Behera S.K., Luo J.J., Masson S., Jury M.R., and Rao S.A. (2004). Coupled ocean–atmosphere variability in the tropical Indian Ocean. In: C. Wang, X.-P. Xie, and J.A. Carton (eds.), *Earth Climate: The Ocean–Atmosphere Interactions*. Springer-Verlag, Berlin, pp. 189–212.

Yu R., Zhang M., and Cess R.D. (1999). Analysis of the atmosphere enrgy budget: A consistency study of available data sets. *J. Geophys. Res.*, **104**(D8), 9655–9661.

Yue T.X., Fan Z.M., and Liu J.Y. (2005). Changes of major terrestrial ecosystems in China since 1960. *Global and Planetary Change*, **48**(4), 287–302.

Zarco-Tejada P.J., Ruweda C.A., and Ustin S.L. (2003). Water content estimation in vegetation with MODIS reflectance data and model inversion methods. *Remote Sensing of Environment,* **85**(1), 109–124.

Zavarzin G. A. (2003). *Lectures on Microbiology Natural History*. Science, Moscow, 403 pp. [in Russian].

Zhan X., Houser P.R., Walker J.P., and Crow W.T. (2006). A method for retrieving high-resolution surface soil moisture from hydros L-band radiometer and radar observations. *IEEE Trans. on Geosci. and Remote Sensing*, **44**(6), 1534–1543.

Zhan X., Sohlberg R.A., Townshend J.R.G., DiMiceli C., Carroll M.L., Eastman J.C., Hansen M.C., and DeFries R.S. (2002). Detection of land cover changes using MODIS 250 m data. *Remote Sensing of Environment*, **83**(1–2), 336–350.

Zhu C. and Anderson G. (2002). *Environmental Applications of Geochemical Modeling*. Cambridge University Press, Cambridge, U.K., 298 pp.

# Index

Printing: Mercedes-Druck, Berlin
Binding: Stein + Lehmann, Berlin